An Introduction to the Mechanics of Machines
S I Units

An Introduction to the Mechanics of Machines

SI UNITS

J. L. M. MORRISON
*Professor of Mechanical Engineering
University of Bristol*

B. CROSSLAND
*Formerly Professor of Mechanical Engineering
The Queen's University of Belfast*

Longman
Scientific &
Technical

Longman Scientific & Technical
Longman Group UK Limited,
Longman House, Burnt Mill, Harlow,
Essex CM20 2JE, England

and Associated Companies throughout the world.

British Units Edition first published 1964
SI Units Edition first published 1970
Reprinted and first Paperback Edition 1971
Tenth impression 1987

ISBN 0-582-44729-1

Produced by Longman Group (FE) Ltd
Printed in Hong Kong

Contents

5. Damped and Forced Vibrations 307

6. Automatic Control 367

7. Friction and Lubrication 422

Preface

Another book on the theory – or mechanics – of machines? Is there any need or justification for another?

As knowledge accumulates, more and more advanced information has to be communicated to the student in the course of his education. Three main factors contribute to making this possible. Specialisation, however much it may be deplored, narrows the field; men like Copernicus or Newton propound fundamental truths which require genius to formulate, but once established are relatively easy to understand, and effect a major simplification for those who follow; and with advancing knowledge the perspective changes, so that detail which at one time seemed of vital importance is seen to be relatively trivial and can be eliminated.

In technological education it is becoming more widely realised that teaching should be directed to the thorough understanding of the few fundamental principles rather than to the transmission of a mass of detail. This approach is in itself economical: more important, it means that if the foundations are well laid the student will be in a much better position to cope with current and future developments in an expanding subject. We have then tried to present the subject with reasonable brevity: let us hope that we shall not have to admit with Horace, 'I labour to be brief, and become obscure'.

Of course, authors of textbooks are inevitably confronted with the problem of defining the boundaries of their endeavour. A treatise to cover even in outline the whole of modern knowledge on the subject of mechanics of machines would occupy many volumes. To indicate that no such purpose was in our minds we have included in the title the phrase 'an introduction to . . .' and correspondingly deleted most of the apologetic statements to the effect that we could add no more on any particular topic. What we have tried to do is to include the material which we feel to be vital in the education of any engineer, and in particular any mechanical engineer, to professional or ordinary degree standard at the present day. We have made no attempt to cover any subject to a level which might be thought to be appropriate to an honours degree candidate who was specialising in it, but instead to give references to a few suitable books which deal each with one such subject in much greater detail, and to which this book might reasonably be regarded as an introduction.

March, 1964   J.L.M.M.
                                                              B.C.

Introduction

Mainly a Revision of Basic Mechanics

§ 0.1 Objective

This chapter is not intended to replace a textbook on mechanics. On the contrary, it is assumed that the principles on which the subject of *Mechanics of Machines* is based have been previously studied. Nevertheless, it is the experience of the authors that a few of the most vital principles have been insufficiently well mastered by the average student, and it is these which have been selected here for revisional treatment. The use of the word 'revisional' is intended to imply that far less attention will be paid to careful introduction and definition of terms and concepts in logical order, and to formal proofs, than would be necessary in a first treatment of the subject, and more attention concentrated on the importance and use of the results.

In the final few paragraphs some topics are discussed which can be regarded as on the borderline between pure and applied mechanics, and which are therefore not always included in textbooks of mechanics. These should not be overlooked.

The student who intends to use this book is urged to read through this introductory chapter with care. If in it he finds nothing novel or ill-understood he will have wasted little time. If, on the other hand, he finds that any of the concepts are new to him, or that he has tended to pass them by as of minor importance, he would be well advised to master them thoroughly before passing on to a study of the subject matter of the later chapters.

§ 0.2 Sign Conventions

Engineers tend to be careless about signs. The reason for this is not far to seek. In many of the problems, particularly in elementary work, with which they have to contend the signs are 'obvious', in the sense that the direction in which, for instance, a displacement occurs or a force acts can be judged intuitively, and so only the numerical value of the result need be calculated. A beam supported at its ends and loaded gravitationally will deflect downwards; a mass on an elastic support will vibrate about its equilibrium position; a rod subjected to a tensile force will increase in length; and so on. In more advanced work this is no longer true: the signs may be far from obvious, and this applies with particular emphasis to many types of problem we have to deal with in dynamics, when, for instance, the elements of a machine are simultaneously subjected to accelerations which involve the application of unbalanced forces to them and required to transmit other forces which must necessarily be balanced by reactions equal in magnitude but opposite in sign. In such instances carelessness leads to hopeless confusion, which can best be avoided by forming the good habit of dealing rigorously with signs even in elementary work.

If there is no good reason for choosing unorthodox axes, it is desirable to follow the standard conventions which have been generally accepted by mathematicians.

0.2.1 One-dimensional problems

In one-dimensional problems the possibility of confusion is, of course, minimal, and it is perfectly permissible to label an axis by any letter which is preferred. Thus, for instance, if we are concerned with the vertical motion only of a mass supported by a spring from a fixed support, there is no objection to choosing, say, the x-axis in the vertical direction. We have an apparently perfectly free choice in deciding to take the upward or downward direction as positive, and because gravity acts downwards there is a slight natural bias in favour of this direction. Provided we are consistent, no immediate trouble will arise: the gravitational force will be positive, the force exerted on the mass by the spring (which is upwards) will be negative, and so on. When we have completed the calculation any quantity, such as the velocity or acceleration of the mass, to which a negative sign may be attached will be in the upward direction, and so on. But at a later date we shall certainly be involved in plotting curves showing how such quantities – force, displacement, velocity, and the like – vary with another variable, time, which we shall want to treat as the independent variable. Do we now plot all such curves with the *x*-axis pointing downwards? Of course we can, but it is all unnecessarily confusing, and sooner or later we, or those to whom we are trying to communicate our results, will make a serious mistake. Surely we should have done better to avoid all this bother by

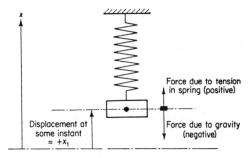

Fig. 0.1 Mass supported by spring

choosing the conventional upward direction as positive in the first place, as indicated in fig. 0.1. 'It is the first step that matters. . . .'

0.2.2 Two-dimensional problems

The positive direction of the *x*-axis is conventionally taken as shown in fig. 0.2, i.e. to the right; and of the *y*-axis (normally, for convenience, perpendicular to it) upwards. Positive rotation is defined as rotation from the (positive direction of the) *x*-axis to the (positive direction of the) *y*-axis: that is, as the 'anti-clockwise' direction.

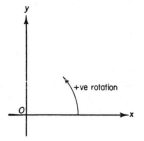

Fig. 0.2 Preferred axes for two-dimensional problems

It may well be desirable to choose other directions for the axes, for instance inclined, so that one may coincide with the axis of a shaft, or even reversed to suit the conditions posed by a particular problem.

Special care should then be exercised, and the difficulties will be minimised if anti-clockwise rotation from x to y is retained as the positive direction of rotation.

0.2.3 Three-dimensional problems

The illustration on paper of three-dimensional problems is complicated by the fact that in the two-dimensional diagram the preferred viewpoint varies, as for instance in fig. 0.3,a,b, and c. Given any such choice of axes, for convenience mutually perpendicular, it is

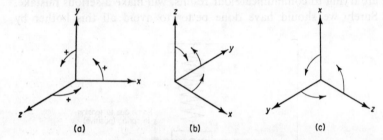

Fig. 0.3 Preferred axes for three-dimensional problems

permissible to label any one axis as the x-axis, but it is then highly desirable to proceed by *anti-clockwise* rotation to the y-axis, and again by *anti-clockwise* rotation to the z-axis. In each of these figures the x-axis has been deliberately chosen in a 'different' direction: the others follow from the rule. It will be noticed that a further rotation anti-clockwise from the z-axis brings us back to the x-axis, so the axes follow in cyclic order. It will further be noticed that if a right-handed screw whose axis coincides with any one of these axes (e.g. the z-axis) were to be rotated in the positive sense (i.e. anti-clockwise, or from the x- to the y-axis) it would be translated in the positive direction along this (z-)axis. This imaginary screw is usually referred to as a corkscrew, presumably for the benefit of non-engineers, and the corresponding rule for determining the direction of a vector (see § 0.5) to represent an angular quantity, as the 'corkscrew rule'.

§ 0.3 D'Alembert's principle

Any particle which is being accelerated must be subject to an unbalanced force in the direction of the acceleration; a body which is being accelerated linearly and rotationally must similarly be subject to an unbalanced force through its centre of gravity and an unbalanced couple, or the equivalent in the form of a force not passing through the centre of gravity. Such accelerations therefore involve for the beginner who has dealt only with problems of static equilibrium a novel concept, and novelty usually implies difficulty. D'Alembert (1717?–83) pointed out that if *imaginary* forces and couples equal and opposite to those required to provide the accelerations were added to the system the bodies would be in *apparent* equilibrium, and that calculations could be carried out exactly as though the system *were* one in static equilibrium. Clearly such an artifice will greatly ease the beginner's difficulties, and it is not surprising that it is widely used. It is, however, on that account all the more necessary to examine carefully the results.

We shall be dealing, a few paragraphs later on, with Newton's Laws of motion and with masses moving in circular paths, but since this is a revisional treatment and we are at the moment concerned with avoiding confusion about signs, let us assume some of the later results to illustrate our present topic.

Consider first the case of a mass *m* (fig. 0.4) constrained by a string *Om* to move with uniform speed in a circle about *O*. (We shall assume for simplicity that there is no gravitational field.) We can prove (§ 0.11) that the mass is being continuously accelerated towards the centre of rotation *O*. This acceleration is given the descriptive adjective *centripetal*, or centre-seeking. It follows at once from Newton's second law of motion (§ 0.6) that the *only* force acting on *m* is in the direction of this acceleration, i.e. an 'inward' or 'centripetal'

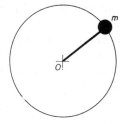

Fig. 0.4 Mass moving in circular path

force; and it is clear enough that the tension in the string provides this force on *m*.

Stated thus, with no explanatory additions, the situation can be readily appreciated. Everyone knows, however, that if he were to take the place of *O*, and to swing a stone on the end of a string in this manner he would experience an *outward* force – the tension in the string. Most of us therefore start off with an 'intuitive' feeling that some outward (or 'centrifugal') force is involved. We have to think quite carefully before we fully understand that this outward force *does not act on the stone at all*, but on our finger, and that it is the 'reaction' (Newton's third law of motion) needed to supply the 'action' on the stone. It does not really help us if at this stage we see diagrams in which non-existent centrifugal forces are inserted to 'reduce the problem to one of statics'.

Or consider now the behaviour of the 'conical pendulum'. If this is presented as in fig. 0.5a, in which are shown *all* the forces acting on the bob, namely its weight w and the tension in the string t, it

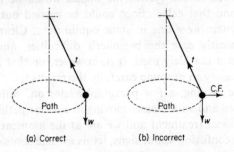

(a) Correct   (b) Incorrect

Fig. 0.5 Forces acting on bob of conical pendulum

can be made clear that this is *not* a problem in statical equilibrium, and that if the bob is to rotate in a horizontal circle, and so be receiving a continuous acceleration towards the centre of this circle, it *must* be subject to an unbalanced force capable of providing this acceleration, that this force is the resultant of t and w . . . and so on.

We have used the phrase 'it can be made clear', but of course this may or may not be altogether easy. This, however, is one of the critical 'moments of truth' when the difficulty (if indeed one exists) must be faced squarely and overcome before proceeding to the next obstacle.

Now consider the alternative presentation using d'Alembert's principle. If an *imaginary* 'centrifugal' force C.F. (fig. 0.5b) is added to the system the problem *can be treated as though* it involved only elementary statical equilibrium. True, and in this elementary case all the right numerical answers can be obtained, and probably at this elementary level full marks can be obtained in the examination question. But at what cost? At the least there is the danger that the pupil will rapidly forget that the force C.F. is purely imaginary and think of it as having a real existence. More probably, he will never give the word 'imaginary' any real thought, but go through life convinced that centrifugal force acts on the mass and with a clear mental picture of figures such as 0.5b to sustain him in his mistaken belief.

The individual to whom d'Alembert's artifice appeals most strongly is the very person least likely to be able to keep clear in his mind the distinction between the real forces acting and the imaginary forces which have been introduced to aid him; it is far better in the long run to use d'Alembert's method only, if at all, when all danger of self-confusion is past, and then as a *conscious* step in the solution.

§ 0.4 Scalar and Vector Quantities

A quantity is described as scalar if it has magnitude but is unassociated with direction, either intrinsically, as in the case of such entities as time, mass, and energy, or because it defines only the magnitude of a vector quantity: examples in the latter category are distance and speed. A quantity is described as vectorial if it has magnitude and direction, and obeys the vector laws of addition, which we shall discuss below. Linear displacement or infinitesimal angular displacement, and velocity, acceleration, force, or momentum, whether these quantities are linear or angular, are vector quantities; angular displacement of finite magnitude, and the moment of inertia of an irregular body, are examples of quantities which though having magnitudes associated with direction are not vectors.

Let us consider first the most elementary example, the difference between distance and displacement. Suppose that a man walks

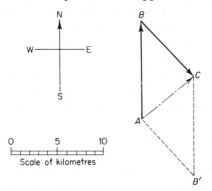

Fig. 0.6 Addition of vector quantities (such as displacement)

north for a distance of 10 km from *A* to *B*, and then south-east for a further distance of 8 km to *C* (fig. 0.6). He has travelled a total *distance* of 18 km: to this the state of his muscles will testify. His *displacement AC* from his starting-point will, however, be roughly 7 km. Clearly the student will be familiar with this situation, with the arithmetical addition of the scalar quantity (i.e. distance), and with the graphical addition of the vector quantity (i.e. displacement). The representation of the vector quantity, in this instance displacement, is by a line whose magnitude represents to some scale (which must be stated) the distance of the displacement and whose direction is the same as (or in this case represents according to a convention so frequently used by map-makers that it need hardly even be stated) the direction of the displacement. Moreover, the rule for addition of such vectors is so 'obvious' that it will probably be accepted without comment: the vector to be added is drawn with its 'tail' at the 'head' of the previous vector. *Exactly the same rule applies to the addition of vectors representing any other quantities.*

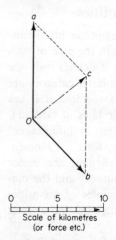

0   5   10

Scale of kilometres
(or force etc.)

Fig. 0.7 Addition of localised vector quantities (such as force)

An alternative method of adding vectors is illustrated in fig. 0.7. Here the first journey of 10 km northwards is represented by the vector *Oa*. The second journey of 8 km south-eastwards, stated in isolation, can be represented by a vector *Ob*, drawn from the same origin *O*.

The addition of these vectors to find the total displacement *Oc* of the traveller is carried out by completing the parallelogram *Oacb*. Clearly the difference between these methods is trivial. The former might be regarded as more 'natural' in the case of displacements, especially if they refer to the translation of a rigid body, and so are not 'localised' or associated with any particular point in the body. The latter might be regarded as more natural in the case of 'localised' vectors, such as two forces which could be represented in fig. 0.7 by *Oa* and *Ob*, acting simultaneously on a particle at *O*.

The final displacement of the traveller will be exactly the same whether he walks first north and then south-east, or first south-east and then north. This illustrates an important characteristic of vector quantities, that the order in which they are added does not affect the result. The dotted lines AB', $B'C$ in fig. 0.6, or the diagram, fig. 0.7, shows that our method of addition of vectors produces the same characteristic, which is satisfactory. But if entities do *not* satisfy this 'commutative law of addition', then their representation by vectors will give incorrect results and must be inadmissible: the entities cannot be called, or treated as, vector quantities. Consider, for instance, angular displacement. In fig. 0.8 is shown a sphere, centre O and radius r, in which a line OA to a

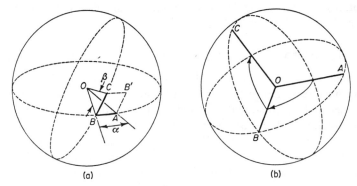

Fig. 0.8 Condition for which angular displacement is a vector quantity (infinitesimal magnitudes only)

point A in the surface is marked. Now suppose the sphere to rotate (fig. 0.8a) through *small* angles α and β successively to bring A to B and then to C. We have $\alpha = AB/r$, $\beta = BC/r$; and the total angular displacement of the sphere is given by the angle $AOC = AC/r$. If the angles are sufficiently small we may regard the lines AB, BC, AC as straight and co-planar, so the total angular displacement bears the same relationship to the component angular displacements as do the linear displacements of the point A. Moreover, the result is the same if the rotations occur in the reverse order, so that A moves to B' and then to C. Angular displacement of an (infinitesimally) small amount is therefore a vector quantity.

If, however, the rotations are large the arguments break down completely. For instance, rotation through 2π radians brings the body back into its original position in space: a vector to represent 2π must therefore have zero length, which is clearly absurd. Again in fig. 0.8b is sketched the situation in which OA turns through two successive right-angles to reach the positions OB, OC. Now, if the sphere were to turn through these angles in the reverse order the radius *OA* would not move during the first rotation, and during the second move only to *OB*. Clearly angular displacement of finite magnitude is not a vector quantity.

Angular velocity is the *rate of change* of angular displacement and is defined at $d\theta/dt$, when these symbols have their usual meaning. However large the velocity, it is defined in terms of an infinitesimally small angle, so we need have no anxieties: angular velocity is always a vector quantity, as are angular acceleration, angular momentum, rate of change of angular momentum, and angular force (or 'moment', or 'couple', or 'torque').

§ 0.5 Vectorial Representation of Angular Quantities

A vector to represent any quantity must represent it unambiguously. We have seen that in the case of a linear quantity a scale must be chosen and stated so that the length of the vector will represent the magnitude; and the direction of the vector can be either in the actual direction of the quantity or in a direction which, to a convention sufficiently well recognised (e.g. a map) or stated, will unambiguously represent it. In the case of an angular quantity, no further difficulty arises about magnitude: the scale can be equally easily stated. However, the direction is slightly less obvious, until it is appreciated that the *only* direction in which a vector can be drawn to indicate a plane of rotation is normal to it. So, for instance, in the case of the angular velocity of a flywheel we *have to* represent it by a vector drawn in a direction perpendicular to the plane of the flywheel. But this is insufficient, as the vector has also got to tell us the sense of the rotation. This is achieved by adopting *purely as a convention* the rule that the vector is drawn in the direction which a right-handed screw (or 'corkscrew') would travel if turned about its axis in the direction of the rotation.

§ 0.6 Newton's 'Axioms, or Laws of Motion'

The philosophical difference between an axiom and a law is profound: too profound to be discussed rationally in a book of this type. If, however, we fail to note that Newton described his postulates as 'axioms, or laws of motion' we do less than justice to his genius. We need not at this stage discuss their scope or limitations, but accept them as utterly satisfactory for our purposes.

Since Newton propounded these axioms in Latin, a certain freedom in translation is permissible. They can be expressed as follows:

1. UNLESS A BODY IS ACTED UPON BY A FORCE, IT MOVES WITH CONSTANT VELOCITY.
2. IF A BODY IS ACTED UPON BY A FORCE, ITS RATE OF CHANGE OF MOMENTUM IS PROPORTIONAL TO, AND IN THE DIRECTION OF, THE FORCE.
3. TO ANY ACTION, THERE IS AN EQUAL AND OPPOSITE REACTION.

And since they are the foundation on which the whole subject of dynamics is based, it is surely unnecessary to stress that they must be thoroughly considered, studied, learnt, and assimilated by every serious student, so that they become part of his conscious and subconscious thought processes.

Of these axioms, the second needs particular care, and special attention will be paid to it in §§ 0.8 and 0.9. For the moment we might note a corollary which can be simply derived from the second and third axioms. Let two bodies act on each other. Over any interval of time, the change in momentum of the first body will be proportional to the time integral of the force applied to it. But this force (or its time integral) is an 'action' to which there must be an equal and opposite 'reaction' on the second body. This second body will therefore suffer a change in momentum equal and opposite to that of the first. It follows that the total change in momentum of the two-body system is inevitably zero; and it is not difficult to see that this argument can be extended to any number of bodies acting on each other. We conclude that

THE MOMENTUM IN A 'CLOSED' SYSTEM IS CONSTANT

the word 'closed' being inserted, perhaps unnecessarily, as a warning not to apply this principle of 'conservation of momentum' to a system on which additional or external forces are acting.

Before passing on we should note that *these axioms, and the principle of conservation of momentum, are just as valid for angular as for linear motion.* For 'force', 'velocity', and 'momentum' we can write 'torque', 'angular velocity', and 'angular momentum' respectively.

§ 0.7 The 'Conservation Principles' – Energy and Momentum

Although every teacher of mechanics must make clear the vital differences between these principles, the student who learns about them at different times is liable to forget and to confuse them, if only because of the use of the word 'conservation' in two different senses. This can be disastrous. It is therefore worth while to spare a moment to compare and contrast them.

Energy, a scalar quantity, can be neither created nor destroyed. *But* it can take many different forms – kinetic, potential, strain, radiant, chemical, electrical, and thermal, for instance (quite apart from the form of mass which as everyone knows can under rather specialised conditions be transformed into energy) – and therefore 'disappear' in one form and turn up in another. Always it tends to be 'degraded': that is, the entropy in a system tends to increase. In certain circumstances – where friction and impact are absent or very small – it may be possible to have energy changing from the kinetic to the potential or strain forms and back again with negligible loss, but such cases are relatively rare, and before writing 'potential energy lost = kinetic energy gained' the student should be very careful indeed to make sure that the appropriate conditions are satisfied; that is, that no energy is being converted to one of the other forms.

We have in the preceding paragraph deduced from Newton's axioms that the momentum, whether linear or angular, in a closed system is constant. Momentum can therefore be neither created nor destroyed, *nor* can it, like energy, take different forms. We need, therefore, have no anxiety about stating that the momentum after any event is equal to the momentum before it – or indeed at any other time – *provided* that we have remembered two things: that we must be discussing a 'closed' system and that momentum is a vector quantity. If we fire a shell from a 'fixed' gun it acquires momentum as it passes through the barrel, and loses momentum again when it strikes the ground, but this is, of course, because some of the earth is included in the system, and as the part of the earth which should be included is incalculable, the principle of conservation of momentum is of no use to us in such an instance. On the other hand, if two billiard balls (or two equal lumps of putty) are, for instance, travelling towards each other with equal speed, we can see that they have equal and opposite momentum, and that the total momentum in the system is zero. When they collide, the billiard balls will rebound, roughly interchanging their speeds, so the total momentum is still zero, and the 'loss' of kinetic energy is comparatively small. The lumps of putty will probably coalesce into a single stationary lump, which makes it even easier to see that the total momentum in the system is still zero, but the 'loss' of kinetic energy, that is, its transformation into heat, is complete.

In many courses of instruction a large proportion of the time is spent on linear motion, and little or nothing is said about angular motion. Many students who are completely at home with the principle of conservation of linear momentum have never considered the parallel principle of conservation of angular momentum. To the mathematically inclined there seems little point in devoting special attention to this, since no novelty is involved. A few minutes spent on convincing the more practical types that the principle is equally valid are not misspent. A qualitative demonstration of the conservation of angular momentum can be obtained by the simple expedient of standing upright with the arms extended horizontally to the sides, spinning round at a reasonable speed, and suddenly bringing the arms in near to the axis of rotation. Ballet dancers and skaters are particularly fond of this simple but effective demonstration: the student may at first be less graceful, but he should nevertheless himself undergo the experience. If he requires an intensified experience, he can hold a kilogram weight in each hand – but this experiment should be performed in the open air or the experience may be a shattering one. He will never again doubt the principle.

§ 0.8 Entities: Fundamental and Derived

In dynamics we have to deal with many related entities: time, length, mass, force (or weight), velocity, acceleration, torque, energy, power, and so on. From purely theoretical considerations it can be shown that all these entities can be expressed in terms of *any three* which are mutually independent and could therefore be regarded as fundamental. Purely from practical considerations of convenience and simplicity there can be no argument but that two of these three must be *time* and *length*. The great majority of scientists choose, as the third, *mass*; some engineers prefer to choose *force* or *weight*. We must examine the reasons for this difference.

Early in life we acquire some appreciation of time, of length, and of force (particularly in the form of weight, or the gravitational attraction of the earth on matter); so early that we could almost regard this appreciation as instinctive. We all 'know' (until we are asked to define them in scientific language) what these words 'mean'. The vast majority of human beings go through life without giving conscious thought to the property of matter we call *mass*. It is hardly surprising that the 'practical man' has a preference for weight as a fundamental unit.

When we come to the scientific study of dynamics the position is very different. We certainly cannot go through life ignoring mass; the whole dynamical behaviour of the universe and of every particle in the universe depends on this property of matter, so one of our first tasks is to gain as clear an appreciation as possible of its meaning. The real difficulty here is that the property is too fundamental to be definable in terms of simpler concepts: to call mass 'inertia' or a 'quantity of matter' is to play unavailingly with words. What we must do is to think carefully about the way in which matter behaves in relation to motion – and to do so (unless we believe that we have a degree of genius comparable with Newton) in the light of his first two axioms. With these to help us, it is relatively easy to appreciate that matter has a property which causes it to respond proportionately to an applied force and that we can call this property 'mass'.

Once we have begun to think on these lines it becomes increasingly obvious that mass is an incomparably more fundamental property of matter than weight. Let us think of a particular lump of matter, for the sake of illustration say a rifle bullet. On the surface of the earth its weight will be roughly, but not exactly, constant, because the earth, which attracts it with this force, is roughly, but not exactly, a sphere with properties constant in any radial direction. If we were to take the bullet to the surface of the moon it would 'weigh' about one-sixth as much; if to outer space, effectively nothing. (Indeed, we need only take it in a satellite to find it 'weightless' in the sense that we should be totally unable to measure any such property as weight, because the bullet would be a part of a tiny 'universe', too small to exert measurable internal gravitational forces of its own. This whole 'universe' is in a condition of 'free fall', and therefore no particle in it has any tendency to fall faster than another and therefore to need support.) But the mass of the bullet would suffer no such changes. The same propellant charge would be required to give it its normal muzzle velocity: when it left the muzzle it would be just as lethal on the moon or in the satellite as on earth.

The pure scientist is concerned with fundamentals, and *weight* would therefore for him be an utterly inappropriate choice for the third basic entity. This does not rule *force* out of court, but it is not difficult to see why he prefers mass. We must remember that he must be able to establish a standard as invariable and/or reproducible as possible. To establish a standard of time he can use the motion of the earth among the stars, or the natural frequencies of vibrations in atoms. To establish a standard of length he can choose a length of a material which he expects to be stable. These standards may not be perfect, but they are very very good. To establish a standard of force he might indeed use the gravitational attraction on a given lump of matter in a given place on the earth's surface, but the difficulties of transferring this standard to another locality are considerable. The substitution of a spring and the measurement of its distortion are theoretically possible but open to grave objections in relation to inconstancy, irreproducibility, and inaccuracy. Why incur all these difficulties rather than choose the obvious candidate, *mass*? He need only select a lump of material which he expects to be stable, and say 'there is my unit of mass'. Again this may not be perfect, but it is in the same category as his standards of time and length. The standard is reproducible by *weighing* the reproduction against the standard, an operation which is fortunately easily carried out in a constant gravitational field.

Should the engineer be driven by the same logic to the same conclusion? This is a controversial matter, and it is seldom that in a controversy all the right is on one side. That the arguments so briefly and therefore inadequately outlined above are nearly as valid for the engineer as for the scientist is obvious: what are the differences, and what are the arguments on the other side?

The engineer must have close and intimate dealings with men who have little or no scientific training. He cannot expect them to understand what mass is, but he knows that they will have a thorough appreciation of force and weight. They cannot talk his language: he must talk theirs. If, for instance, he labels a lump of cast iron '1 kg' they will inevitably regard this as a standard of weight or force, and use it in this fashion. Again, the engineer is seldom if ever concerned with the order of accuracy which is so important to the pure scientist. If he manufactures in Great Britain a bridge which is to be erected in Rhodesia he will not worry about its collapsing because the gravitational field differs in strength in the two places.* As long as he is on the surface of the earth he can regard the gravitational field as constant, and the same lump of cast iron can be taken as a standard of force. For these and similar reasons, allied perhaps to the possibility that some have never given the opposing arguments a hearing, or perhaps rid themselves of the elementary confusion, many engineers continue to prefer to regard force, or its manifestation in the form of weight, as the third basic entity, and mass as the derived entity.

The deplorable practical consequences of this failure on the part of engineers to agree on the choice of the basic entities will be discussed in the next paragraph.

§ 0.9 Units for Dynamical Calculations

Fundamental entities, and the relationships between them, appear to exist independently of man. While we are working in general terms, therefore, and provided we do not violate the natural laws, we can discover and deduce such relationships without reference to any particular system of units. These relationships are usually described as 'physical', and we shall discuss them more fully in the next paragraph.

Engineers, however, cannot progress very far with their proper function of 'ordering the forces of nature for the use of man', for instance by designing and building structures and machines which will serve his purposes, without descending from such lofty generalised ideas to mundane calculations; and for this purpose it is essential to choose a system of units. These units would in an ideal world be universally agreed; unfortunately we have not yet reached such an agreement. We need not waste time in discussing the merits of the alternatives: in this book we shall as far as possible work in general terms independent of units, but when it comes to the point we shall adopt the metre–kilogram–second system.

But this statement is far from sufficient. The metre and the second give us no trouble: the former is derived from the supposed size of

* As a matter of interest, it may be noted that the value of g varies over the earth's surface from about 9·77 to 9·83 m/s^2.

the earth but it is now defined in terms of a number of wavelengths of a particular radiation of light; the latter from the mean sidereal year; and we are familiar with them and their multiples and sub-multiples. But the kilogram? We know that there is a particular lump of platinum in Sèvres, near Paris, that its mass is accepted by all scientists and many engineers as a basic unit of mass, and that its weight (strictly, at a place where the value of g is 9·80665 m/s^2) is accepted by many engineers as the basic unit of force. Which unit shall we choose? Can we accept both?

Let us go back to fundamentals. Newton's second law of motion, as applied to a body of constant mass, states that,

Force $\propto$ Mass $\times$ Acceleration

Now we cannot proceed to specific calculations until we replace the proportionality sign by something more definite. According to our choice of basic unit we can write

$$1 \text{ derived unit of force} = 1 \text{ kilogram mass} \times 1 \text{ m/s}^2 \quad (0.1)$$
$$(= 1 \text{ newton})$$

or $1 \text{ kilogram force} = 1 \text{ derived unit of mass} \times 1 \text{ m/s}^2 \quad (0.2)$
(for which there is no generally accepted name)

or $1 \text{ kilogram force} = \text{a derived } \textit{numerical constant}^*$
$$\times 1 \text{ kilogram mass} \times 1 \text{ m/s}^2 \quad (0.3)$$

Each of these equations leads to a usable system, but it does not follow that all are equally satisfactory. In a moment we shall consider them one by one, but let us first discuss briefly the meaning of the word 'coherent' as applied to a system of units, and its importance.

A system of units is coherent if the product or quotient of any two unit quantities in the system is the unit of the resultant quantity. Thus if we decide to adopt a coherent system and have chosen a metre as the unit of length we do not need to discuss the basic units of area or volume: they will be a square metre and a cubic metre respectively, and hectares and litres, though used in everyday life for convenience – after all to ask for $0·5 \times 10^{-3}$ m^3 of beer might cause the barman a certain degree of perplexity or annoyance – must be relegated to a strictly subsidiary role with no place in technical calculations. Going one step further, when we choose a second as the unit of time our basic unit of speed automatically becomes one

* $= g_0$, whose value is 9·80665 $\simeq$ 9·81.

completely

metre/second, of acceleration one metre/second^2, and so on. These derived basic units are not very complicated, so no very compelling reason exists to coin a simpler name for use in calculations, and none has been chosen ... but sailors who are concerned with speeds maintained for long periods of time will no doubt continue to discuss it in terms of knots, and airmen and designers of high-speed machinery will continue to use g as a unit of acceleration. And why not? The sensible engineer should appreciate the immense advantages *to him* of carrying out all his calculations in a coherent system of units with its avoidance of unnecessary conversion constants in every equation, and the convenience of non-standard units to other people, and be prepared to accept the trivial chore of converting his data from, and his final answer into, units which may be more convenient to, or readily understandable by, those to whom he is talking.

Now let us look back at the first system of units (0.1) we were discussing. Clearly it has been based on length, time and mass; moreover, it has been chosen to be coherent, so it will have all the relevant virtues and conveniences when used in calculations. The unit of force is inescapably one kg m/s^2 or (if one tried to pronounce it) one kilogram (mass) metre per second squared: logical, but quite a mouthful, even for scientists. So the sensible thing to do is to invent a new simple name for this unit, and this (internationally agreed) name is the 'newton'. It is important to realise that this typical procedure *is only* a matter of convenience. At any time we come across a force in the form of kg m/s^2 we can replace this complex by the single word 'newton', or the single-letter abbreviation N, and vice-versa.

What are the disadvantages of such a system? To the scientist isolated from contact with non-scientists, none. He uses it and cannot understand why there should be any argument about it. To the engineer, the disadvantage is obvious and apparently serious, and has already been elaborated; he is talking a language which the man in the street, with whom he must be able to converse frequently, does not and *cannot* understand. This is no exaggeration. It would of course be perfectly easy to cast little lumps of metal – they would have to have a mass of roughly 1/9·81 kg or 102 g – and label them '1 newton'. It might even be possible to explain that their weight was an approximate standard measure of force ... but then one would have to explain why they were labelled with the word newton while the almost identical lumps labelled 100 g, already used as units of *weight* in every grocer's shop, were not really units of weight at all, but units of mass. . . . The way out of this difficulty has already been discussed: it is simply to translate newtons into 'kilograms-force' when talking to the man in the street; but *only* when talking to the man in the street; and to translate back into newtons before attempting any calculations, however simple.

As has already been said, the other two equations 0.2 and 0.3 lead to usable systems. About this there can be no argument: they have been used ever since the metric system was invented, and similar systems with other basic units for very much longer. Their merit is that they use only terms which are familiar to, whether or not they are understood by, the general public; and above all that the vitally important and constantly-used unit for force is one which is both familiar to and understood by everyone. This merit should not be underestimated: it will probably suffice to prolong the use of the systems for many years. But as soon as a wider view of the situation is taken their grave disadvantages become apparent. The system 0.3 is not even coherent, so every argument and calculation is complicated by the intrusion of the numerical constant g_0 (which to make matters infinitely worse is often confused with the numerically-equal gravitational constant); moreover this system, using as it does both kilogram-mass and kilogram-force, offers the maximum opportunity for confusion between them. The system 0.2 is at least coherent so far as it goes, and in a sense avoids in simple dynamical calculations the gravest dangers of confusion by giving no name at all to the unit of mass. But in a wider context coherence breaks down, and a partial system built on the kilogram force comes into direct conflict with an all-embracing system based on the kilogram mass. This latter system (of which 0.1 is a small but vital part) is known as SI, the letters standing for *Système International (d'Unités)*. It is based on six primary units (the metre, kilogram, second, ampere, degree Kelvin, and candela) and has after long and careful consideration, and after many years of discussion, been accepted internationally. To the authors it no longer seems to make sense to perpetuate the difficulties inherent in the use of 0.2 or 0.3, and this book will be written in terms of the metre, second, kilogram-mass*, and equation 0.1.

---

* One distressing illogicality (at the time of writing) remains: the use of a complex *word*, kilogram, for the basic *unit* of mass. It would be immensely helpful if a

§ 0.10 The Manipulation of Units

The difficulties involved in the manipulation of units are minimised, if not eliminated, by the consistent use of a coherent system. We must, of course, be familiar with the names given to complex units, of which we have, in § 0.9, come across one example: 1 newton (N) $\equiv$ 1 kg m/s^2. Others we shall mention below. But apart from this, and *provided* that every time we write down a number for a quantity we append the (standard) units in which it is expressed,* we can apply the rules of arithmetic to the unit symbols as to the numbers, whereupon the units in which our answer has been obtained are automatically revealed. Thus, for instance, if we are told that a body of mass m kg has a constant acceleration a m/s^2 we know that the force F acting is

$$F = m \text{ kg} \times a \text{ m/s}^2$$
$$= m\, a \text{ kg m/s}^2$$

or, from the identity above, $= m\, a$ newtons

If this force acts through a distance S metres the work W done by it is: $W = F.S$ newton metres or joules since the name joule (J) has been accepted for this unit of energy or work; 1 Nm $\equiv$ 1 kg m^2/s^2 $\equiv$ 1 J.†

If this operation has occurred during a time of t seconds the rate of working, or power P is:

$$P = W/t \text{ Nm/s or J/s or kg m}^2/\text{s}^3 \text{ or watts}$$

again the name watt (W) having been accepted for this basic unit of

---

suitable simple name, for which many reasonable possibilities have been suggested, were internationally adopted as an alternative to kilogram, *not* to supplement it in everyday use when it would be misunderstood and misused, but in scientific use *only*.

* Perhaps a rather elementary but nevertheless important corollary may be added. The units must be expressed in reasonable algebraic form, otherwise it is unnecessarily difficult to carry out the arithmetical processes on them. Thus, for instance, a volume might be expressed as m^3 but *not* cu m; a pressure as N/m^2 or N m^{-2} but *not* newtons per sq m; an acceleration as m/s^2 or m s^{-2} but *not* m per s per s *nor* the equally ambiguous m/s/s. In particular, angular quantities should *always* be expressed in radians, since this measure is assumed in all theoretical work: to use any other measures, such as revolutions or degrees, introduces awkward numerical constants, and the likelihood of avoidable errors.

† A word of caution: the complex kg m^2/s^2 or Nm is not always a measure of work; it can equally be a measure of the moment of a force about a point.

power.* These three coined words for complex *basic* units (N, J, W) are the only ones of importance to mechanical engineers; others (coulomb, volt, farad, ohm, and the like) apply mainly to electrical affairs. But there are several cases in which the magnitude of the basic unit is very awkward. Thus, for instance, the basic unit of pressure, N/m^2, has a value which is most easily appreciated when it is expressed as about one hundred-thousandth of atmospheric pressure: if one is interested in high-vacuum technology this is convenient enough, but seldom otherwise. So a name whose derivation is obvious enough, bar, has been coined for 10^5 N/m^2 rather than one for 1 N/m^2. High pressure technologists can then discuss their work in terms of, say, kilobars, and those interested in the strength of materials in terms of, say, hectobars without the inconvenience of using 'astronomical' numbers. And, of course, hundreds of words for non-basic units will rightly remain in common use for the foreseeable future – litres and hectares and hours and years and degrees and revolutions and light-years and parsecs and so on almost *ad infinitum*.

Now when we come to the manipulation of units we must similarly preserve our common sense. There are many occasions when we can carry out calculations in non-standard units without the slightest danger of introducing errors. It would therefore be stupid to contend that all quantities should be converted into standard units before calculations are carried out: all that is necessary is that the units associated with each number should in every case accompany that number and be dealt with as outlined above. But where the danger exists (and this is particularly true of dynamical calculations) conversion of every quantity into standard units is without doubt the safest procedure.

As a special case which is of no very great importance but which illustrates a safe method of dealing with non-standard units, let us suppose that we are running a trial on an engine and wish to determine the shaft power developed for a series of conditions. Power P is in this case the product of torque T and angular speed ω, so the physical equation is simply:

$$P = T\omega$$

and if the dynamometer has an arm L calibrated in metres, the load

---

* To those accustomed to the requirement for conversion factors from units of 'mechanical' power to units of 'electrical' power it may come as a surprise – and a relief – to find that these units, joule and watt, are identical in both branches.

W is measured on a spring balance calibrated in newtons, and our speed-measuring device is calibrated in rad/s, $T = WL$ and

$$P = WL\omega \qquad \text{Nm/s or watts}$$

But suppose the load W_1 is supplied as pound weights, the arm is calibrated in feet, L_1, and the speed-measuring device gives us rev/min, N_1. We can convert each quantity as it is measured into standard units, but if we have a large number of readings this procedure is somewhat wasteful, and it may be more economical to proceed as follows:

$$\text{Power} = W_1 \text{ lbf} \times L_1 \text{ ft} \times N_1 \text{ rev/min}$$

Now to get rid of the unwanted units we use the (approximate) identities

$$1 \text{ lbf} \equiv 0\cdot453 \text{ kgf} \equiv 0\cdot453 \times 9\cdot81 \text{ N} \equiv 4\cdot448 \text{ N}$$
$$1 \text{ ft} \equiv 0\cdot3048 \text{ m}$$
$$1 \text{ rev/min} \equiv 2\pi/60 \text{ rad/s} \equiv 0\cdot1047 \text{ rad/s}$$

and the technique of multiplying the answer above and below by identical quantities, which of course does not alter its value. Thus,

$$\text{Power} = W_1 L_1 N_1 \frac{\text{lbf ft rev}}{\text{min}} \times \frac{4\cdot448 \text{ N}}{1 \text{ lbf}}$$
$$\times \frac{0\cdot3048 \text{ m}}{1 \text{ ft}} \times \frac{0\cdot1047 \text{ rad min}}{1 \quad \text{s} \quad \text{rev}}$$

and after doing the arithmetic and cancelling units we obtain

$$\text{Power} = 0\cdot143 \ W_1 L_1 N_1 \text{ Nm/s or watts}$$

This type of equation, which is correct only if the quantities are measured in the units specified, is called a 'numerical' equation, in distinction from the 'physical' equation $P = T\omega$ which is not expressed in terms of specified units.

§ 0.11 Motion in a Circular Path

Let a particle P (fig. 0.9) move round a circle, centre O and radius r, with *uniform* linear speed $v = \omega r$, where ω is the angular velocity (needless to say in *radians* per unit time) of the radius vector OP. We wish to find its acceleration. At some instant in time let it be at P_1, and a short time δt later at P_2, where angle $P_1 O P_2 = \omega.\delta t$. The direction of motion is at all times perpendicular to the radius: we can, therefore, represent the velocity at the two instants by the two vectors op_1 and op_2, each of length v, respectively perpendicular to OP_1 and OP_2, and therefore at an angle $\omega.\delta t$ to each other. The change in velocity in time δt is found by subtracting the vectors to be p_1p_2. Since δt is small, the chord $p_1p_2 \simeq$ the arc $p_1p_2 = v(\omega.\delta t)$. The magnitude of the acceleration is the rate of change of velocity or $v.\omega.\delta t \div \delta t$ or $v.\omega$, which may be written as $\omega^2 r$ or v^2/r; and its direction is towards the centre of the circle. This acceleration is given the name 'centripetal', which means 'centre-seeking'.

If the particle is not moving with uniform speed it is 'obvious' that it will have in addition an acceleration in the direction of its path; this is called a 'tangential' acceleration, and the total acceleration of the particle will be the vectorial sum of the centripetal and

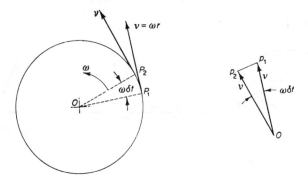

Fig. 0.9 Uniform motion in a circle

tangential components. (We cannot determine the magnitude of the tangential acceleration from the facts so far provided.) This deduction is not only 'obvious' but also true, which is more fortunate than might appear, since we might have made an equally 'obvious' deduction about what would happen if the curvature of the path were not circular, and this would have been quite untrue (see § 1.8). This may serve as a warning in such matters: for the moment it may encourage us to deal with the present problem by another method, which is very helpful when we come to a more complicated case.

Again (fig. 0.10) let the particle P be moving round the circle centre O and radius r, with an angular speed ω, which need *not* be uniform. At some instant let P be in the position shown relative to the co-ordinate axes xOy. We can write down the co-ordinates of P,

$$x = r \cos \theta \qquad\qquad y = r \sin \theta$$

If we now differentiate with respect to time, since r is constant, we obtain,

$$\dot{x} = -r \sin \theta . \dot{\theta} \qquad\qquad \dot{y} = r \cos \theta . \dot{\theta}*$$

and since $\dot{\theta} = \omega$,

$$\dot{x} = -\omega r \sin \theta \qquad\qquad \dot{y} = \omega r \cos \theta$$

And by differentiating again (remembering that both ω and θ are variable),

$$\ddot{x} = -r(\omega \cos \theta . \dot{\theta} + \dot{\omega} \sin \theta)$$
$$= -\omega^2 r \cos \theta - \dot{\omega} r \sin \theta \qquad\qquad (0.4)$$

$$\ddot{y} = r(-\omega \sin \theta . \dot{\theta} + \dot{\omega} \cos \theta)$$
$$= -\omega^2 r \sin \theta + \dot{\omega} r \cos \theta \qquad\qquad (0.5)$$

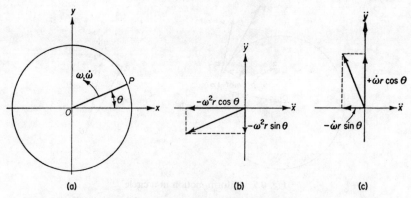

Fig. 0.10 Non-uniform motion in a circle

We can now obtain the physical interpretation of this result in either of two ways. If we look at the first term in each of these expressions we can see that it is a component acceleration, and that the two together add up to give a total acceleration $\omega^2 r$ in the direction opposite to OP (see fig. 0.10b); this is the centripetal acceleration as before. Similarly, the second terms add up to give an acceleration $\dot{\omega} r$ in a direction perpendicular to OP: if $\dot{\omega}$ is positive the acceleration is in the direction indicated by the arrow in fig. 0.10c.

* Note the use of the Newtonian 'fluxional' notation of a dot above a letter to indicate differentiation with respect to time; e.g. $dx/dt = \dot{x}$, $d^2x/dt^2 = \ddot{x}$, and so on.

Alternatively, we can say that if an *x*-axis were chosen to coincide with the direction *OP* at the instant considered, the value of θ would be zero and we should, by writing in equations (0.4) and (0.5) 1 for cos θ and 0 for sin θ, obtain immediately the component accelerations along and perpendicular to *OP* as $\ddot{x} = -\omega^2 r$ and $\ddot{y}$ as $\dot{\omega}r$ respectively. This is a slightly more sophisticated approach.

§ 0.12 The Simple Engine Governor

Engines are frequently required to run at an approximately constant speed irrespective of the power they are being asked to produce.

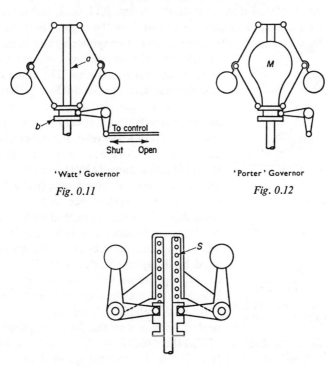

'Watt' Governor

Fig. 0.11

'Porter' Governor

Fig. 0.12

'Hartnell' Governor

Fig. 0.13

The control could be a valve in the case of a steam engine, a throttle for a petrol engine, a fuel-pump setting for an oil engine, and so on: the device to set the control at the appropriate level is called a governor. Governors can take many forms, but several of the most common types depend essentially on the characteristics of motion in a circle (and often consist of an adaptation of the simple conical pendulum), which we have just been discussing. And so, although the major problems in effective governing of an engine are much more complex than will immediately appear, and will be discussed in chapter 6, and although the minor problems are usually discussed as part of the subject of 'machines', it is by no means inappropriate to consider these minor problems here as an illustration of basic mechanics.

We may take three basic types as representative of many others. They are sketched diagrammatically in figs. 0.11, 0.12, and 0.13. In each case a vertical spindle *a* is driven from the engine shaft. In the first, fig. 0.11, this spindle supports a pair (for symmetry and balance)

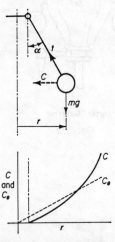

Fig. 0.14 Analysis of behaviour of Watt governor

of balls on rigid pivoted links, which act exactly as conical pendulums. The faster the spindle rotates, the farther do these balls tend to move outwards, and the simple linkage shown converts this outward motion into displacement of a 'sleeve' *b*, and hence of the control member. In the second case the sleeve is loaded gravitationally by the weight of *M*, whose effect we shall investigate. In the third case the direct resemblance to the conical pendulum has gone, but instead we have the rotating balls supported on bell-crank levers, whose other ends are subjected to a force supplied by the spring *S*: clearly the behaviour of this type of governor will be similar in principle, but may be very different in detail.

In this preliminary treatment we shall neglect friction and the forces required to operate the control, as well as the conditions which arise when the engine speed *changes*, and shall consider only the speed at which the governor will be *stable* in a given configuration.

In the 'Watt' governor, as we have seen, each ball will behave as a conical pendulum. In fig. 0.14 let the *mass* be *m*, and the radius of rotation at any instant *r*. Then the *only* forces acting on *m* are its weight *mg* and the tension *t* in the supporting link. This position can be one of equilibrium only if these forces are supplying to the ball a horizontal force C called the *controlling force* sufficient to give m its centripetal acceleration $\omega^2 r$, where ω is the speed of rotation in radians per unit time.

For equilibrium, then, vertically

$$t \cos \alpha = mg$$

and horizontally

$$C = t \sin \alpha = m \cdot \omega^2 r$$

Hence it follows that $\omega = \sqrt{\left(\dfrac{g \tan \alpha}{r}\right)}$, so we have found *an* expression for the equilibrium speed, though not in a very convenient form, since r and α are interdependent. It is, however, much more informative to work in terms of the parametric equations:

From geometry the controlling force

$$C = mg \tan \alpha$$

and, defining C_e as the force required to give the ball its centripetal acceleration at a speed ω_1,

$$C_e = m \omega_1^2 r$$

For equilibrium at speed ω_1

$$C_e = C$$

If we now plot the curve of C against a base of r we obtain a result such as that shown by the full line in fig. 0.14 for the controlling force which the device will supply to the ball in any configuration. For any given speed ω_1 we can now plot the force C_e required by the ball for equilibrium, which will, of course, be a straight line passing through the origin. The crossing point gives the equilibrium position. But the diagram tells us much more than this. We can see at a glance the excess force $(C - C_e)$ available to alter the configuration if this does not correspond to the correct speed ω_1. Even more important, we can deduce with fair ease how the governor will respond to different speeds, since the curve C is invariant with speed, and each C_e curve will be a straight line passing through the origin, the gradient varying as ω^2. Inspection shows that there will be one, and only one, equilibrium position, and that r will increase monotonically* as the speed rises (rapidly at first and then more and more slowly). From this elementary viewpoint we can therefore say that this governor is inherently 'stable'.

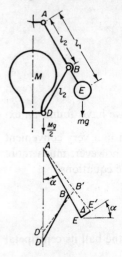

Simple calculation, however, shows that it is really suitable only for very low spindle speeds: for instance, if $r = 150$ mm and $\alpha = 45°$, $\omega = \sqrt{(9\cdot81/0\cdot15)} \simeq 8$ rad/s or roughly 80 rev/min. The Porter governor (fig. 0.12) represents a modification which alters this characteristic. If we assume, with the notation shown in fig. 0.15, that $AE = l_1$, $AB = l_2 = BD$, that there are two balls each of mass m (and weight mg), and a sleeve of mass M (and weight Mg), we can show that

$$C = \left(m + \frac{l_2}{l_1} M \right) g \tan \alpha.$$

The easiest proof is probably provided by applying the 'principle of virtual work'.† Let the linkage be displaced by a small amount from the configuration $ABED$ to the configuration $AB'E'D'$, EE' having, say, the value

Fig. 0.15 Analysis of behaviour of Porter governor

Δ. The horizontal motion of m will be $\Delta \cos \alpha$, the vertical motion $\Delta \sin \alpha$, the vertical motion of B $\Delta \dfrac{l_2}{l_1} \sin \alpha$ and of D (since ABD remains an isosceles triangle) $2\Delta \dfrac{l_2}{l_1} \sin \alpha$. Equating the 'virtual work' done against C to that done against mg and $Mg/2$, we have

$$C \cdot \Delta \cos \alpha = mg \, \Delta \sin \alpha + \frac{Mg}{2} \cdot 2\Delta \cdot \frac{l_2}{l_1} \cdot \sin \alpha$$

whence $C = \left(m + \dfrac{l_2}{l_1} M \right) g \tan \alpha$

---

* A useful word in mathematics, meaning that there is no change of sign in the gradient of the ω–r curve.

† *Principle of virtual work.* From the principle of conservation of energy it follows immediately that if a frictionless system is in equilibrium the net work done by external forces in altering (slightly) its configuration is zero. The extension of this principle to say that the work done by a system of forces is equal to the work done by their resultant offers no difficulty. The principle is frequently useful in problems such as this. The student might care to tackle it by 'statics': he will find the solution relatively tedious.

It follows immediately that the characteristic curve of C for a Porter governor will be identical in shape to that for a Watt governor: only the scale will be different. The C_e curves are unaffected, so the behaviour of the governor will be similar to that of the Watt, but the equilibrium speeds will be higher, as will be the forces available to alter the control.

A spring-controlled governor has significantly different characteristics which we can very quickly appreciate. If, in the example shown in fig. 0.13, we neglect the *weight* of m (which will have a small moment about the pivot when the arm carrying it is not vertical) and assume two balls and the dimensions shown in fig. 0.16a, we can by taking moments about the pivot write down the value of the controlling force C as $\frac{S}{2} \cdot \frac{h}{v} = \text{const} \times S$.

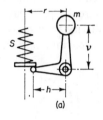

(a)

S, the force exerted by the spring on the sleeve, is, of course, not in itself a constant, but will be proportional to the compression of the spring, which will vary with the radius at which the balls are situated. In principle, then, the curve of controlling force will be a straight line as shown in fig. 0.16b, the height and gradient depending on the particulars of the design. As before, the C_e curves of force required on each ball for equilibrium will be straight lines passing through the origin. In the case shown, in which the C curve produced

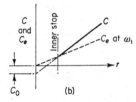

(b)

Fig. 0.16 Analysis of behaviour of Hartnell governor

backwards to the C-axis gives a negative intercept C_0, the governor will be inoperative, the balls (or sleeve) resting against the 'inner stop', until the speed rises to some value ω_1 which gives equilibrium at this configuration: as the speed rises so will the equilibrium configuration change monotonically until the outer stop is reached. Clearly the characteristics of the governor are under the designer's control. The smaller the numerical value of C_0, the smaller will be the change of speed required to give full movement to the controlling member – *but* the smaller, too, will be the force available to move it. In the limit, if $C_0 = 0$ the governor would 'want' to move from inner to outer stop at a critical speed, but could exert *no* force on the control member until the speed had risen above the critical value, when the full movement would take place and give rise to disastrous 'instability'. Such an (impractical) governor is described as 'isochronous'. If C_0 is positive the position is even worse. For reasons which we shall discuss in chapter 6, engine–governor combinations in practice may well behave in an unstable fashion, even though the elementary considerations we have been discussing would predict stability for the governor as an isolated unit.

§ 0.13 The Simple Gyroscope

To the average student tackling for the first time the subject we have treated in § 0.11, the fact that a particle moving with constant *speed* in a circular path has a centripetal *acceleration* comes as a surprise: familiarity soon makes him feel that the result is natural and right. The behaviour of a simple gyroscope is similarly at first surprising,

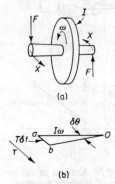

(a)

(b)

Fig. 0.17 Behaviour of simple gyroscope

but essentially no more difficult to understand: it is in effect the rotational analogue of the same problem, and may therefore be treated here, though in the first instance with a slightly different method of attack.

Suppose (fig. 0.17) that we have a stationary flywheel, of moment of inertia I, on a shaft which is so mounted (on 'gimbal bearings', see fig. 4.27) that it is supported but free to rotate in any sense, and that a couple represented by the two vertical forces F is applied to the shaft. Under these conditions the whole assembly shown will, in accordance with Newton's second law of motion, accelerate in the sense of the applied couple, the shaft 'toppling' end over end with ever-increasing speed.

Now let the flywheel be spinning about its axis with angular speed ω when the couple is applied. Under these conditions the behaviour of the assembly is utterly different. It rotates with constant angular speed in a horizontal plane at right angles to the forces F as indicated by the arrows XX. Why? Have we found an exception to Newton's second law?

Of course the answer to the last question is 'no'. Indeed, we can, by applying this law with due care, explain quite easily why the flywheel behaves as it does.

Let us restate the law (§ 0.6) as it applies to rotational motion, '. . . the rate of change of angular momentum is proportional to, and in the direction of, the applied couple'. When the flywheel is stationary the initial angular momentum is zero, so no special care is needed in interpreting the law. But when the flywheel is rotating we start with a quantity of angular momentum, to which the change must be *added*, and momentum is a vector quantity (§ 0.4), so the addition must be in accordance with the rules.

Initially, then, the angular momentum of the system (i.e. the flywheel) is $I\omega$, and must be represented (§ 0.5) by a vector perpendicular to the plane of rotation, i.e. parallel to the axis of the shaft, and in the direction in which a right-handed screw would travel. This vector is shown as *oa* in fig. 0.17b.

The applied couple – let us call its magnitude T – *must* be similarly represented by a vector perpendicular to the plane of the forces F, F: this vector is shown as T in fig. 0.17b, and it is at right angles to the vector *oa*.

If the couple T acts for a short time δt we can write down,

Change of angular momentum, in direction of $T = T\delta t$

So we can add this change to *oa* (§ 0.4) and find the angular momentum *ob* after δt. The result (of course an approximation which becomes exact as $\delta t \to 0$) is clearly to alter the *direction* of the angular momentum vector by an amount $\delta\theta$ (say) without altering its magnitude: and the physical interpretation of this is equally clearly that the axis of the shaft rotates with the vector but that the flywheel speed is unaltered. From simple geometry (again taking the approximation which becomes exact as $\delta t \to 0$)

$$T\,\delta t \simeq I\omega \,.\, \delta\theta$$

or in the limit $\qquad T = I\omega\dot\theta$

The angular motion θ is called 'precession'. Strictly speaking, this precessional motion implies another angular momentum term whose vector is in the direction of the forces F, but as long as the imposed conditions are constant, so also is this angular momentum, and we need not concern ourselves with it. This is not the appropriate stage at which to consider the problems which arise when the imposed conditions are transient.

We could, of course, equally easily have dealt with the converse case: that is, we could have determined the torque necessary to keep the axis of spin horizontal when the gyroscope precessed at an angular speed $\dot{\theta}$. The answer would be the same, but the method less illuminating. This viewpoint, however, is completely analogous to the linear momentum problem of a particle being constrained to rotate in a circle, where it is necessary to apply a centripetal force.

If we replace the flywheel by a single point mass at radius r we have a situation which requires rather more careful consideration and which it would be unreasonable to deal with in a chapter described as revisional. This problem is, however, instructive, and in the *Bulletin of Mechanical Engineering Education* for January 1955 it will be found treated both from this point of view and from the point of view of the forces required to give the mass its acceleration.

§ 0.14 Simple Harmonic Motion

As we shall see in later chapters, any elastic structure which is deformed and released tends to vibrate about a mean position, and, if the frequencies of vibration are in the audible range, to give out musical tones. This type of motion is therefore described as harmonic, and in its simplest form, when only one frequency occurs, as simple harmonic motion.

From the engineer's point of view there are advantages in adopting in the first instance a physical approach to the consideration of this type of motion. We can define simple harmonic motion by saying that 'if a point moves round a circle with uniform speed, its projection on any diameter of the circle moves with simple harmonic

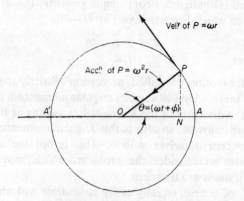

Fig. 0.18 Simple harmonic motion

motion'. In fig. 0.18 P is the point moving round the circle, centre O and radius r, with uniform speed $v = \omega r$, and AA' the diameter (taken for convenience in the direction of the x-axis) along which its projection N moves. The angle AOP, of course, varies with time, and if we measure time from some arbitrary datum can be written down in its most general form as $(\omega t + \phi)$. In § 0.11 we have discussed motion in a circle, so without difficulty we can write down as in col. 2 of the table below expressions for the displacement, velocity, and acceleration of P.

	POINT	
PROPERTY	P	N
Displacement from O	r along OP	$r \cos(\omega t + \phi)$
Velocity	ωr perpendicular to OP	$-\omega r \sin(\omega t + \phi)$
Acceleration	$\omega^2 r$ along PO	$-\omega^2 r \cos(\omega t + \phi)$

We now wish, in col. 3, to write down similar expressions for the displacement, velocity, and acceleration of N, and this we can do (since N has the same horizontal motion as P but no vertical motion) by taking the horizontal components of the corresponding terms. Alternatively, of course, having written down the displacement of N as $r \cos(\omega t + \phi)$, we can differentiate this twice with respect to time to obtain the velocity and acceleration of N.

By inspection we see that the acceleration of N is a negative constant $(-\omega^2)$ times its displacement. *Whenever we can show that the acceleration of a particle is a negative constant, say $-k$, times its displacement we can immediately state that its motion is simple harmonic, and that $\omega = \sqrt{k}$, and we can picture ω as the angular speed of the radius vector OP, whether the point P has a real existence or not.*

Also from inspection of fig. 0.18 we see that N will execute one complete vibration while P rotates once, so we can write down the 'period', or time for one complete vibration, as $2\pi/\omega$, and the 'frequency' or number of complete vibrations in unit time, as $\omega/2\pi$. These are values which we may want to quote for practical purposes, but at all intermediate stages of manipulation and calculation (and two whole chapters of this book are taken up in discussing vibrations) the introduction of the constant 2π is simply an inconvenience. It is far easier to work in terms of ω than of period or frequency, and the only hindrance to doing so is that there is no commonly accepted name for it. *In this book we shall use the term* 'RADIANCY' * *for this quantity* ω.

We have so far confined our discussion to the rectilinear motion of a particle, but we have no need so to restrict the meaning of the variable which we have called displacement. Any quantity u which varies so that

$$\ddot{u} = -k \cdot u$$

where k is a positive, and $-k$ therefore a negative, constant, is said to vary in a simple harmonic manner; and either by solving this differential equation or by analogy we can write down

$$u = U \cos (\omega t + \phi)$$

where U, the maximum value of u, corresponds to the radius r, ω is, of course, the radiancy of the motion, and ϕ is a constant depending on the instant which we choose to take as zero time. U and ϕ will normally have to be found from our starting conditions: mathematically speaking, they are arbitrary constants.

The visualisation of a simple harmonic quantity by means of a rotating vector, and its representation by the projection of the vector, frequently give very easy solutions to problems which would otherwise be difficult. As a simple example, let us suppose we want to add two harmonic quantities having the same frequency, say $U_1 \cos (\omega t + \phi_1)$ and $U_2 \cos (\omega t + \phi_2)$. All we need do is to draw, say for the instant $t = 0$, two radius vectors as in fig. 0.19, U_1 at angle ϕ_1 and U_2 at angle ϕ_2, and sum these vectors by the parallelogram (or triangle) rule to give U. Clearly the projections of U_1 and U_2 on any axis OA drawn at $-\omega t$ give the components, and the projection of U on this same axis their sum. Note that we *cannot* usefully sum two harmonic quantities of differing frequencies in this way.

---

* An angular speed measured in radians per unit time, i.e. normally in rad/s, is a quantity which inevitably recurs in all calculations involving harmonic motion No word for this quantity has gained general acceptance. The word 'pulsatance' was coined, but has dropped entirely out of favour. Sometimes the word 'frequency' is used, but this should be avoided at all cost, since frequency should properly be reserved for $\omega/2\pi$. Electrical engineers commonly use the term 'angular frequency', and this has been accepted by the British Standards Institution. Others use the term 'circular frequency', but both these terms are clumsy and might be regarded as misleading. The authors suggest and hope that the word 'radiancy' (as a portmanteau word incorporating radians and frequency) might provide an acceptable term.

When we use a rotating vector to represent a harmonically varying quantity we are interested only in its 'horizontal' projection, and no physical meaning need be attached to its 'vertical' component. In order to facilitate the arithmetical and algebraical manipulation of such vectors we may usefully employ the concepts of complex numbers (which are made up of real and imaginary parts, e.g.

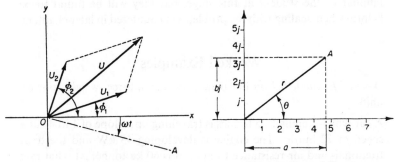

Fig. 0.19 Vectorial addition of harmonic quantities

Fig. 0.20 Argand diagram

$a + jb$, where $j = \sqrt{-1}$), and the Argand diagram (in which the real part is represented by the appropriate length, a units, measured along the x-axis, and the imaginary part by the corresponding length, b units, measured along the y-axis). Thus, the vector OA in fig. 0.20 can be defined as in the preceding paragraphs by its length r and its inclination θ to the x-axis; equally well it can be expressed in the complex number form, $a + jb$, where $a = r \cos \theta$ and $b = r \sin \theta$, that is, as

$$r(\cos \theta + j \sin \theta)$$

Again, the Maclaurin series for e^x, where $x = j\theta$ is

$$e^{j\theta} = 1 + j\theta + \frac{j^2\theta^2}{2!} + \frac{j^3\theta^3}{3!} + \frac{j^4\theta^4}{4!} + \frac{j^5\theta^5}{5!} + \cdots$$

$$= 1 + j\theta - \frac{\theta^2}{2!} - \frac{j\theta^3}{3!} + \frac{\theta^4}{4!} + \frac{j\theta^5}{5!} - \cdots$$

and the exponential series for $\cos \theta$ and $j \sin \theta$ are respectively

$$\cos \theta = 1 - \frac{\theta^2}{2!} + \frac{\theta^4}{4!} - \cdots$$

$$j \sin \theta = j\theta - j\frac{\theta^3}{3!} + j\frac{\theta^5}{5!} - \cdots$$

from which we see that

$$e^{j\theta} = \cos\theta + j\sin\theta$$

Thus, the vector can be expressed in the form

$$r \cdot e^{j\theta}$$

These mathematical techniques for dealing with vectors may not be familiar to the student at this stage, but they will be found to be helpful when dealing with the problems encountered in later chapters.

Worked Examples

Worked Example 0.1 on linear motion: mass, force, weight, power, units

A car which weighs 1·25 tonnes is travelling up a slope of 1 in 40 at a speed of 72 km/hr. The engine is developing 45 kW and the total frictional- and air-resistance to motion is 60 kg/tonne. At what rate is the car accelerating?

SOLUTION

Clearly this question has been posed by someone who is not using SI units. The car is said to 'weigh 1·25 tonnes', which is, of course, non-sense unless by tonnes he means tonnes force or tf. We shall need to know *both* the mass (to deal with the acceleration) *and* the weight (to deal with the gradient) so let us start by translating into our agreed language:

mass of car = 1·25 t = = 1 250 kg

weight of car = 1·25 tf = 1 250 kgf = 1 250 × 9·81 = 12 260 N

Similarly we note that the resistance to motion, stated as 60 kg/tonne, must mean 60 kgf/tf (or /t: it does not matter which, but the former is more logical physically). Anyway the resisting force R_1 has a magnitude of:

$$R_1 = 1\cdot25 \times 60 \text{ kgf} = 1\cdot25 \times 60 \times 9\cdot81 \text{ N} = 736 \text{ N}$$

The retarding force R_2 due to the gradient is:

$$R_2 = 12\ 260/40 = 307 \text{ N}$$

From the figures for the power (45 kW) being developed by the engine and the speed of 72 km/hr or 20 m/s we know that the total forward force F_1 is:

$$F_1 = 45\,000 \text{ W} \div 20 \text{ m/s}$$
$$= 45\,000 \text{ Nm/s} \div 20 \text{ m/s} = 2\,250 \text{ N}$$

so the net accelerating force F_2 is:

$$F_2 = F_1 - R_1 - R_2$$
$$= 2\,250 - 736 - 307 \text{ N}$$
$$= 1\,207 \text{ N}$$

and the acceleration is

$$1\,207 \text{ N} \div 1\,250 \text{ kg} = 1\,207 \text{ kg m/s}^2 \div 1\,250 \text{ kg}$$
$$= 0\cdot965 \text{ m/s}^2$$

Worked Example 0.2 on units, manipulation of units, motion in a circle

In a machine built to test the effect of acceleration on human physiology a man whose weight is 80 kg sits on a chair at the end of an arm 4 metres long which rotates 20 times a minute in a horizontal plane.
Calculate:

(1) his angular speed and momentum
(2) his linear speed and momentum
(3) his kinetic energy
(4) the total force exerted on his body by the chair

SOLUTION
Here again we note that one of the data is expressed in the usual 'popular' form which does not conform to approved technical usage. The man's mass m is 80 kg: his weight W is 80 kgf or 785 N.

Angular speed $= \omega = \dfrac{2\pi}{60} \times 20 = 2\cdot095$ rad/s

Moment of inertia $= I = \Sigma mr^2 = mr^2$
$$= 80 \times 16 = 1\,280 \text{ kg m}^2$$

Angular momentum $= I\omega = 1\,280$ kg m^2 $\times$ 2·095 rad/s
$$= 2\,680 \text{ kg m}^2/\text{s (or Nm s)}$$

Linear speed $= v = \omega r = 2\cdot095 \times 4 = 8\cdot38$ m/s

Linear momentum $= mv = 80$ kg $\times$ 8·38 m/s $= 670$ kg m/s (or Ns)

Kinetic energy (working from 'angular' information) $= \frac{1}{2}I\omega^2$
$$= \frac{1}{2} \times 1\ 280 \text{ kg m}^2 \times 2·095^2 \text{ rad}^2/\text{s}^2$$
$$= 2\ 810 \text{ kg m}^2/\text{s}^2 \text{ or Nm}$$

(*or*, working from 'linear' information) $= \frac{1}{2}mv^2$

$$= \frac{1}{2} \times 80 \text{ kg} \times 8·38^2 \text{ m}^2/\text{s}^2$$
$$= 2\ 810 \text{ kg m}^2/\text{s}^2 \text{ or Nm}$$

Centripetal acceleration $= -\omega^2 r = -2·095^2 \text{ rad}^2/\text{s}^2 \times 4 \text{ m}$
$$= -17·5 \text{ m/s}^2 \text{ (i.e. inwards)}$$

Horizontal force on man $= 80$ kg $\times$ 17·5 m/s^2
$$= 1\ 400 \text{ kg m/s}^2 \text{ or N (inwards)}$$

Vertical force on man $= 785$ N (upwards)

Resultant force on man $= \sqrt{[(1\ 400)^2 + (785)^2]} \text{ N}$
$$= 1\ 600 \text{ N inwards and upwards at}$$
$$\tan^{-1} 785/1\ 400 \text{ to horizontal.}$$

NOTE:
If we are giving information to a carpenter who has to build a suitable chair, he might like to know that the force it has to withstand is about 160 kgf ($\simeq 1\ 600/9·81$).

Worked Example 0.3 on the 'conservation' principles

A pile weighing 1·75 tf is being driven into the ground by blows from a tup of weight 220 kgf, which falls freely a distance of 2 m on to the pile without rebounding. The pile is driven in 10 cm by each blow. If the resistance of the ground is uniform find its value.

SOLUTION
This question has three distinct parts, corresponding to three phases in the occurrence.

(1) *The tup falls freely*
Here we may safely assume that no energy is lost but that all the potential energy is converted into kinetic energy, so:

$$Wh = \frac{1}{2}mv^2$$

$\therefore$ $(220 \times 9\cdot81)$ N $\times$ 2 m $= \frac{1}{2} \times 220$ kg $\times v^2$
and, writing kg m/s^2 for N,

$$v^2 = 9\cdot81 \times 2 \times 2 \text{ kg m}^2/\text{s}^2 \text{ kg}$$
$$v = 6\cdot26 \text{ m/s}.$$

(2) *Impact occurs*
Here energy will certainly be 'lost' (i.e. converted into heat). On the other hand, it is reasonable to assume that during the duration of the impact the tup and pile form a 'closed system', whose momentum will be constant.

$\therefore$ 220 kg $\times$ 6·26 m/s $= (1\ 750 + 220)$ kg $\times v_2$

$$\therefore v_2 = 0\cdot70 \text{ m/s}$$
$$= \text{velocity of (pile + tup) after impact.}$$

(3) *The (pile + tup) move downwards into the ground*
Here we are *told to assume* that the resistance to motion is uniform, so we can write down:

net force F' causing deceleration $\times$ distance $=$ KE to be 'destroyed'
or $F' \times 0\cdot10$ m $= \frac{1}{2} \times 1\ 970$ kg $\times 0\cdot70^2$ m^2/s^2

$\therefore F' = 4\ 820$ kg m/s^2 or N
    $= 492$ kgf   (The calculation being finished)

But the earth is supplying this force in addition to supporting the weight of the pile and tup, so the total upward force F is:

$$F = 1\ 970 \text{ kgf} + 492 \text{ kgf} = 2\ 460 \text{ kgf}$$

Worked Example 0.4 on rotational harmonic motion

Two gearwheels A and B are mounted with their axes horizontal as shown in the sketch, fig. 0.21. A has a radius r and moment of

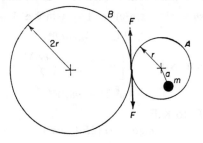

Fig. 0.21

inertia I, B a radius $2r$ and moment of inertia $4I$. An out-of balance mass m is attached to A at radius a. Show that if m is moved slightly from its position of stable equilibrium the resulting motion will be simple harmonic, and find its frequency.

SOLUTION

(a) From consideration of forces:

Let the angular displacement of m at any instant be θ_A.
Let the tangential tooth load between the wheels be F.

Torque on $A = -mga \sin\theta_A + Fr \simeq -mga\,\theta_A + Fr$
Torque on $B = +F \cdot 2r$

$$\therefore \ddot{\theta}_A = -\frac{(mga\,\theta_A - Fr)}{I + ma^2} = -2\,\ddot{\theta}_B = -2\frac{F \cdot 2r}{4I} = -\frac{Fr}{I}$$

$$\therefore Imga\,\theta_A - IFr = IFr + Fr \cdot ma^2$$

$$\therefore Fr(2I + ma^2) = Imga\,\theta_A$$

$$\therefore \ddot{\theta}_A = -\frac{Fr}{I} = -\frac{mga}{2I + ma^2} \cdot \theta_A$$

$$\therefore \text{Radiancy} = \omega = \sqrt{\left(\frac{mga}{2I + ma^2}\right)}$$

$$\text{frequency} = \frac{\omega}{2\pi} = \frac{1}{2\pi}\sqrt{\left(\frac{mga}{2I + ma^2}\right)}$$

SOLUTION

(b) By energy method:

Let θ = max. value of θ_A
then max. angular velocity of $A = \omega\theta$
In displaced position P.E. $= mga\,(1 - \cos\theta)$ and since θ is small

$$\cos\theta = (1 - \sin^2\theta)^{\frac{1}{2}} \simeq 1 - \tfrac{1}{2}\sin^2\theta \simeq 1 - \tfrac{1}{2}\theta^2$$

$$\therefore \text{P.E.} = \tfrac{1}{2}mga\,\theta^2$$

In mid position K.E. $= \tfrac{1}{2}\left[I\omega^2\theta^2 + 4I\left(\frac{\omega\theta}{2}\right)^2 + m(a\omega\theta)^2\right]$

$$= \tfrac{1}{2}\omega^2(2I + ma^2)\theta^2$$

Equating P.E. to K.E.

$$mga = \omega^2(2I + ma^2)$$

hence ω as before.

Reference Books

There are many good textbooks on mechanics. One which may be recommended is

D. HUMPHREY and J. TOPPING. *Shorter Intermediate Mechanics*, 3rd edn. Longmans, 1963.

A useful reference to the SI system of units is provided by *Using the S.I. System*, HMSO, 1969.

Examples

0.1 The pilot of a light aeroplane notes that the smoke from a ship travelling south-west lies 10° to the east of north, while the smoke from a ship travelling due south lies 20° to the west of north. He estimates the speed of each ship as 20 km/hr. What can he deduce about the direction and speed of the wind?

0.2 A flywheel weighs 2tf and has a radius of gyration of 1·2 m. Find the torque necessary to make it reach a speed of 250 rev/min in $\frac{1}{2}$ min and the work done in getting up speed.

0.3 A flywheel has an axle 25 mm in diameter which rests on a pair of parallel rails having a gradient of 1 in 20. Starting from rest, it rolls down 0·6 m in 5 s.

Calculate the radius of gyration of the flywheel-and-axle about the centre line of the axle

(*a*) by finding the angular acceleration
(*b*) by an energy method
(*c*) by equating the time integral of the torque to the gain of angular momentum.

0.4 Two spheres, of masses 3 and 5 kg, are moving towards each other along their line of centres with speeds of 25 and 35 m/s respectively. Determine their velocities after impact, and also the loss of kinetic energy, the coefficient of restitution being $\frac{2}{3}$.

0.5 Two bodies of mass 10 kg and 20 kg and velocities 5 and 10 m/s respectively, are moving on straight paths, in the same plane, inclined at 60° to one another.

They collide and subsequently move on together. Determine the common velocity after the impact and its direction with respect to that initially of the smaller mass. What fraction of the initial kinetic energy of the masses is lost during the collision, and what happens to this energy?

0.6 What are the dimensions of force, momentum, kinetic energy, and horsepower respectively,

(*a*) in length, time, and mass units
(*b*) in length, time, and force units?

A mass of *M* tonnes, initially moving due east along a smooth horizontal plane at a uniform speed of *v* kilometres an hour, meets a constant resisting force of *P* tonnes force, which continues to act on the mass in a constant direction, from an angle of α south of east, for *t* min. Prove that the resulting change of kinetic energy of the mass in this time in newton metres is

$$Pt(1{\cdot}73 \times 10^8\, Pt/M - 1{\cdot}63 \times 10^5 v \cos \alpha)$$

0.7 The SI unit of pressure, 1 N/m², is one of the more awkward basic units in that it is for most purposes inconveniently small. The 'bar' ≡ 10⁵ N/m² and its derivatives are accepted as secondary units. Many other units for pressure are likely to be with us for a long time to come, and it is useful to be able to convert them readily to SI units.

Working to slide-rule accuracy, express in N/m², and in bars, the following pressures: 760 mm of mercury*; 30 in of mercury*; 1 000 lbf/in²; 1 tonf/in²; 1 kgf/mm²;

0.8 If two parallel surfaces are separated by a layer of fluid and one moves parallel to itself at not too great a speed relative to the other it experiences a resistance *R* due to the 'shearing' of the fluid, which is found to be directly proportional to the area *a* being sheared, and to the relative speed *v*, and inversely proportional to the thickness *t* of the film; i.e. $R \propto av/t$. Alternatively, we can write $R = \eta \, . \, av/t$, and so define the 'coefficient of viscosity, η', of the fluid.

In certain cases it is found that the property of the fluid which controls its behaviour is the ratio of its viscosity to its density; this ratio, say $\nu = \eta/\rho$, is called the 'kinematic viscosity' of the fluid.

* Take *g* = 9·81 m/s²; density of mercury = 13 590 kg/m³; and note that 1 in ≡ 0·0254 m; 1 'atmosphere' ≡ 101 325 N/m²; 1 lb = 0·4536 kg; 1 ton ≡ 2 240 lb.

Again in many problems of fluid flow it is found that a parameter called the 'Reynolds number', say *Re*, is of vital importance. This can be defined as $V \cdot d/\nu$, where V is some representative velocity, d some representative dimension or length, and ν (as above) the kinematic viscosity of the fluid.

Given this information, state the basic SI units in which η, ν, and *Re* should be measured.

0.9 A motor-cyclist has to go round a curve of 15 m radius, and the track is banked the 'wrong' way at an angle of 5° to the horizontal. If the coefficient of friction between track and tyres is 0·6 what is his maximum possible speed, and what will be the included angle between track and motor-cycle at this speed?

0.10 An engine governor of the Hartnell type (fig. 0.16) is provided with balls each weighing 2 kg. The vertical and horizontal arms of the bell-cranks are 10 cm and 7·5 cm long respectively, and the distance of the fulcrum from the axis of rotation is 12 cm. The spring stiffness is 65 N/cm, and the initial compression is adjusted so that the balls are vertically above the fulcrum when the speed is 250 rev/min. What is the spring force in this configuration?

Draw the controlling-force curve for a range of movement of the balls of ±2 cm, and so determine whether or not the governor will in itself be stable.

0.11 A bicycle is being ridden 'hands off' and the rider wishes to turn a corner to the left. What does he do, and what is the consequent effect of the gyroscopic action of the front wheel?

0.12 A child's spinning top consists of a brass disk 8 cm in diameter and 6 mm thick, through which a light spindle with a sharp point protrudes 4·7 cm. It is spinning at 2 500 rev/min, in a clockwise direction when viewed from above, and the spindle is inclined at 15° to the vertical. In which direction and at what speed is the spindle precessing?

This top would not set itself upright; on the other hand, a top in which the end of the spindle is blunt would do so. Can you work out why?

0.13 An aircraft travelling at 1 000 km/hr turns to the right in a curve of radius 1 500 m. What is the centripetal acceleration? Express the result in terms of g.

If the moment of inertia of the rotating parts of the engine is 6·5 kg m², the speed 9 000 rev/min, and the direction of rotation clockwise when viewed from the tail, what, in magnitude and sense, is the gyroscopic torque *on the rotor bearings*?

0.14 The tension of a spring is proportional to its extension beyond its unstretched length. The spring is hung up vertically with a mass attached to its lower end. Show that, if the mass be set oscillating vertically, its motion will be simple harmonic, and that its 'radiancy' will be $\sqrt{(g/\delta)}$, and its period $2\pi\sqrt{(\delta/g)}$, where δ is the extension in the spring due to the weight of the oscillating mass when the latter is hanging at rest.

0.15 A particle of mass 5 kg moving with simple harmonic motion has a maximum velocity of 3 m/s, and it performs a complete oscillation in $2\frac{1}{2}$ s. Calculate the complete range of its oscillation and the maximum value of the force applied to the particle.

1. Mechanisms

§ 1.1 Introduction: Preparatory Definitions and Terminology

A *machine*, according to one of the definitions given in the *Oxford English Dictionary*, is 'an apparatus for applying mechanical power, consisting of a number of interrelated parts, each having a definite function', and this is the sense in which the word 'machine' is used in the title of this book.

The parts of the machine are usually called *elements*, and two elements which are in contact, and between which there is relative motion, are known as a *pair*. An element joining two pairs is called a *link* or *bar*. If a group of links and pairs is capable of relative motion but can be made rigid by joining any two elements it is called a *kinematic chain*. If one of the links of a kinematic chain is 'fixed' the chain becomes a *mechanism*, and if a mechanism is used to transmit force it becomes a *machine*.

The term 'fixed' has, of course, no meaning in the absolute sense, but may be used for convenience in dealing with terrestrial affairs, when it means fixed relative to the earth or to some specified frame of reference. It is possible to obtain from one kinematic chain a number of different mechanisms; if one mechanism is converted into another by fixing an alternative link the process is described as *inversion*.

Some of these terms may be illustrated by considering a well-known machine such as a motor-car engine. A particular example is shown in cross-section in fig. 1.1. The crankshaft *A* rotates in bearings in the frame; these two elements are said to be coupled by a *turning pair*. In the cylinders *C* (which are kinematically part of

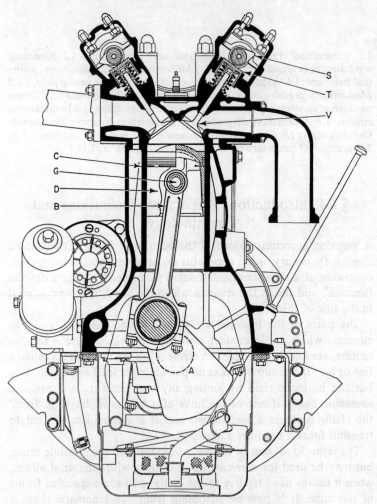

Fig. 1.1 Cross-section of 3·8-litre motor car engine
(By permission of Jaguar Cars Ltd.)

the frame) the pistons *D* reciprocate: these elements are said to be coupled by *sliding pairs*. These two types of pair (together with the relatively uncommon *screw pair*, which may be exemplified by the lead screw and nut in a lathe) allow contact to be made over a surface and are sometimes classified as *lower pairs*. Each connecting-rod *B* couples the gudgeon pin *G* of a piston to the crankshaft by a turning pair at each end (though, of course, the former does not rotate but only oscillates). Many other examples of such turning and sliding pairs in the complete engine will be apparent. The valves *V* (of the poppet type) are, however, driven through tappets *T* from a cam-shaft *S*, and between these latter elements there is only 'line' contact; again, between the gear-wheels in engine or gearbox only line contact occurs: such pairs are called *higher pairs*.

A drawing such as fig. 1.1 is intended to illustrate the general arrangement of the engine, and omits many details which have to be completely specified in the 'working' drawings. Even so, it is far too complicated for most of the purposes with which we are here concerned, and it is customary to make further drastic simplifications. Thus, what we might regard as the main unit in such an engine consists of the frame, crankshaft, connecting-rod, and piston, and it can be represented conventionally but adequately as in fig. 1.2, where the frame, crank, and connecting-rod appear merely as straight lines; each turning pair is represented by a dot or small circle, and the piston sliding in the cylinder by a rectangle

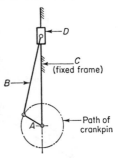

Fig. 1.2 Simple engine mechanism

over the line representing the frame. A 'fixed' point is usually indicated by a vertical cross (derived from the draughtsman's positioning lines), as at the centre of the crankshaft, while a fixed member is often indicated by hatching as shown on the 'frame'.

This unit is in itself an example of a machine, and if we are not, for the moment, particularly interested in the forces which are being transmitted we can call it a mechanism: it is in fact usually de-scribed as a 'simple engine mechanism'. The kinematic chain from which it is derived has four elements, three turning pairs and one sliding pair, and is often called the 'four-bar single-slider chain'.

This same kinematic chain can, by inversion, give several other mechanisms. Thus, for instance (fig. 1.3), if we fix the crank *A* and leave the rest of the mechanism (cylinder, piston, and connecting-rod) free to rotate, we do not alter the relative motions, so the piston will reciprocate in the cylinder as before. This arrangement

was in fact used in some early 'rotary' aero-engines; not of course
with only one cylinder nor with the cylinders arranged in a bank
along the crankshaft, but with several cylinders disposed radially
round the crankshaft, and all the connecting-rods coupled to the
one crank-pin. More commonly, the mechanism shown in fig. 1.3 is
used in the form indicated, that is with a single slider in the shape
of a block *D* sliding in a slotted member *C*. It can readily be appreciated that if the member *B* is driven at a constant angular speed
the angular speed of *C* will vary cyclically (and vice versa), since
while *B* is rotating through half a revolution (from $\theta = 0$ through
$\theta = +90°$ to $\theta = +180°$) *C* will execute less than 180°, and during
the next half-revolution of *B*, *C* will do more than 180°. The amount
of this speed variation will vary with the length of the member *A*.

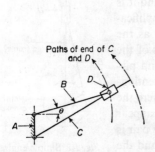

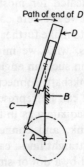

Fig. 1.3 Rotary engine (or   *Fig. 1.4* Oscillating-cylinder
quick-return) mechanism   engine mechanism

Such a mechanism can be used in various machines, for instance to
cut a moving sheet of board or metal into different lengths, or to
cause a cutting tool to move slowly in one direction while it is
cutting, and return rapidly to start another cut: it is often called a
'quick-return mechanism'. Its appearance can vary greatly if the
lengths of the various links are altered: compare, for instance,
fig. 1.50a.

Returning to the simple engine mechanism in fig. 1.2, we can see
that if the connecting-rod *B* were made the frame of the engine and
fixed in space and *A* were allowed to rotate about its end, we should
have an engine in which the cylinder reciprocated and oscillated.
This would be too awkward altogether to be of practical use, but if
as in fig. 1.4 the cylinder is pivoted to the upper end of *B*, the
reciprocating motion can be restored to the piston, while the cylinder

oscillates about the pivot. Unlikely as this arrangement might appear, it was in fact used in some early steam engines, and survives in some oil pumps, not to mention toy steam engines. Kinematically, this is still a mechanism derived from the single-slider four-bar chain.

Books have been written about mechanisms, their analysis, and, more recently, their synthesis, and one or two are mentioned in the references at the end of this chapter, but the subject is somewhat specialised and of limited importance, and this brief discussion of one of the most elementary mechanisms is intended primarily to illustrate the terms used.

§ 1.2 Kinematic Constraint: Principles and Practice

The position of a *point* in space must be defined in relation to a *frame of reference*, and for this definition three measurements or *co-ordinates* are required. These co-ordinates may be of a number of kinds, such as spherical, cylindrical, or Cartesian. The most commonly used system is that of Cartesian co-ordinates in which the frame of reference consists of three, frequently but not necessarily mutually perpendicular, axes Ox, Oy, Oz, passing through a point O described as the origin, the distance along each of these axes being specified. Similarly, the position of a *body* in space can be defined by specifying the position of one point in the body with reference to the three co-ordinate axes and the *orientation* of the body in relation to the three axes. When a body is free to move in space its motion can be analysed into translation along the x, y, and z axes and rotation about each of these axes. The ability of the body to move in each of these senses is described as a *degree of freedom*, and if the motion of a body in any one of these senses is prevented the body is said to be subjected to a *constraint*.

If a 'rigid' body has to be fixed relative to a frame of reference six appropriate constraints must therefore be applied, and such a condition is called complete kinematic constraint. If fewer than six appropriate constraints are applied the body will be capable of motion. If an attempt is made to apply more than the six constraints either some of the constraints will be ineffective or, since no real body is truly rigid, the body itself may be distorted.

In some cases, notably in the design of instruments, it may be desirable to apply to an element pure kinematic constraint, either complete or partial, in order to fix its position or allow it the desired degree of freedom. Such considerations are paramount when accuracy of positioning or motion is essential, when the forces applied are so small in relation to the stiffness of the element that it can justifiably be assumed to be rigid, and when the forces and motions are such that wear is not an important factor. In other cases, and these will include the majority of machines, an attempt to apply *pure* kinematic constraint would be disastrous. Compromise, in the sense of modifying the design to provide mating surfaces instead of point contacts, and in the sense of applying more than the minimum number of constraints to provide greater stability, or to allow for the fact that no real element is truly rigid, is often essential.

An 'ideal' method of applying a single translational constraint to a rigid body is cause a point on its surface to make contact with a surface which may be taken as 'fixed'. For simplicity, let us assume

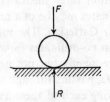

Fig. 1.5 Single constraint with force closure

Fig. 1.6 Double constraint

that the body has a spherical surface – or ball foot – which makes contact with a plane fixed surface (fig. 1.5). Then inward motion in the direction of the normal to the plane is prevented. To prevent outward motion without applying a force to the ball it would be necessary to provide another fixed surface touching the ball at a point diametrically opposite the first point of contact, and this implies that two fixed surfaces, *exactly* parallel to each other and *exactly* the diameter of the ball apart, have to be provided. Such a provision is unrealistic, and its necessity can be obviated by providing instead that a force shall always act on the ball towards the first surface. This will ensure that the ball cannot move in either direction along the normal, and the arrangement is described as *force closure*. Translation in either of the other directions and rotation about any of the axes are not prevented, though there would be a frictional force opposing such motions. This force, though often important, may be neglected in the present discussion.

If we wish to apply two translational constraints to this ball we can provide two surfaces as shown in fig. 1.6, with force closure. The ball can now move in the direction of the 'groove' formed by the surfaces and rotate in any sense.

If we wish to prevent all translational motion of the ball we must provide three surfaces against which it is forced. These need not be mutually perpendicular. Fig. 1.7 shows a plan view of a ball resting (in this instance under gravitational force closure) in a trihedral depression in a flat plate. This is a particularly useful form of such a system frequently found in instrument design: the ball is completely positioned but able to rotate about any axis.

If we wish to eliminate any of these rotational degrees of freedom we must apply a constraint to another part of the body than the single ball foot we have so far been considering. Let us suppose that the body is provided with a second ball foot: since the body is

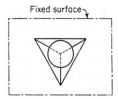

Fig. 1.7 Triple constraint positioning ball foot

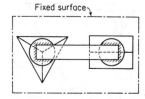

Fig. 1.8 Five constraints giving single degree of rotational freedom

rigid, these two feet will necessarily be at a fixed distance apart. If then we provide a plane fixed surface and cause the second foot to remain in contact with it, as in fig. 1.5, the body will be capable of rotation in two senses. If instead the ball is forced against two plane surfaces in the form of a 'groove' (fig. 1.8), pointing in any direction other than the normal to the line of centres of the ball feet but preferably roughly parallel to this line of centres, we eliminate five of the six degrees of freedom, and the body is now capable only of rotation about the line joining the centres of the ball feet. This again is a system very frequently used in instrument design where an element is required to have one degree of rotational freedom, and it provides 'ideal' constraint provided that the friction and possible wear associated with rotation can be tolerated. *Note that we do* NOT *provide* a trihedral depression for both ball feet: to do so would amount to attempting to provide six constraints to eliminate five degrees of freedom. The result would be that unless the distance between the balls and that between the depressions were at great trouble and expense made to be identical, and remained so under, for instance, fluctuating temperatures – a most improbable occurence – the position of the body would not be defined, since each of the feet would be attempting to locate it differently, and either or neither might be successful. To go to great trouble and expense to obtain an unsatisfactory result instead of obtaining easily and cheaply a satisfactory result is unwise, yet examples of this sort of lack of thought are not infrequently met in practice.

To eliminate the remaining rotational freedom, that is, to provide complete kinematic constraint for the body, the sixth constraint is preferably provided at a surface remote from the present axis of rotation. To satisfy this requirement, the body may have a third foot, which need only be held in contact with, for instance, the plane which already contains the trihedral hole and the groove, or another at a different level (fig. 1.9). A body so positioned by three ball feet has complete kinematic constraint. This particular solution, which is due to Lord Kelvin and is usually called a 'Kelvin clamp', is frequently used, for instance in siting surveying instruments. It is not, of course, the only solution, but it is probably the most elegant. The student is

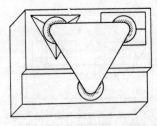

Fig. 1.9 (With force closure) complete kinematic constraint

advised to consider for himself some possible alternatives, as well as alternative methods of providing for various defined degrees of freedom.

When we turn to the problem of machine design it soon appears that radical modifications must be made in our approach. Let us suppose that we are faced with the problem of mounting a shaft so that it is free to rotate, but not to move in any other sense. Kinematically, the problem is solved in fig. 1.8; we should, of course, have to replace the bar shown in that diagram by a shaft passing through the centres of the balls as in fig. 1.10a, but then we should have achieved our aim. Practically, such a solution would be ludicrous, because (even if force closure introduced no practical difficulties) the constraints are made by point contacts which would accept no reasonable load without wearing and which would be impossible to lubricate effectively. We therefore have to replace them by appropriate surfaces. The normal solution might consist of two cylindrical bearings, and thrust bearings to provide axial location, as sketched in fig. 1.10b. We must, however, realise that these bearings introduce unnecessary additional constraints which are described as kinematically 'redundant', and we have to be careful to ensure that these constraints will not introduce intolerable conditions. For instance, if our bearings are a good fit on the shaft but are not accurately aligned, the shaft will not enter them, and if sufficient clearance is provided to allow the shaft to enter it will touch them, not over the whole surface as intended, but over relatively tiny areas at the edges of the bearings, so causing excessive pressure there, and consequent wear. Even if the bearings are accurately aligned, any bending in the shaft will immediately produce similar effects, so

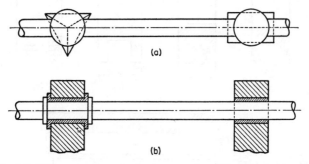

Fig. 1.10 Kinematic constraint in theory and modifications in practice

we must be sure that the shaft is stiff enough to prevent this sort of trouble, and this often gives difficulty. Compare this situation with the kinematic one in fig. 1.10a, where bending of the shaft would produce no change in the constraints, and no alignment is needed.

If our solution were to envisage ball or roller bearings we should have to give at least as careful thought to their selection and use. A few of the more common types of these bearings are illustrated in fig. 1.11; (*a*) shows the 'deep-groove' type in which (to increase the allowable load) the races fit the balls with very little clearance: this means that the bearing provides not only a radial constraint but also axial and directional constraints; (*b*) shows the 'self-aligning' type, in which the outer race is spherical: this eliminates the directional but not the axial constraint; (*c*) is a roller bearing which eliminates the axial but not the directional constraint; (*d*) shows a special form of 'needle-roller' bearing to which we shall have

occasion to refer later on, and in fig. 1.51c is shown the mounting
for the front wheel of a motor car which incorporates two bearings
of yet another type, the taper-roller bearing.

It is not difficult to see why we have to be careful. Fig. 1.11e
shows a common arrangement of two bearings to provide a short
shaft with one degree of rotational freedom: note the use of one
bearing of a type which provides an axial constraint, and the second

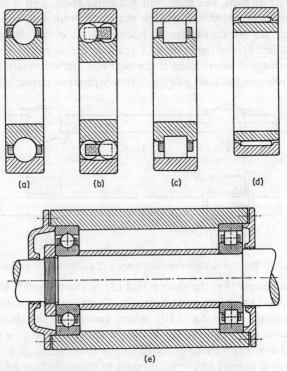

Fig. 1.11 Types of ball and roller bearings (cf. also figs. 1.10 and 1.51)

which allows axial freedom; this is the exact analogue of Fig. 1.10b.
Two *ball*-races so mounted would almost inevitably give rise to
serious trouble, since, even if they were initially positioned with
sufficient accuracy to avoid a 'locked-up' load between them,
temperature changes would soon give rise to such a load, which
would immediately increase the friction in the races and so increase
the heat generation, temperature differences, and locked-up load,
and so on in a vicious spiral until failure occurred. (The mounting

for two ball races would have to be modified to allow one of the races axial freedom as a whole; but again this can be done only if the conditions of loading permit it.) Note that even the arrangement shown in fig. 1.11e is satisfactory only if the bearings are accurately aligned and the shaft between them is stiff.

Kinematically, two bearings are, as we have seen, always suffi-cient to give a shaft the single degree of rotational freedom which it should have. Needless to say, it is often necessary to provide many more bearings in practice. In car engines such as that illustrated in fig. 1.1, for instance, only the smallest are ever provided with only two bearings for the crankshaft: many have an additional bearing between each cylinder, so that for a shaft only a couple of feet long there may be seven or more bearings. These are necessary because to make a large crankshaft sufficiently rigid and strong to run in only two bearings and yet to withstand the forces applied by the pistons would be virtually impossible. This may be taken as another illustration of the sort of compromise required in practical engineer-ing.

Again, it is often necessary to give a part of a machine one degree of translational freedom. This is a common requirement in a machine tool, and the bed of a planing machine may be chosen as an example. A kinematic solution would be achieved by modifying the Kelvin clamp in fig. 1.9 by eliminating the trihedral depression and allowing two of the ball feet to slide in an extended groove. For instruments in which the travel of the moving member is very re-stricted such a design can be modified by allowing the balls freedom to roll between the fixed surface and a similar member inverted and loaded to rest on top of the three balls. For a planing machine, of which the moving bed may be many feet in length, may weigh many tons and must have a relatively long travel, neither solution approaches the practicable. A reasonable compromise may be reached by replacing two of the balls by a V-shaped projection which fits a corresponding groove in the foundation and supports the bed over the whole or most of its length, and replacing the single ball by a flat surface (fig. 1.12a) which similarly supports the bed over its whole length. Such a design provides adequate surface area for carrying the loads and gives the far-from-rigid moving bed adequate support from the foundations. But what is to be said in favour of the very common design sketched in fig. 1.12b?

The purpose of these few paragraphs is, however, not to provide guidance in the design of shaft bearings, or machine tools, but merely to point out that even where strict kinematic constraint is inadvisable it is important to keep the principles in mind, and to

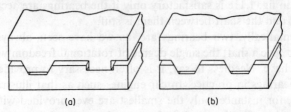

(a)   (b)

Fig. 1.12 Approximate kinematic constraint and over-constraint

depart from them only when and to the extent that there is good and sufficient reason for doing so. This type of intelligent compromise between conflicting ideologies or requirements is the stuff of which good engineering design is made.

§ 1.3 Motion of Elements in Machines: Analytical Treatment

Having considered, even so briefly, the elements of a mechanism or machine, we must now direct our attention to the methods available for analysing the motion of, and forces on, these elements. The determination of the various velocities is often a useful aid to the analysis of the forces which the elements must be designed to transmit, and is an essential preliminary to the determination of accelerations; the higher the speed of the machines, the more important do inertia forces become (sometimes indeed swamping the forces required to do useful work), and to find these inertia forces it is necessary to determine the acceleration of the elements.

It is sometimes desirable to obtain the results in the form of mathematical expressions, and some of the simpler mechanisms are amenable to analytical treatment. In the more complicated mechanisms when such expressions, even if found, would usually be too clumsy to be useful, graphical methods may provide the answers we require more quickly, and in a more illuminating way. We must therefore consider examples of both methods.

To illustrate the application of analytical methods we shall choose the most widely used of all mechanisms, the so-called 'simple engine mechanism', which was referred to in § 1.1. It is again illustrated diagrammatically in fig. 1.13.

Let the dimensions of the mechanism be as shown (i.e. crank length $= r$, connecting-rod $= l$, or q times r, where q is a numerical constant) and let the crankshaft speed be constant and have a value ω (in radians per unit time, normally rad/s). Our problem is to obtain analytical expressions for the velocity and acceleration of the piston in terms of these quantities and the angle θ between the crank and the line joining the centre of the crankshaft to the piston.*

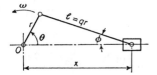

Fig. 1.13 Simple engine mechanism

Let us call the acute angle between the connecting-rod and this line ϕ, and the distance between the crankshaft centre and the piston x. Then by inspection we can write

$$x = r \cos \theta + l \cos \phi$$
$$= r \cos \theta + qr \cos \phi$$

To find the velocity of the piston we have to differentiate x with respect to time. In this expression $\cos \phi$ gives rise to awkwardness, since ϕ is an angle varying in an irregular manner with time. We can eliminate ϕ by noting that

$$l \sin \phi = r \sin \theta$$
$$\therefore \cos \phi = \sqrt{(1 - \sin^2 \phi)} = \left(1 - \frac{r^2}{l^2} \sin^2 \theta\right)^{\frac{1}{2}} = \left(1 - \frac{1}{q^2} \sin^2 \theta\right)^{\frac{1}{2}}$$

Further, although we could differentiate this quantity, the result would be in a less useful form than is obtained by first expanding it as a binomial. We note that, with any reasonable dimensions of the elements, $\frac{1}{q^2} \sin^2 \theta \ll 1$, so the expansion will be rapidly convergent.

---

* The position of the crank $\theta = 0$ is often referred to as 'inner' or 'top' dead centre. 'Dead centre' refers to the fact that the piston at this instant has no velocity; 'inner' means that the crank is pointing towards the 'inside' of the engine; 'top' implies that the most common orientation of an engine, with the cylinders above the crankshaft as in fig. 1.1, is in mind. The term 'top dead centre' will be used in all subsequent sections of this book, even though we shall, to conform with normal mathematical conventions, most frequently be considering horizontal arrangements of engines so that the crank is in the positive direction of the x-axis when $\theta = 0$.

$$\cos \phi = 1 - \frac{1}{2q^2} \sin^2 \theta - \frac{1}{8q^4} \sin^4 \theta - \cdots$$

$$\therefore x = r \left[\cos \theta + q - \frac{1}{2q} \sin^2 \theta - \frac{1}{8q^3} \sin^4 \theta - \cdots \right]$$

We can now differentiate with respect to time

$$\dot{x} = r \left[-\sin \theta . \dot{\theta} + 0 - \frac{1}{2q} . 2 \sin \theta \cos \theta . \dot{\theta} \right.$$
$$\left. - \frac{1}{8q^3} . 4 \sin^3 \theta . \cos \theta . \dot{\theta} \cdots \right]$$

and $\dot{\theta} = \omega$ so

$$\dot{x} = - \omega r \left[\sin \theta + \frac{1}{2q} . \sin 2\theta + \frac{1}{8q^3} (\sin 2\theta - \tfrac{1}{2} \sin 4\theta) + \cdots \right]$$

and to find the acceleration of the piston we differentiate again:

$$\ddot{x} = - \omega^2 r \left[\cos \theta + \frac{1}{q} \cos 2\theta + \frac{1}{4q^3} (\cos 2\theta - \cos 4\theta) + \cdots \right]$$

We are thus able to approximate to the velocity $\dot{x}$ and the acceleration $\ddot{x}$ of the piston to any desired degree of accuracy by taking the appropriate number of terms of this infinite series. We must now decide how many terms to take into consideration, and the answer clearly depends on the magnitude of q. In the majority of engines q has a value of 3·5 to 4 or more, and a few seconds' inspection indicates that the third term in the expression for velocity is of the order $\frac{1}{2}$ per cent, and in the acceleration of the order 1 per cent, of the first term. For many purposes such magnitudes may well be neglected, and we obtain the delightfully simple further approximations

$$\dot{x} \simeq - \omega r \left[\sin \theta + \frac{1}{2q} \sin 2\theta \right] \tag{1.1}$$

$$\ddot{x} \simeq - \omega^2 r \left[\cos \theta + \frac{1}{q} \cos 2\theta \right] \tag{1.2}$$

It will be observed that if the connecting-rod were 'infinitely' long we should, of course, obtain expressions showing that the piston had simple harmonic motion (cf. § 0.14).

This, then, is clearly a case in which the analytical approach is immensely helpful, and use will be made in later chapters of the expressions we have just obtained. For the time being, however, we must consider the graphical methods, which are more suitable for complex mechanisms.

§ 1.4 Velocity Diagrams

1.4.1 Mechanisms with turning pairs

It has been pointed out (§ 0.4) that velocity is a vector quantity and that it can be specified only by quoting *speed* and *direction* of motion relative to a given *frame of reference*. In dealing with terrestrial matters it may be convenient, and is indeed customary, tacitly to choose the local surface of the earth as a frame of reference and loosely to regard and speak of this as 'fixed'. Such a custom should not be allowed to become too ingrained as a habit, since it causes many problems which should be easily solved to appear unnecessarily difficult.

Thus, for instance, one may be given the information: 'two particles, A and B, have velocities V_A and V_B as shown in fig. 1.14'; and be required to find the velocity of B relative to A. A standard method of solution is to 'add' a velocity $-V_A$ to each (an operation which does not alter their *relative* velocity), whereupon it is seen that the total velocity of A is zero and that of B is the vector sum of V_B and $-V_A$, which can be obtained by drawing the two vectors

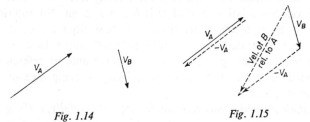

Fig. 1.14   Fig. 1.15

in succession tail to head (fig. 1.15), or, of course, by drawing a parallelogram instead of a triangle of velocities (fig. 1.16, and see § 0.4).

This is a perfectly satisfactory method of obtaining the *required* answer; the objections to it are simply that because the frame of reference has been tacitly understood throughout, the generality of the problem has been obscured, restrictive arrow-heads have been unnecessarily introduced, and other equally valid facts, which should be apparent, can be deduced only by the expenditure of further thought.

It is more efficient to argue in general terms from the start. Thus (if the problem has been presented as above) the first step in solving it is to note that we are dealing with the translational velocities of *A* and *B* relative to some fixed frame of reference whose origin we may denote by *O*. We can now choose on the paper (fig. 1.17) a point *o* to represent the (zero) velocity of *O*, and from *o* draw the vectors *oa* and *ob* to represent the velocities V_A and V_B relative to *O*. Note that no arrow-heads are inserted. The resulting diagram (which we may complete by drawing the line *ab*, and which is called a 'velocity diagram') now gives information about the

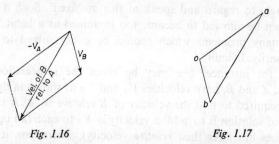

Fig. 1.16   Fig. 1.17

relative velocity of *any* particle in the system with respect to *any* other – in particular, for instance, the velocity relative to *A* of *B* is *ab*, the velocity relative to *B* of *O* is *bo*, and so on. No additional 'proof' of this statement is required, since basically the procedure is exactly that illustrated in fig. 1.15. Note that to make the directions appear 'natural' the particle serving as the frame of reference is mentioned first, the particle whose velocity is being measured is mentioned second.

The student should study this simple diagram carefully, making sure that he can visualise and understand the situation not only from the original viewpoint of an observer at rest with respect to O, but with equal facility from the viewpoint of an observer moving with A, when a becomes the 'origin' or 'pole' of the diagram and the velocities of O and B are seen to be ao, ab respectively, and so on.

In an example as easy as this the advantages of the method advocated are, of course, minimal: in a complicated case, such as we shall consider presently, in which many particles or points are involved, they are very great.

The form in which such problems usually appear in machines and mechanisms is in relation to an element, which for simplicity we may reduce conceptually to a rigid bar or 'link', such as *AB* in fig. 1.18; the velocity of one end *A*, and the direction of motion but *not* the speed of the other end, *B*, being, for instance, specified, and the speed of *B* being required.

Again, the first step in the solution is to realise that the velocities must have been specified in relation to a frame of reference, which

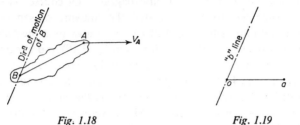

Fig. 1.18   Fig. 1.19

we may denote as before by *O*. We therefore choose a 'pole' *o* (fig. 1.19) to represent the velocity of *O*, and as before we draw a vector *oa* to represent the velocity of *A* relative to *O* and start to draw a vector *ob* to represent the velocity of *B* relative to *O*. We know the direction of *ob*, but not its length: in other words we can draw through *o* a line parallel to the direction of motion of *B* on which the point *b* must lie; if it seems helpful we can label this a '*b*' line.

The second step in the solution is taken by regarding the situation

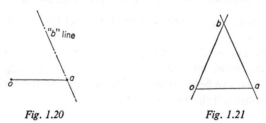

Fig. 1.20   Fig. 1.21

from the viewpoint of an observer moving with *A*. He will regard himself as 'fixed', he will choose the point *a* as the 'pole' of the diagram and consider *O* to be moving with velocity *ao* (fig. 1.20). But he also knows that the link *AB* is rigid; in other words, that whatever the velocity of *B*, it *has no component in his direction and can therefore be only in a direction at right angles to this. This fact he can record in fig. 1.20 by starting to draw a vector ab in a direction at right angles to the link AB*, and, since he has no information about the magnitude of this velocity, leaving the line of indefinite length and labelling it a '*b*' line.

The third and final step in the solution is taken by regarding the situation in an impartial manner. To the detached observer, there is no virtue in labelling either *o* or *a* a 'pole'. Figs. 1.19 and 1.20 are partial representations of the same situation, and when combined yield fig. 1.21, the 'velocity diagram'. Of course, as soon as the principle has been appreciated the drawing of separate diagrams is without virtue: fig. 1.21 can be drawn by representing the given facts (*oa*, incomplete *ob* line as in fig. 1.19) and adding a line through *a* perpendicular to *AB* to determine *b*. *No* points should be labelled 'poles'. In fact, the process is easily reduced to a 'rule of thumb': *if AB is a rigid link in a mechanism, the line ab in the velocity diagram is perpendicular to it.* Many velocity diagrams can be constructed by the repeated application of this rule alone. It need hardly be added that to adopt such a rule without a full understanding of the simple principle behind it would later lead to disaster.

Two simple corollaries should be considered and if necessary proved by each student for himself at this stage:

1. If *ABC* is a rigid triangular link in a mechanism, the velocity diagram will contain a similar triangle, *abc*, of which each side is perpendicular to the corresponding side in the mechanism.

2. If *ABC* are three points on a straight link the velocity diagram will contain a straight line *abc* such that $ab : bc = AB : BC$.

(These similar figures are often referred to as the 'velocity images' of the links.)

With the aid of this single principle the velocity diagram for many mechanisms, involving rigid elements and turning pairs or pin joints only, can be constructed. The student should, for instance, attempt for himself the example 1.1 below.

Worked Example 1.1

In the mechanism shown to scale in fig. 1.22a, *O*, *P*, and *Q* are 'fixed' points. The crank *OA* rotates at 85 rev/min clockwise. *ACB* is a rigid link (note the convention here adopted to indicate this).

DATA (with lengths in mm): Co-ordinates of *O* are (0, 0), of *P* (−450, +50), of *Q* (75, 200), $OA = 75$, $PB = 250$, $AB = 500$, $AC = 300$, $CD = 250$, $QD = QE = DE = 150$, $\angle AOX = 30°$.

Find, for the configuration given, the instantaneous velocity of the point *E*.

SOLUTION: fig. 1.22b

NOTES:

Since O, P, and Q are 'fixed' points and have therefore zero relative velocity, the positions of o, p, and q in the velocity diagram will coincide. Relative to O (and therefore for the moment considering o as the 'pole' of the diagram to be drawn): the angular

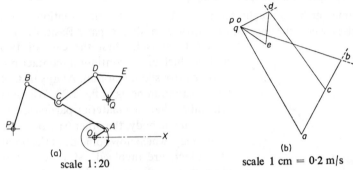

(a)
scale 1:20

(b)
scale 1 cm = 0·2 m/s

Fig. 1.22 Mechanism and velocity diagram

velocity of OA is 85 rev/min = $85 \times 2\pi/60$ rad/s $\simeq$ 8·9 rad/s, so the linear velocity of A is $8·9 \times 0·075 \simeq 0·667$ m/s perpendicular to OA. This is represented in fig. 1.22b by the vector oa.

The next point to be considered must be B, since its motion is controlled by its attachment to P and A, two points whose velocities are known.

BP is a rigid link. Relative to P (and therefore considering p (or o) as the pole) B's motion can be only perpendicular to BP. A 'b line' can therefore be drawn perpendicular to BP through p.

AB is a rigid link. Relative to A (and therefore considering for the moment a as the pole) B's motion can be only perpendicular to AB. A 'b line' can therefore be drawn perpendicular to AB through a.

Hence b is found.

Still considering motion relative to A, C's motion can only be perpendicular to AC: in other words, c lies on the line ab. Moreover, relative to A, C's velocity must be smaller than B's, in the ratio AC/AB. Hence (i.e. by applying corollary 2 above) we find c, and so on.

Finally, from first principles or by applying corollary 1, we find the velocity of E to be 0·17 m/s, at right angles to QE, in the sense shown by qe.

Of course, every other dimension in this diagram gives the velocity of one point relative to another; e.g.

dc is the velocity relative to *D* of *C*
eb is the velocity relative to *E* of *B*

and so on.

1.4.2 Mechanisms with sliding pairs

Some further thought must now be given to the situation which arises when the mechanism involves a sliding pair. Basically it is easiest to consider first the case of two bodies which are in continuous contact but have relative sliding motion. At any instant let the situation be as in fig. 1.23. We must remember that *two* material particles, one (say *A*) in one body, the other (say *B*) in the other body, which now coincide at the point of contact, are involved. Moreover, it is clear that at this point there must be a common tangent to the two surfaces, and therefore, perpendicular to this, a common normal. Now, whatever the motion, the point *B* can, at this instant, have *no* velocity relative to *A* along the common normal, since otherwise, at an infinitesimally short time earlier or later, the two surfaces would require to penetrate each other: it follows that *the velocity of B relative to A is wholly along the common tangent and that in the velocity diagram the vector ab will have this direction.*

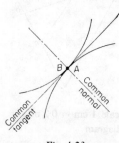

Fig. 1.23

When drawing a velocity diagram for a mechanism which involves a sliding pair, therefore, no difficulty arises *provided* these basic facts are kept in mind: first, that two separate points or material particles at the point of contact are involved (and should, to avoid confusion, *always* be separately lettered), and second, that the vector representing their relative velocity must be drawn parallel to the common tangent at the point of contact.

Worked Example 1.2: fig. 1.24a

Note that this is basically a quick-return mechanism as illustrated in fig. 1.3, but with very different proportions for the links. The crank *OB* of length 150 mm carries at its end a block *B* which slides on the link *PC*. *O* and *P* are fixed centres, 300 mm apart; *PC* is 525 mm in length. *OB* is rotating at 300 rev/min anti-clockwise, and at the instant shown $\angle POB = 135°$. What in this configuration is the velocity of C?

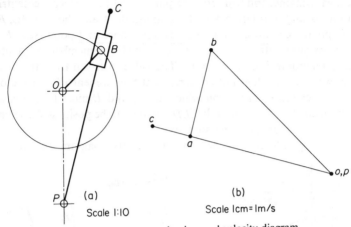

Fig. 1.24 Quick-return mechanism and velocity diagram

SOLUTION: fig. 1.24b

NOTES:

Velocity of $B = \left(\dfrac{2\pi}{60} \cdot 300\right) \cdot 0.15 = 4.71$ m/s $\perp OB$; hence ob.

Let A be the point on PC which at this instant coincides with B.
$ab \parallel PC$; $pa \perp PA$; hence a.

$$\frac{pc}{pa} = \frac{PC}{PA} = \frac{52\cdot5}{41\cdot4}\;;\; \text{hence } c.$$

ANSWER: Velocity of C is 5.1 m/s in direction pc.

1.4.3 Mechanisms with 'floating links'

Finally, we must consider the awkwardness which arises when the mechanism involves a 'floating link', i.e. a link (straight or curved) such as the member CED in fig. 1.25a, whose position is determined by constraints applied at more than two points along its length. In setting out on a drawing board the configuration of such a mechanism from its given dimensions difficulty arises because no individual point on the link can be located by the use of the usual simple tools, compasses and rules: the easiest way is to define the possible position or locus of each of the constraining points and then to set out the link on a piece of tracing paper and adjust its position until all the necessary conditions are satisfied. This is best illustrated by an example: fig. 1.25a shows the well-known Stephenson valve gear for a steam engine, and will serve the purpose. OA and OB are eccentrics fixed to, and so rotating with, the crankshaft O, and the rods AC, BD couple the eccentrics to a (curved) link whose mid-point (say) E is, for any given setting of the gear, positioned by a member PE pivoted at a temporarily fixed point P. (The valve is driven from the block shown, but this does not concern us at the moment.) In setting out this mechanism, curves on which C, D, and E must lie are easily obtained; the 'floating' link CED has then to be positioned by some method such as that suggested above.

Correspondingly, in drawing a velocity diagram for such a

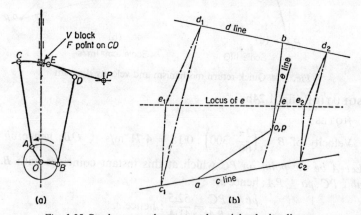

Fig. 1.25 Stephenson valve gear and partial velocity diagram

mechanism, no difficulty arises in obtaining the 'c', 'd', and 'e' *lines* as shown in fig. 1.25b, but none of these *points* can be located directly. The method of solution is to choose a number of possible positions for (say) c, calling them c_1, c_2, . . . and, ignoring for the moment the e line, to deduce corresponding positions of d and e (d_1, d_2, . . .; e_1, e_2, . . .). From inspection or from elementary geometry, it is clear that the locus of e so obtained is a straight line which intersects the original e line, giving the correct position of e The rest of the diagram can therefore be completed in a straightforward fashion: in Fig. 1.25 it has been omitted for clarity. The correct positions of c and d are found by drawing $ec \perp EC$ and $ed \perp ED$. The velocity image ced of CED is constructed by drawing a circular arc through the points c, e, and d. f is found from the relationship $cf:fd = CF:FD$. Relative to O, V has a vertical velocity, and relative to F, a velocity tangential to the link at F.

§ 1.5 Determination of Accelerations from Velocity Diagrams: Hodograph

Before we leave velocity diagrams it should be noted that the acceleration of any point in a mechanism can be estimated with reasonable accuracy by the simple expedient of drawing two super-imposed velocity diagrams, one for an instant preferably shortly before the mechanism reaches the configuration in which we are interested, the second for an instant preferably an equal interval of time after it reaches this configuration. The change in position of any point, say x, in the velocity diagram, is, of course, a measure of the change in the velocity of the corresponding point X in the mechanism during the time interval which has elapsed: if this velocity change is divided by the elapsed time the quotient is the average acceleration during this time. Clearly a limitation in this method is that if the time interval chosen is too long the average acceleration may differ materially from the acceleration we wish to know; if short enough to make this error small it may be difficult to measure the velocity change with sufficient accuracy. Nevertheless, the method has the advantages of simplicity and ease of comprehension. For this reason an example of its use will be given in § 1.8, where these advantages are particularly important.

If a vector drawn from a fixed origin continuously represents the velocity of a point it follows from the above argument that the velocity of the end of the vector is a measure of the acceleration of the point. The locus of the end of the vector is called a *hodograph*.

§ 1.6 Instantaneous Centres

The use of the concept of instantaneous centres may be regarded as an alternative to that of the velocity diagram as a means of determining the relative velocities in a mechanism. The major advantage is that it frequently provides a clearer insight into the reasons for the behaviour of a mechanism, and so assists a designer to improve his initial design. The major disadvantage is that, as the mechanisms considered become more complicated, the difficulties involved increase disproportionately.

We may conveniently introduce the concept by again considering (fig. 1.26a; cf. fig. 1.18) a rigid body or element containing two points A and B. Let us again assume that we know the velocity of A in magnitude and direction, and the direction of motion, but not the speed, of B; and take as our objective the determination of the speed of B. We note that the data can, in general, apply only to an instant in time, since at any earlier or later instant the position of the body, and in all probability the velocities, will be different.

Consider, then, the situation at this instant. We know the velocity, and hence the direction of motion of A. The body *may* therefore be rotating about some point in the line aa, which has been drawn

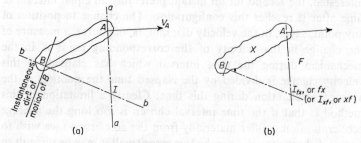

Fig. 1.26 Instantaneous centre relative to 'fixed' frame

through A perpendicular to V_A, but cannot possibly be rotating about any other point. (It is strictly unnecessary to use the word 'may', since the only other possibility is that the body is moving *without* rotation, and this is mathematically equivalent to saying that the point in aa about which the rotation is occurring is infinitely far away.) Similarly, since we know the direction of motion of B, we can draw perpendicularly to it, through B, the line bb, and state that the body 'may', or 'must', be at this instant rotating about some point in bb. It follows that at this instant the body is rotating about the point I in which aa and bb coincide, and I is called the 'instantaneous centre of rotation' of the body. It is then clear that the angular velocity, ω, of the body about I is V_A/IA; and that the velocity of B is $\omega.IB$.

$$\text{hence } \frac{V_B}{V_A} = \frac{IB}{IA}$$

Although we have achieved our objective, it is worth while to pause for a moment and compare the situation with that outlined in § 1.4. Again we have treated the frame of reference rather scurvily: we have, in fact, ignored it completely, and must therefore have been tacitly assuming that it was 'fixed'. This is a wholly unnecessary restriction which may appear to simplify consideration of a problem as elementary as the one we have chosen, but it is one which would stultify any attempt to deal with a more complicated mechanism. If we amend our ways now we shall avoid serious difficulties later on. In fact, none of the arguments has to be modified: all we need to do is to add 'relative to a frame of reference F' where appropriate. What may usefully be changed is our system of notation. As soon as we realise that the instantaneous centre we have found is that of the whole body or element, which we may call X, relative to the frame of reference, which we have called F, we can agree to call this instantaneous centre I_{fx}, as in fig. 1.26b (or fx if we are short of space in a diagram, as often happens), thus specifying both and so allowing for the consideration, without confusion, of other elements in the system or mechanism.

Before we leave this elementary example we must take the opportunity to understand fully a few corollaries. The first is straightforward. At the instant considered the body X is rotating, relative to F, about the point I_{fx}. It is equally true to say that the frame or 'body' F is rotating, relative to X, round this same point, which we might therefore equally well designate I_{xf}. The points I_{fx} and I_{xf} are essentially identical; there is therefore no need to distinguish between them.

To appreciate the second corollary, we may imagine that the bodies F and X are, if necessary, enlarged, so that they overlap at the instantaneous centre. We can further imagine that they could be momentarily linked by a pin joint at this point, the pin passing through both bodies. This may help us to realise that we are dealing with *two* points, one 'in' (or to be considered in relation to) each body. These points at this instant coincide, and have no relative velocity. Now let a short interval of time elapse. Again at this second instant it will be possible to find the instantaneous centre of one body relative to the other, and this instantaneous centre will define a new point in, or relative to, each body. We may repeat this consideration as often as we wish, and so trace out a series of successive instantaneous centres in each of the bodies. The locus of each is called a 'centrode'. Now we have just seen that at successive instants successive points on these two centrodes move into coincidence, coincide without relative velocity, and separate; in other words, *as the bodies move, the two centrodes roll round each other.*

In such general terms, centrodes are not easy to visualise, but a specific example or two will help to clarify the situation. Take first the case (fig. 1.27) of a wheel W rolling without slipping along the ground G. By inspection we can see that at the instant depicted the instantaneous centre I_{gw} is at the point of contact.* This, then, is a point on the wheel centrode *and* on the ground centrode. If we now consider a series of instants in time separated by short intervals we, of course, find that the instantaneous centre is always, as far as the wheel is concerned, a point on its rim; and as far as the ground is concerned, a point on its surface. In other words, the rim and the surface *are* the wheel centrode and the ground

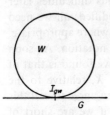

Fig. 1.27 Centrodes: rolling wheel

centrode respectively, and we started by assuming that one was rolling on the other. The wheel need not be round, nor the ground flat. Neither of the bodies need be regarded as fixed in space: thus, the argument applies equally well, for instance, to a pair of gear wheels, whose relative instantaneous centre is the 'pitch point'† and whose relative centrodes are the 'pitch circles'.† When it is desirable to emphasise that one of the bodies is to be regarded as fixed in space the centrodes may be called body and space centrodes.

The example taken as a first illustration was chosen because the centrodes coincide with physical surfaces and are therefore easy to appreciate. In contrast, we may take the case of the kinematic chain shown in fig. 1.28, in which the ends of a bar B are constrained to move in rectangular directions by means of blocks in slots in the disk D.

By regarding temporarily the member D as our frame of reference, we see that the instantaneous centre I_{bd} is at the position shown.

* If the point on the wheel corresponding to I_{gw} had any velocity normal to the ground it would, at an infinitesimally short time earlier or later, have to penetrate the ground; if it had any velocity tangential to the ground the wheel would be slipping. The student should be careful to see that he really understands this argument, that having accepted it he can prove to himself that the centre of the wheel has – with the usual notation – a horizontal velocity of ωr, that the top of the wheel has a horizontal velocity of $2\omega r$, and that he can state immediately the direction of motion, relative to the ground, of any point on the wheel.

† For a definition of these terms see § 2.6.

To find the two centrodes it is necessary to specify I_{bd} *as a point relative to each of the bodies B and D in turn.* Relative to B we note that the angle $YI_{bd}X$ is a right angle: it follows that I_{bd} is always a point on a circle with diameter XY, i.e. the bar itself. This circle is therefore the bar centrode. Relative to D, we note that the figure $(XO\,YI_{bd})$ is a rectangle: since the diameters of a rectangle are equal, it follows that I_{bd} is a point at a constant distance XY from the centre O of the disk. The disk centrode is thus defined as a circle with O as centre and the length of the

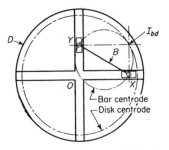

Fig. 1.28 Centrodes: 'elliptical trammels'

bar as radius. As the ends of the bar move in the slots, the circle which is the bar centrode rolls inside the circle of twice its diameter which is the disk centrode.

(This may be regarded simply as an exercise in finding centrodes which correspond to no physical surfaces. Certain other aspects may, however, be mentioned briefly in passing. To those familiar with the properties of cycloids,* it will be evident that each slot in the disk may be regarded as the hypocycloid traced out by a point on the smaller circle rolling inside the larger circle. Again, those familiar with the device known as 'elliptical trammels' or the 'ellipsograph' will recognise it here, and be aware that any point on the bar B or its extension traces out an ellipse relative to D. Again, another application of this kinematic chain is obtained by 'fixing in space' the centres of the disk and of the bar, but allowing both to rotate. This provides a mechanism whereby two shafts may be coupled to provide without gears a change in angular speed of $2:1$, a fact which is easily understood in the light of knowledge of the centrodes, but would hardly otherwise be evident. See example 1.13 at the end of this chapter.)

After this somewhat lengthy discussion of centrodes, we must return to the second corollary which led up to it. It will be remembered that we were considering two bodies 'overlapping' at their relative instantaneous centre and momentarily linked in imagination by a pin passing through each. We had agreed that two points, one in each body, had to be considered, and that their relative velocity was at this instant zero. It follows that they must have the same

* See also §§ 2.6.3 and 2.6.4.

velocity relative to any third body which we may care to consider as a new frame of reference.

Now let A, B, and C (fig. 1.29) be three bodies in motion, and let us assume that we have found the three instantaneous centres, I_{ab}, I_{bc}, I_{ac}, according to the notation we have chosen. I_{ab}, the instantaneous centre of B relative to A (or of A relative to B) defines a point in each of these bodies, linked, we may imagine, by a pin. At this same instant the centre of rotation of A relative to C is I_{ac}:

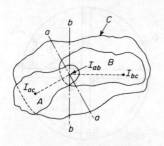

Fig. 1.29 Relative instantaneous centres

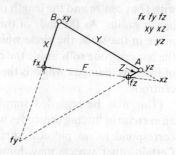

Fig. 1.30 Mechanism and instantaneous centres

relative to C, the pin at I_{ab} must therefore be moving in the direction *aa*. Similarly, the centre of rotation of B relative to C is I_{bc}: relative to C, the pin must therefore be moving in the direction *bb*. But this is clearly impossible unless the directions *aa* and *bb* coincide, and this will be so only if the three instantaneous centres we originally assumed are in a straight line.

This is an important result, which in fact provides a method which can be used to find all but the most obvious instantaneous centres in a complicated mechanism. It is worthy of repetition in an explicit form. *If three bodies, A, B, and C have relative instantaneous centres I_{ab}, I_{bc}, I_{ac}, these three instantaneous centres must lie in a straight line.*

We can best illustrate the application of this theorem by means of an example, and we shall deliberately, so that methods and results can be compared, choose first a part of, and then the whole, mechanism illustrated in fig. 1.22. The part is shown again in fig. 1.30, though it will be noticed that letters have been added, each of which represents, *not* a joint or pair, but an element or member – not forgetting the frame which has been denoted by F.

It is worth while, certainly until the methods have been mastered, to note down in logical sequence, as shown inset in fig. 1.30, a list of all the possible instantaneous centres relating to the mechanism, that is of each member relative to each of the others. (The total is the number of combinations of n members taken two at a time, or $^nC_2$, i.e. in this case $^4C_2$ or $\dfrac{4 \times 3}{1 \times 2}$ or 6.) It is usually unnecessary to locate all these centres to solve a particular problem, but at this stage it will be instructive to do so.

Two of the centres are immediately obvious, in the sense that, relative to the frame F, members Z and X are constrained to rotate about the points which have been labelled $_{fz}$ and $_{fx}$ (shorthand for I_{fz} and I_{fx}) respectively. It requires little more imagination to regard the member Y as fixed, whereupon I_{xy} and I_{yz} are located as shown. If we regard the members X and Z in turn as fixed, we merely confirm the centres already found. Two centres, I_{fy} and I_{xz}, remain to be found. Taking I_{fy} first, we can use our theorem to say that: (1) I_{xy}, I_{fx}, and I_{fy} must be collinear, and (2) I_{yz}, I_{fz}, and I_{fy} must be collinear. These two facts give us the I_{fy} lines which converge at the point indicated, and similar reasoning gives the location of I_{xz}.

If our problem were now to be particularised by assuming, for instance, that the angular velocity of the crank Z, as in example 1.1, fig. 1.22, was known and the angular velocity of the member X had to be found, we should have to note that the member linking these was the connecting-rod Y, which at this instant is rotating about I_{fy}. The ratio of the linear velocity of the ends of this rod (points A and B) is therefore given by $V_B/V_A = BI_{fy}/AI_{fy}$: the rest of the calculation is trivial.

The intelligent student will note at this stage that he has gone a very long way round to reach this result. The only instantaneous centre needed for this particular purpose is I_{fy}. What was to prevent him saying to himself, 'The end A of member Y is moving at right angles to the crank Z. I_{fy} must therefore lie on the line of Z. The end B is moving at right angles to the link X: I_{fy} must therefore lie on the line of X. Produce the lines of X and Z to meet in I_{fy}'? The answer is 'nothing', and this illustrates the peculiar virtue of the method of instantaneous centres; that in many simple mechanisms the centres required for a particular purpose can be located very easily indeed. Often the centres relative to the frame of reference are sufficient in themselves.

These lessons, that the method of instantaneous centres used indiscriminately or for general purposes can be extremely laborious but used with discrimination and intelligence for particular purposes can be quick and neat, may be further illustrated by taking now the whole mechanism of the Worked Example 1.1 and finding (*a*) all the relative instantaneous centres, and (*b*) only those required

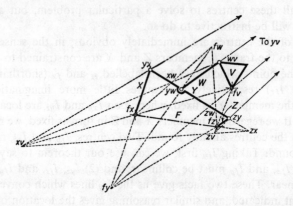

Fig. 1.31 Mechanism of fig. 1.22 and all instantaneous centres

for the purpose of determining the velocity of the point *E*. The first is illustrated in fig. 1.31, and all the centres have been found by the use of the theorem proved above. The second is illustrated in fig.

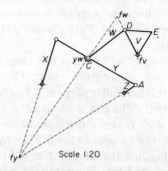

Scale 1:20

Fig. 1.32 Mechanism of fig. 1.22 and important instantaneous centres

1.32, where attention has been concentrated on those centres, I_{fy}, I_{fw}, and I_{fv}, which are required. (Of these, I_{fv} is obvious, I_{fy} has been already found, and I_{fw} follows from exactly analagous reasoning, or from a more judicious application of the theorem.) The student should carry out each of these exercises for himself, but the former is hardly to be recommended as an efficient method of solving problems.

The required answer, the speed of E, is found from fig. 1.32 thus:

$$\frac{V_E}{V_A} = \frac{V_E}{V_D} \cdot \frac{V_D}{V_C} \cdot \frac{V_C}{V_A} = \frac{I_{fv}E}{I_{fv}D} \cdot \frac{I_{fw}D}{I_{fw}C} \cdot \frac{I_{fy}C}{I_{fy}A}$$

$$= \frac{15}{15} \times \frac{8}{27} \times \frac{66}{75} = 0 \cdot 26$$

$\therefore V_E \simeq 0 \cdot 17$ m/s as before,* perpendicular to $I_{fv}E$.

§ 1.7 Acceleration Diagrams

Like velocity, acceleration is a vector quantity and can be specified only by quoting its *magnitude* and *direction* relative to a given *frame of reference*, and illustrated by a vector drawn to a known scale from a point representing the frame of reference. It is, however, much more difficult to visualise accelerations than velocities. The reason is fairly simple. If a particle is moving in a known path its velocity is at all times tangent to that path: that may be regarded as 'obvious'. Its acceleration, however, may have any direction (through an arc of 180° towards the concave side of the path) relative to this tangent, and coincides with the tangent only if the path happens to be straight. Even the direction of the acceleration is therefore *not* 'obvious'.

Before attempting to draw acceleration diagrams for mechanisms we must be sure that the basic principles outlined in § 0.11 are understood. In this paragraph it is shown that a particle P travelling in a circular arc of radius r at an angular speed ω *must* have an acceleration of $\omega^2 r$ ($= \omega.v = v^2/r$, where $v = \omega r$ is the linear speed of the particle) and that this acceleration is 'centripetal' or directed towards the centre of curvature of the path. It *may* also have an acceleration along its path, called a 'tangential' acceleration. This acceleration will, of course, be in the same direction as the motion of P if its speed is increasing, or in the opposite sense if its speed is decreasing; it will be zero only if the speed happens to be constant. Thus, the *total* acceleration of P will be the vector sum of the centripetal and tangential components, as sketched in fig. 1.33.

These principles can now be applied to elements in mechanisms. As in the case of velocity diagrams, it is well to start with a single

* Worked Example 1.1.

rigid link, *AB*, and it may be assumed that we are given complete information (relative to a frame of reference) about the velocity and acceleration of *A*, but that only the directions of the velocity and acceleration of *B* are given, as sketched in fig. 1.34. We must first

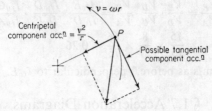

Fig. 1.33 Acceleration of point moving in a circle

determine the velocity of *B* relative to *A*, and this offers no difficulty: the method is described in § 1.4 and the result is *ab* as indicated in fig. 1.35.

We can now apply exactly analogous reasoning to the acceleration problem. We start by choosing a pole *o′* to represent the frame of

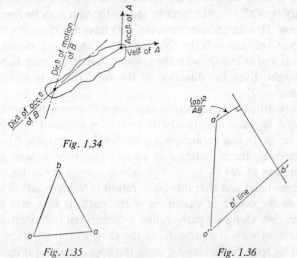

Fig. 1.34

Fig. 1.35   *Fig. 1.36*

reference relative to which the velocities and accelerations have been measured (fig. 1.36). From *o′* we draw a vector *o′a′* to represent in magnitude and direction the acceleration of *A*. Also from *o′* we start to draw a vector *o′b′* to represent the acceleration of *B*, but since we do not know the magnitude, we have to leave the length open: the line we can describe as a '*b′* line'.

We now have to look at the situation as it will appear to our observer moving with A. To him, at the centre of his universe, a and a' are, of course, the poles from which the vectors ao and $a'o'$, representing the velocity and acceleration of O, and the vector ab, representing the velocity of B, have been drawn. From his point of view (since the link AB is rigid), B is a point moving round him with speed ab at a constant distance, that is, in a circle. He therefore knows that B *must* have a component acceleration towards him $(= v^2/r = (ab)^2/AB)$ and *may* have a tangential component acceleration also. These facts he can illustrate in fig. 1.36 by drawing a vector from a' in the direction BA of length $(ab)^2/AB$ to represent the centripetal component, and adding to it a line drawn in the tangential (i.e. $\perp^{ar}$ to BA) direction of unknown length on which b' must, from his knowledge, lie. Thus, again we have two separate indications of the position of b', which is therefore defined.

(The beginner may have a little anxiety about the scales to be used in this operation. There is, in fact, no difficulty. On a drawing board the quantities ab and AB will normally be measured in convenient units, for instance mm. Each must, of course, be translated, ab into m/s and AB into m, when the calculation $(ab)^2/AB$ gives m/s² automatically.)

It is perfectly possible (though certainly not initially advisable) to reduce the process just described to a rule of thumb. If AB is a rigid link in a mechanism and ab is its velocity image the centripetal component acceleration will be of length $(ab)^2/AB$ and its direction will be 'opposite' to that of the link: the tangential component will be perpendicular to it.

Corollaries corresponding closely to those in § 1.4.1 are valid and should be proved by each student for himself at this stage.

1. If ABC is a rigid triangular link in a mechanism, the acceleration diagram will contain a similar triangle $a'b'c'$.

2. If ABC are three points on a straight link the acceleration diagram will contain a straight line $a'b'c'$ such that $a'b': b'c' = AB: BC$.

(These similar figures are often referred to as the acceleration images of the links.)

With the aid of the single principle outlined above, many acceleration diagrams may be constructed, in particular those involving only turning pairs, or in which a sliding member moves only in relation to a fixed member. The very important but more difficult case of a mechanism involving a moving sliding pair will be dealt with in the next paragraph.

Examples involving only the principles above should be tackled by the student at this stage: one worked example follows.

Worked Example 1.3

The heavy lines in fig. 1.37a show diagrammatically to scale part of the mechanism of an opposed-piston engine which should need little explanation. The frame of the engine carries the cylinder Q, the

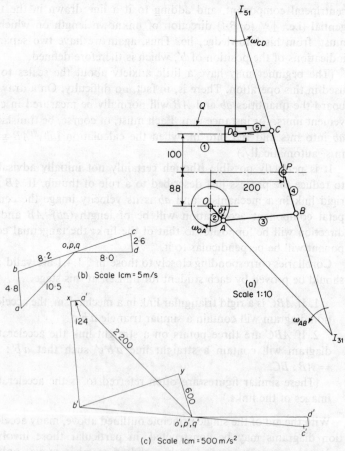

(b) Scale 1cm = 5m/s

(a) Scale 1:10

(c) Scale 1cm = 500 m/s²

Fig. 1.37 Engine mechanism and acceleration diagram

main crankshaft bearings O, and the pivot P. One of the pistons, D, drives the crank OA through the linkage DC, CPB, BA. CPB is a rigid link. $OA = 50$ mm, $AB = 200$ mm, $BP = 112$ mm, $BC = 224$ mm, $CD = 100$ mm. The crankshaft speed is 2 000 rev/min clockwise. Find the (velocity and) acceleration of the piston when the mechanism is in the configuration shown, i.e. when $\theta = 45°$.

Before we tackle the acceleration diagram, we require to know various velocities, and perhaps the most straightforward procedure is to draw the velocity diagram as in fig. 1.37b; this should give no difficulty. Alternatively, since in fact our strict needs are limited to the angular velocities of the links, we may prefer to find the instantaneous centres of the links, which are numbered, relative to the frame. These can all be determined by inspection (or by applying the 'rules' outlined in § 1.6), and the only two which are needed and are not immediately apparent are shown in fig. 1.37a, the construction lines being drawn faintly relative to the mechanism.

From the data we know that

$$V_A = \omega_{OA} \cdot OA = \frac{2\,000 \times 2\pi}{60} \times \frac{50}{1\,000} = 10\cdot5 \text{ m/s}$$

From the velocity diagram we find that the *velocity of the piston*, *od*, *is* 8 m/s; and in addition obtain the information that $ab = 4\cdot8$ $pb = 8\cdot2$, $dc = 2\cdot6$ m/s.

Alternatively, from the instantaneous centre diagram we deduce that:

$$\omega_{AB} = V_A/AI_{31} = \omega_{OA} \cdot OA/AI_{31} = 24\cdot8 \text{ rad/s}$$

Similarly:

$$\omega_{BC} = 73 \text{ rad/s}$$

and $\quad \omega_{DC} = 25\cdot5 \text{ rad/s}$

and in particular $V_D = \omega_{DC} \cdot DI_{51} = 8$ m/s.

We start the acceleration diagram (fig. 1.37c) by choosing a pole $(o'p'q')$ to represent the frame of the machine. Relative to this frame we know A's (total) acceleration $(-\omega_{OA}^2 \cdot OA = -V_A^2/OA = -2\,200 \text{ m/s}^2)$ which we represent by an appropriate vector $o'a'$. Just as in drawing the velocity diagram, we have to consider next the point B, whose motion is clearly controlled by the links PB, AB. P is a fixed point; PB is a rigid link; B is therefore rotating about P with a linear velocity pb (or angular velocity ω_{BC}), and must have a centripetal acceleration of V^2/r (or $\omega^2 r) = (bp)^2/BP = 600 \text{ m/s}^2$. Thus, we can represent this centripetal component of acceleration by the appropriate vector $p'y$ (y is a letter chosen at random to have no ulterior significance). But of course B's speed is not likely to be constant: we therefore accept the fact that B may have in addition a tangential acceleration. Through y, therefore, we draw a line perpendicular to $p'y$ on which b' must lie.

We now have to consider the acceleration of B relative to A, and so have to regard the situation from the viewpoint of an observer moving with A, to whom a' is the pole of the acceleration diagram. Apart from this the method is identical. The centripetal acceleration of B relative to A is $(ab)^2/AB$ or $\omega^2_{AB} . AB$; this we represent on the diagram by the vector $a'x$, and through x we draw a b' line perpendicular to $a'x$ to represent the unknown tangential component acceleration of B relative to A. The crossing point of these two b' lines gives the point b'.

The link BPC is continuous and $BP = PC$, so relative to P the acceleration of C will be in the opposite sense to that of B, and equal in magnitude. We therefore find the point c', and so on.

d' is found by: (1) considering the acceleration of D relative to C, and (2) appreciating that D is constrained to accelerate in the horizontal direction.

The magnitudes given in figs. 1.37b and 1.37c should be more than sufficient in the way of assistance, but other magnitudes can be scaled off if required. The acceleration of D is found to be 1 620 m/s² to the right.

§ 1.8 Acceleration Diagrams (continued): Coriolis's Term

A sliding pair in a mechanism for which the velocity diagram had to be constructed gave rise to little difficulty, and was dealt with in §1.4.2. The situation is more awkward when we come to acceleration, and must be considered rather carefully. We can deal with it either by using the hodograph or by using a mathematical approach (compare the treatment of motion in a circle in § 0.11). The choice is a matter of taste; but it may reasonably be said that the former gives a better physical insight into the problem, and the latter a greater certainty that no factors have been overlooked. It is worth while to consider both.

Hodograph method

Let us suppose (fig. 1.38a) that a block B is sliding with uniform speed $\dot{r}$ along a link OL which, relative to 'fixed' axes xOy, is rotating with uniform angular speed ω about O. Our problem is to determine the acceleration of the block.

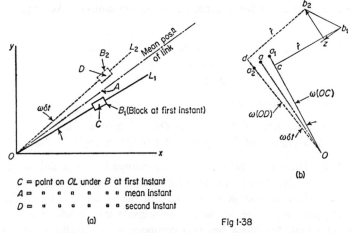

C = point on OL under B at first instant
A = " " " " " " " mean instant
D = " " " " " " " second instant

(a)   Fig 1·38

Fig. 1.38 Block sliding on rotating link: hodograph

Following the method outlined in § 1.5, we sketch the mechanism at two instants in time δt apart just before and just after the instant at which we wish to know the acceleration. The first position of the mechanism is shown in full lines; OL_1 is the link, B_1 the block, and C the *point on the link OL* with which B_1 momentarily coincides. The second position, OL_2, B_2, and D, the point on OL with which B_2 now momentarily coincides, is shown in dotted lines. The mean position, in which B coincides with A, is indicated. The angle L_1OL_2 is, of course, $\omega . \delta t$.

There is no difficulty in drawing the corresponding velocity diagrams as in fig. 1.38b. In the first case the velocity of C is $\omega . OC$ perpendicular to OL_1, and is represented by oc. The velocity of B relative to C is $\dot{r}$ parallel to OL_1, so we arrive at the total velocity of B_1, ob_1. The velocity of A at this instant is indicated by oa_1. In the second case the components are $\omega . OD$ perpendicular to OL_2 and $\dot{r}$ parallel to OL_2, so we arrive at the velocity of B_2, ob_2. The total change in velocity of B is b_1b_2, and the mean acceleration therefore $b_1b_2/\delta t$.

This acceleration can be split into two components $b_1z/_{\partial t}$ parallel, and $zb_2/_{\partial t}$ perpendicular, to the mean direction of the link OL. To a good approximation (which will become exact as $\delta t \rightarrow o$)$b_1z = a_1a_2$, so we see that B has the same centripetal acceleration as A, namely $-\omega^2 r$. Similarly, we can see that the total magnitude of zb_2 is derived from two factors, first that od is longer than oc, and second that cb_1 and db_2 differ in direction by an amount $\omega\delta t$. Adding,

$$
\begin{aligned}
zb_2 &\simeq (od - oc) + \dot{r}(\omega\delta t) \\
&\simeq \omega(OD - OC) + \dot{r}(\omega\delta t) \\
&\simeq \omega(\dot{r}\delta t) + \dot{r}(\omega\delta t) \\
&\simeq 2\omega\dot{r}\delta t
\end{aligned}
$$

so in the limit as $\delta t \rightarrow 0$

Acceleration of B parallel to $OL \simeq b_1z/\delta t = -\omega^2 r$
Acceleration of B perpendicular to $OL \simeq zb_2/\delta t = 2\omega\dot{r}$

half this latter term being due to the change in tangential velocity caused by the change in effective radius from OB_1 to OB_2, the other half to the change in direction of the radial velocity $\dot{r}$. Thus, in the particular case we have been considering, in which ω and $\dot{r}$ are constant, the block has two component accelerations, one the centripetal acceleration of the point A underneath it, the second an acceleration relative to this point in a direction perpendicular to the link and of magnitude $2\omega\dot{r}$. Moreover, the hodograph makes the physical explanation of this latter acceleration (which it is very easy to overlook) clear. Its existence was first discovered by G. G. Coriolis (1792–1843), and it is usually called by his name.

It is not difficult to see that if ω and $\dot{r}$ are not constant two more terms will arise, one the tangential acceleration of the point A on the link, of magnitude $\dot{\omega}r$, and the second the radial acceleration of B relative to A of amount $\ddot{r}$. Thus, the block has four component accelerations. It is, however, probably advisable to tackle this general problem by normal mathematical methods, exactly as in § 0.11.

Again, let us suppose (fig. 1.39) that a block B is sliding along a link OL, which is itself rotating at angular speed ω, which need not be constant, about its end O relative to 'fixed' axes xOy, and that we have to find the acceleration (relative to these axes) of B. Let the Cartesian and polar co-ordinates of B be (x, y) and (r, θ). Then $x = r\cos\theta$ (all these quantities being variables with respect to time).

Differentiating twice with respect to time

$$\dot{x} = \dot{r}\cos\theta - r\sin\theta \,.\, \dot{\theta}$$
$$\ddot{x} = \ddot{r}\cos\theta - 2\dot{r}\sin\theta \,.\, \dot{\theta} - r\cos\theta\,(\dot{\theta})^2 - r\sin\theta \,.\, \ddot{\theta}$$

Similarly

$$y = r\sin\theta$$
$$\dot{y} = \dot{r}\sin\theta + r\cos\theta \,.\, \dot{\theta}$$
$$\ddot{y} = \ddot{r}\sin\theta + 2\dot{r}\cos\theta \,.\, \dot{\theta} - r\sin\theta\,(\dot{\theta})^2 + r\cos\theta \,.\, \ddot{\theta}$$

By inspection of these expressions for $\ddot{x}$ and $\ddot{y}$ we can see that they combine to give four component accelerations, of which two are parallel to the link and two perpendicular to it (compare § 0.11 and fig. 0.10). Thus, for instance, the first terms combine to give the

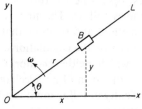

Fig. 1.39 Block sliding on rotating link: calculation of acceleration

component acceleration $\ddot{r}$ along the link, the second term $2\dot{r}\dot{\theta}$ at an angle $+\pi/2$ to it, and so on. The same result can be obtained rather more simply by *choosing* the direction of the axis OX to coincide with *OL at the instant considered*. At this instant, then, $\theta = 0$, so $\sin\theta = 0$, $\cos\theta = 1$, and the expressions for the component accelerations reduce to

$$\ddot{x} = \ddot{r} - r(\dot{\theta})^2 \qquad \text{(along the link)}$$
$$\ddot{y} = 2\dot{r}\dot{\theta} + r\ddot{\theta} \qquad \text{(perpendicular to the link)}$$

and we might now adopt our more normal terminology of writing ω for the instantaneous angular speed of OL instead of $\dot{\theta}$, and $\dot{\omega}$ instead of $\ddot{\theta}$, and rearrange the order to give

$$\ddot{x} = -\omega^2 r + \ddot{r}$$
$$\ddot{y} = \dot{\omega} r + 2\omega\dot{r} \tag{1.3}$$

which are the required component accelerations of the block B in the direction of, and perpendicular to, the link.

If we again call A the point on OL which at the instant considered coincides with the centre of the block, we shall have no difficulty in writing down the component accelerations of A. A is merely a point moving round O in a circular path with varying speed, and has the usual centripetal and tangential accelerations.

For A, then

$$\ddot{x} = -\omega^2 r$$
$$\ddot{y} = \dot{\omega} r$$

By comparing these accelerations of B and A we can see at a glance that the component accelerations of B relative to A must be

$$\ddot{x} = \ddot{r}$$
$$\ddot{y} = 2\omega\dot{r}$$

Thus, we have obtained the same results as before, by a method which to some will appear less satisfying because providing less physical insight into the problem, and to others will appeal as being much quicker and less open to error through an oversight.

Two worked examples follow. The first is intended to illustrate only the hodograph method and to show that the numerical results are those which are given by the calculations. The second is a more practical example in which neither ω nor $\dot{r}$ is constant, and the solution provided is in the form of an acceleration diagram, but the student who finds difficulty in following it is advised to construct the hodograph for this case also.

Worked Example 1.4

A link OL is rotating about O with a constant speed of 20 rev/s, and a block B is sliding outwards along OL with a constant speed relative to the link of 30 m/s. At a particular instant the configuration is that shown in full in fig. 1.40a, when the block is at a distance of 0·25 m from O. Find relative to 'fixed' axes the acceleration of the block.

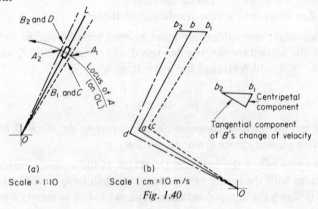

B_2 and D
L
A_1
A_2
B_1 and C
Locus of A (on OL)
O

b_2 b b_1
b_2 b_1 Centripetal component
Tangential component of B's change of velocity
d a c
O

(a)
Scale = 1:10

(b)
Scale 1 cm = 10 m/s

Fig. 1.40

A convenient time interval is 0·0005 s. In this time the link will rotate through 0·01 rev and the block will move along it a distance of 0·015 m. The configurations 0·0005 s before and after the instant are shown in dotted and chain-dotted lines respectively, and the subscripts 1 and 2 refer to these configurations. At the earlier time the block will coincide with a point C on the link, where $OC = 0·235$ m. The speed of C is $40\pi \times 0·235$ m/s, the speed of B relative to C is 30 m/s, and there is no difficulty in drawing the velocity diagram ocb_1. Similarly, we can draw the velocity diagram odb_2 for the later time (and oab, if we wish, for the instant itself). The change in velocity of B during the total time interval of 0·001 s is therefore given in magnitude and direction by the vector b_1b_2, and the mean acceleration during this time is easily found. The small inset shows how this acceleration can be split up into centripetal and tangential components, and these will be found to agree with the calculated values of $\omega^2 r$ (3 950 m/s^2) and $2\omega\dot{r}$ (7 550 m/s^2) respectively, to the order of accuracy with which such a drawing can be made.

Worked Example 1.5
A crank OB, of length 500 mm, rotating clockwise at a constant speed of 10 rad/s about a fixed centre O, carries at its end a block B which slides on a link PL of length 1 000 mm. P is another fixed centre distant 250 mm from O. At the instant when the configuration is as shown in fig. 1.41a find the acceleration of L.

METHOD

Let the point in PL at this instant coinciding with the centre of the block be A. Let ω, $\dot{\omega}$ refer to the link PL. Draw the velocity diagram (a perfectly straightforward operation, see § 1.4) as in fig. 1.41b. From these two diagrams we can obtain by measurement the quantities we require to draw the acceleration diagram. Thus with the notation we have established

$r = PB = PA = 670$ mm

$\dot{r} = ab = -1·65$ m/s   (N.B. negative because r is *decreasing* in magnitude)

$\omega = \dfrac{pa}{PA} = -7·1$ rad/s   (N.B. negative because the link is rotating in the clockwise direction, i.e. from the x-axis *away* from the y-axis)

(Particular care should be given to these signs: it should be noted that we are adapting the standard mathematical conventions (see § 0.2) throughout. We take the *x*-axis along the link *PL*, and therefore the *y*-axis in the direction indicated in fig. 1.41a; this establishes

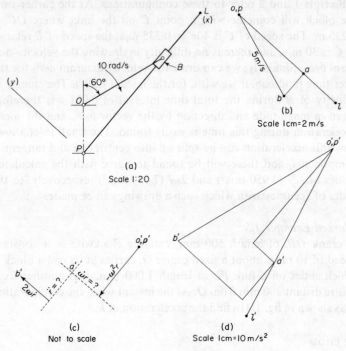

(a)
Scale 1:20

(b)
Scale 1cm=2 m/s

(c)
Not to scale

(d)
Scale 1cm=10 m/s²

Fig. 1.41

anti-clockwise rotation, from the *x*- to the *y*-axis, as positive. The radius vector *r* is always positive; inspection of fig. 1.41a *or* consideration of the direction of *ab* in fig. 1.41b shows that *r* is decreasing in magnitude, so $\dot{r}$ must be negative.)

It is now desirable, to avoid confusion, to direct our attention to the acceleration of *B*, considered first as a block sliding on the rotating link. We can write down the four terms in the two component accelerations thus:

in *x* direction $-\omega^2 r = -(-7\cdot1)^2 \times 0\cdot67$

$\quad\quad\quad\quad\quad\quad\quad\quad = -33\cdot8$ m/s² and these give the acceleration of *A*

in *y* direction $\dot{\omega}r$ is unknown in magnitude and sign in x direction $\ddot{r}$ is unknown in magnitude and sign $\Bigg\}$ and these give the acceleration of B relative to A.

in y direction $+2\omega\dot{r} = 2(-7\cdot1)(-1\cdot65)$
$$= +23\cdot5 \text{ m/s}^2$$

And it is helpful if we *sketch* their vectorial addition in that order, inserting merely dotted lines for the second and third components whose magnitude and signs are unknown: see fig. 1.41c. The components of whose magnitudes and directions we are certain have been decorated (for once) with arrow-heads because we are considering specifically the acceleration of A and B relative to O.

Of course we have not thus found the correct location of a' or b': we have merely indicated the route to be followed in locating them.

We now consider B as the end of the crank OB, from which it appears that we *know* the correct location of b'. Its total acceleration is centripetal, of magnitude $(ob)^2/OB$ or 50 m/s². We can now draw the acceleration diagram properly (fig. 1.41d) by starting with *this* vector $o'b'$, drawing the $-\omega^2 r$ component 'forwards' from o' (or p'), drawing the $2\omega r$ component 'backwards' from b', and filling in the two remaining components in their only possible places.

So we find the point a': the rest of the problem should give no difficulty.

The method indicated may at first sight seem laborious and unnecessarily complicated. The reason for adopting it is simply that if care is taken no trouble arises over signs. It is, however, quite possible to adopt a more straightforward method if one has appreciated to the full the physical significance of the various component accelerations.

§ 1.9 Forces Required to Accelerate Elements of Machines

If the three 'Laws of Motion' enunciated by Newton (see § 0.6) have been thoroughly understood it is unnecessary to devote much time to detailed consideration of the forces required to accelerate elements in a machine. Each such element can be regarded as a body, of which the translational acceleration of the centre of gravity, and the rotational acceleration, can be determined by the methods we have been discussing. Let the mass of the body be M, its moment of inertia about its centre of gravity G be $I = Mk^2$, the translational acceleration of its C.G. be a_G, and its rotational acceleration be $\dot{\omega}$.

We can then write down the magnitude of the force required to accelerate the body translationally as $F_G = Ma_G$ and we know that the direction of this force is the direction of a_G. This force, if acting through the C.G. of the body, will, of course, produce no angular acceleration, for which we must have in addition a torque or couple $T = I\dot{\omega}$. It does not matter in the least whether we regard these two components F_G and T of the total force separately, or whether we combine them to give a total force $F = F_G$ acting on the body in a direction parallel to F_G but at a distance d such that $F \times d = T$ (see fig. 1.42).

An alternative method of determining this total force F depends essentially upon 'replacing' the body by two 'dynamically equivalent' point masses. For many purposes* this method has considerable

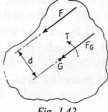

Fig. 1.42

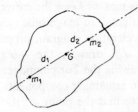

Fig. 1.43 Replacing distributed mass by two point masses

advantages: we shall therefore consider it here, and begin by revising the principles under which the 'replacement' is carried out. From the preceding discussion it is clear that the attributes of the body which determine its dynamical properties are its mass, the position of its centre of gravity, and its moment of inertia. As far as the external forces such as F on the body are concerned (but not, of course, in relation to the internal stresses caused by the acceleration of its distributed mass) we are entitled to replace it by two point masses, if, and only if, they together have these three properties identical with those of the body.

Let the body be as sketched in fig. 1.43, let the notation be as indicated above, and let the two point masses to replace the element be m_1 and m_2, distant respectively d_1 and d_2 from G. Note that in sketching the positions of m_1 and m_2 as collinear with, and on opposite sides of, G, we have used our common sense, since the position of G could not otherwise be unaltered. We have then still to satisfy the following conditions:

* See, for example, § 3.3.

(1) Total mass unaltered, i.e.

$$m_1 + m_2 = M \qquad (1.4a)$$

(2) Position of C.G. unaltered, i.e.

$$m_1 d_1 = m_2 d_2 \qquad (1.4b)$$

(3) Moment of inertia unaltered, i.e.

$$m_1 d_1^2 + m_2 d_2^2 = I = Mk^2 \qquad (1.4c)$$

The third of these conditions can be more simply expressed. Substituting $m_2 d_2$ for $m_1 d_1$ (from b) in the first term, and the inverse relationship in the second, we obtain

$$m_2 d_2 \cdot d_1 + m_1 d_1 \cdot d_2 = Mk^2$$

from which, using (a) we obtain

$$d_1 d_2 = k^2 \qquad (1.4d)$$

and (a), (b), and (d) express the necessary conditions.

We have four quantities ($m_1 m_2 d_1 d_2$) to find and only three conditions to satisfy: it follows that we can make an arbitrary (within reason: e.g. $m_1 \not> M$) convenient choice of any one, and that the remaining three will then be determinate.*

We may now tackle by either of the methods we have been discussing the problem of determining the total force required to accelerate a distributed mass, and since no new principles are involved, we can most effectively do so by taking a practical example. A convenient case is the connecting-rod in a simple engine mechanism such as that sketched in fig. 1.44a. Let the mass of the rod AB be M, its moment of inertia about its centre of gravity G in the

---

* It is, perhaps, worth emphasising the fact that no novel principles are involved in this procedure; examples commonly dealt with in mechanics include the 'centre of percussion' and the compound pendulum. Thus, if a distributed body is subjected to an impulsive force at a point P (other than the C.G.) it begins to rotate about a centre O, and vice versa, and O and P may be described as conjugate centres of rotation and percussion. It is not difficult to identify these as conjugate positions of m_1 and m_2 respectively, whereupon another appreciation of the reason for the behaviour of the body may become evident: the immediate effect of the force is to accelerate the point mass to which it is applied, and to leave the other mass at rest. (In the limiting case where P coincides with G, O is, of course, at infinity and the body accelerates without rotation.) Again, it is commonly pointed out in dealing with the compound pendulum that its period of oscillation is the same as that of an 'equivalent simple pendulum' (see § 4.3). It is easy to show that if the body treated above is mounted as a pendulum oscillating about a centre which coincides with the position of m_1 or m_2 the length of the equivalent simple pendulum is the distance between these masses: the reason will be immediately apparent in the light of the present treatment.

plane of motion $I = Mk^2$, and its length between centres l. The crank OA is of length r, the crankshaft speed is ω rad/s, and our

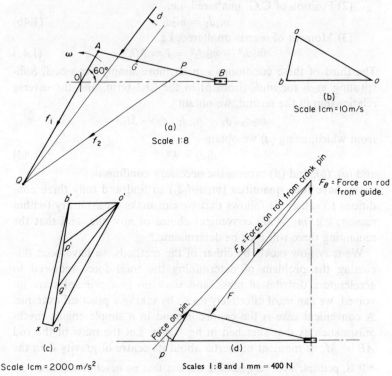

Fig. 1.44 Inertia forces on connecting rod

problem is to find the force required to give the rod its acceleration in the configuration shown.

We start by drawing the velocity and acceleration diagrams for the mechanism as shown in fig. 1.44b and 1.44c respectively.* In the acceleration diagram we can draw the 'acceleration image' $a'b'$ for the connecting-rod: to find the acceleration of any point, e.g. G on the rod, we need only divide $a'b'$ in g' so that $a'g' : g'b' = AG : GB$,

---

* There are special methods, such as Klein's construction, for doing this particular part of this particular problem which save a little labour, but these are deliberately not included here, since the labour is saved only at the cost of a considerable complication of the diagram; and of inverting the acceleration diagram and so needlessly confusing the question of signs.

and so obtain the required acceleration $o'g'$. Applying Newton's second law of motion, we can now state that the force F required to give the connecting-rod its translational acceleration is $M.o'g'$, and that this force is in the direction $o'g'$.

We can now proceed to find the *position* of this force F by either method. The rotational acceleration of the rod, $\dot{\omega}$, is, of course (from fig. 1.44c), xb'/AB, since xb' is the tangential acceleration of B relative to A; so the torque T which causes this acceleration can be calculated as $I\dot{\omega}$. Let the distance from the C.G. to the line of action of the force F be d; then

$$T = I\dot{\omega} = Fd$$

so
$$d = \frac{T}{F} = \frac{I\dot{\omega}}{M \cdot o'g'} = \frac{k^2\dot{\omega}}{o'g'} = \frac{k^2 \cdot xb'}{AB \cdot o'g'}$$

No difficulty should arise over signs. For instance, in the example shown the direction of xb' indicates an anti-clockwise rotational acceleration which can be caused only by an anti-clockwise torque due to the downward force F being displaced to the left-hand side of G.

The alternative method involves replacing the rod AB by two point masses, and we can conveniently exercise our 'free' choice by placing one of these masses at either A or B. Let us choose A. Then $d_1 = AG$, so $d_2 = k^2/AG = GP$ (say). Now we can obtain immediately from the acceleration diagram the *direction* – we are uninterested in the magnitude – of acceleration of each of these masses ($o'a'$ and $o'p'$ respectively), and hence deduce the direction of the force on each, and, since each is a point mass, the position of each force, since it must act through the mass. These two component forces, say f_1 and f_2, are as shown in fig. 1.44a. But the total force on the rod is the resultant of these two component forces f_1 and f_2 and must be concurrent with them. It must therefore act through the point Q where they meet – and, of course, we already know its magnitude and direction, so we have no need to add f_1 and f_2 to find them.

In the example outlined there is little to choose between the two methods. It is as well to solve one problem by both, and if desired with an alternative choice of d_1, to make sure that each is well understood. With this in mind, the example given has been drawn to suit the following data, and may therefore be regarded as a worked example.

Worked Example 1.6

DATA

Fig. 1.44a: Scale 1:8, crank $OA = 76$ mm; connecting-rod AB length 280 mm, mass 1·81 kg, I about G in the plane of motion 0·0164 kg m²; $AG = 89$ mm. Crankshaft speed 3 000 rev/min.

$$GP = \frac{k^2}{d_1} = \frac{0\cdot0164}{1\cdot81} \times \frac{1}{0\cdot089} = 0\cdot102 \text{ m} = 102 \text{ mm}.$$

Fig. 1.44b: $oa = 314 \times 0\cdot076 = 23\cdot9$ m/s. Choose scale 1 cm $= 10$ m/s. $ab = 12\cdot3$ m/s.

Fig. 1.44c: $o'a' = 314^2 \times 0\cdot076 = 7\ 500$ m/s². Choose scale 1 cm $= 2\ 000$ m/s². $o'g' = 5\ 620$ m/s². Therefore $F = 1\cdot81 \times 5\ 620 = 10\ 200$ N: $xb' = 6\ 550$ m/s², therefore

$$d = \frac{0\cdot00906 \times 6550}{0\cdot28 \times 5620} = 0\cdot038 \text{ m} = 38 \text{ mm}.$$

We have found the resultant force which must be applied to the connecting-rod to give it its acceleration, but this force *can* be exerted *only* as the resultant of two forces at the points of attachment to other parts of the machine, i.e. at A and B. Irrespective of any additional forces which may be being transmitted through the mechanism and which may be superimposed on those we are considering, we can say that the guides in which B moves can be regarded as frictionless and therefore capable of applying to the connecting-rod only a force at right angles to them. (If the friction were not negligible the force could as easily be taken as acting at the appropriate angle of friction to this normal.) To avoid confusion fig. 1.44a has been redrawn in fig. 1.44d, the data which are now irrelevant being omitted. The total force F is retained, and this component applied at B, F_B, has been inserted. It is now clear that the component force F_A applied at A must be concurrent with these: thus, a triangle of forces may be drawn to find the magnitudes of F_A and F_B.

If we wish to know the effect of the inertia of the connecting-rod on the turning moment of the crankshaft we first note that *the force on the crankpin is equal and opposite*[*] *to the force* F_A, and that the arm of this force about O is (as shown) p. Thus, at the instant shown

[*] See Worked Example 1.7, and in particular fig. 1.50f, for an illustration of a helpful means of avoiding confusion in describing these 'actions and reactions'.

the effect is a clockwise moment of amount $F_A \cdot p$ (or numerically 8 900 N × 18 mm = 160 Nm.

Both methods given are perfectly general and may be applied to any element in a machine in any specific configuration. It may, however, be worth while to point out that a particular virtue of the method of replacing a distributed mass by two point masses has not been given adequate prominence. It is, of course, obvious that these masses are independent of machine configuration and speed. If now we have in a machine an element which is oscillating about a fixed centre we can replace it by two point masses, of which one can be chosen to be at this fixed centre. It then becomes apparent that we can deal not only with the forces which have to be applied to give the element its motion but also with the inertia forces which will have to be balanced* or taken by the foundations, by considering the element as effectively *a single point mass* NOT *at the centre of gravity but at the centre of percussion* in relation to the fixed axis. This is a simplification of major importance.

It is desirable, before leaving this specific example of a connecting-rod in an engine mechanism, to add that for many purposes a sufficiently good approximation to the effect of its inertia can be obtained by replacing it by two point masses, m_1, at the centre of the big end and m_2 at the centre of the small end, where m_1 and m_2 satisfy equations 1·4(a) and (b) at the beginning of this paragraph. We can now regard m_1, as part of the rotating mass associated with the crankpin, and m_2 as part of the reciprocating mass associated with the piston; and the usefulness of this approximation will be appreciated when we deal with the topic of balancing of engines.† The error introduced by our failure to satisfy equation (c), which is usually very small, can be calculated for any particular configuration by determining by the methods given in this paragraph the exact force required to accelerate the rod and comparing it with the sum of these approximate forces.

§ 1.10 Transmission of Forces through Machines: Torque Diagrams

The determination of the forces transmitted through a mechanism or machine in all configurations involves considerable labour, but

---

* See chapter 3, and in particular § 3.4.

† See chapter 3, and in particular § 3.3.2.

no fundamental difficulties. The methods most commonly used include elementary statics or the principles of virtual work or virtual power, and these are dealt with in mechanics.

A particular case which, though straightforward, is of sufficient practical importance to merit mention as an example is the deduction of the torque or turning moment exerted on the crankshaft of a simple engine mechanism due to the forces applied through the piston and connecting-rod by the working fluid and those arising from the inertia of the parts. We may assume that the pressure exerted on the piston at any stage in the cycle is obtainable from an indicator diagram. The gas force F is then the product of this pressure and the area of the piston. The reciprocating parts must be accelerated: we can find their acceleration either graphically or, in this case more sensibly, by calculation as indicated in § 1.3. The force required to provide this acceleration is, of course, the product of the mass of these parts and their acceleration, and

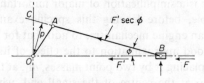

Fig. 1.45 Force transmitted through connecting rod

this must be deducted from the gas force F to find the net force, say F', which the piston applies to the gudgeon pin. The force acting along the connecting-rod is the resultant of F' and the side thrust from the cylinder wall (or crosshead slide) and has a magnitude F' sec ϕ (see fig. 1.45), where ϕ is the acute angle between the rod and the line of stroke. The torque T or moment of this force about the centre of rotation of the crankshaft O is then

$$T = F' \sec \phi \,.\, p$$

Let the line of the connecting-rod cut in C the perpendicular through O to the line of stroke. Then $OC = p \sec \phi$, so $T = F'.OC$, and OC can be obtained by calculation or, more easily, by drawing the mechanism to scale. There is, therefore, no difficulty in determining, to a reasonable degree of accuracy, the torque which is exerted on the crankshaft at any point in the cycle, and this procedure can be repeated for different configurations.

The simplified calculation outlined makes no allowance for

relatively minor effects such as friction, and the true moment of inertia of the connecting-rod, but if greater accuracy is required, and the data are available, there is no theoretical difficulty in making the appropriate corrections. The exact effect of the inertia of the connecting-rod has been dealt with in § 1.9; friction between the piston and cylinder *could* be allowed for by taking the side-thrust

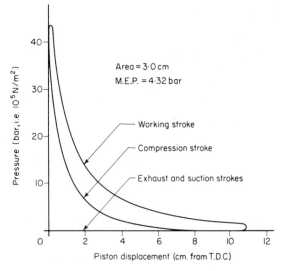

Fig. 1.46

as acting at the angle of friction to the normal, and friction in each turning pair by using the device of the friction circle and drawing the line of thrust tangent to such a circle instead of through the geometric centre of the pair; but these are matters of elementary mechanics. In fact, the 'coefficients of friction', with satisfactory conditions of lubrication (see chapter 7), will be very small indeed, and are unlikely to be known with sufficient accuracy to make such corrections worthwhile.

As an illustration, the straightforward calculations outlined above have been carried out for a Ricardo E.6 single-cylinder, four-stroke oil engine running at a speed of 1 500 rev/min, under a load corresponding to the indicator diagram in fig. 1.46. Sufficient particulars are given to allow the method of computation and the calculations to be checked: the arithmetic for one particular configuration is given as an example.

DATA

> Stroke 111 mm, bore 76·2 mm
>
> | Mass of piston and gudgeon pin | 0·70 kg |
> |---|---|
> | Mass of connecting-rod | 1·90 kg |
> | Length of connecting-rod between centres | 238 mm |
> | C.G. of rod from big-end centre | 73·5 mm |

Arithmetical example

Configuration chosen, crank $30°$ past top dead centre on firing stroke. In this configuration displacement of piston from top dead centre (see § 1.3)

$$= r(1 - \cos \theta) + l(1 - \cos \phi)$$
$$= r(1 - \cos \theta) + l \sin^2 \theta / 2q^2 = 9\cdot1 \text{ mm}$$

And at this displacement gas pressure (from indicator diagram)

$$= 23 \text{ bar}$$

and piston acceleration $= -\omega^2 r \left(\cos \theta + \dfrac{1}{q} \cos 2\theta \right) = -1\,370$

$$\times 0\cdot982$$
$$= -1\,345 \text{ m/s}^2$$

Reciprocating part of connecting-rod (approx) $= 1\cdot9 \times \dfrac{73\cdot5}{238}$

$$= 0\cdot59 \text{ kg}$$

∴ Force required to accelerate reciprocating parts

$$= (0\cdot70 + 0\cdot59) \times (-1\,345)$$
$$= -1\,738 \text{ N (i.e. force of 1 738 N to the left)}$$

Gas force on piston

$$= -\frac{\pi}{4} \times 76\cdot2^2 \times 10^{-6} \times 23 \times 10^5 = -10\,500 \text{ N (i.e. to the left)}$$

∴ Net force on gudgeon pin

$$= -10\,500 - (-1\,738) = -8\,760 \text{ N (i.e. to the left)}$$

OC (from drawing) $= 33\cdot5$ mm

∴ Torque on crank $= +294$ Nm (i.e. anti-clockwise)

and this gives us one point ($\theta = 30° = \pi/6$; $T = 294$) on the curve drawn in fig. 1.47. The student is recommended to check one or two other points for himself.

If the whole torque diagram is required the calculation is, of course, carried out in tabular form, a method which eases considerably the labour involved. Whether arithmetical or graphical methods are used is largely a matter of personal preference. The

result of such a computation is displayed in fig. 1.47. The gas
pressures in this engine during the exhaust and suction strokes are
negligibly different from atmospheric, so the torque diagram shows
in the region $\pi <\theta<3\pi$ the effect of the reciprocating masses;
elsewhere the gas pressure is the dominating feature.

From such a diagram it is a very easy matter to build up diagrams
from which forecasts of the behaviour of multi-cylinder engines

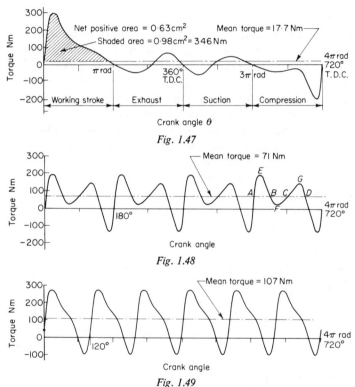

Fig. 1.47

Fig. 1.48

Fig. 1.49

can be made. For instance, a six-cylinder engine with evenly spaced
firing strokes will give a total torque which is the algebraic sum of
six such individual torques each displaced 120° from the preceding
one, and the summation can be effected graphically or, even more
simply, arithmetically. Again in illustration the results for four- and
six-cylinder engines are shown in figs. 1.48 and 1.49.

Torque diagrams are of use for two purposes: in the avoidance
of serious torsional vibration troubles and in the design of fly-

wheels. Vibrations are discussed in later chapters (4 and 5) of this book; flywheels in the next paragraph. For the time being, it will suffice to add that to use torque diagrams for vibration analysis, it is usually necessary to express the torque not as a graph but in mathematical form as a Fourier series; the methods of doing so are outside the scope of this book, but are readily available;[7,8] the subject will be given more attention in § 5.11.

Worked Example 1.7

As a further example consider the quick-return mechanism shown in fig. 1.50a, which is employed in a shaping machine. The cutting tool is attached to the ram shown at D, the total mass being 70 kg, and the cutting force is 270 kgf, resisting the motion of the ram. The machine is driven by the crank OB, which has a speed of $+300$ rev/min. For the position shown we wish to know the forces (additional to those due to gravity) which are transmitted through the various turning and sliding pairs and the instantaneous value of the torque which the crank OB is exerting on the machine. We shall assume that friction and the inertia of the elements other than the ram are negligible.

DATA (with lengths in mm): $QO = 330$, $QZ = 660$, $OB = 135$, $QC = 600$, $CD = 435$; $\angle ZOB = 45°$.

No assistance should at this stage be required in drawing the velocity diagram: in relation to the acceleration diagram it is perhaps advisable to give the warning that the Coriolis term should not be forgotten. These diagrams should preferably be constructed *without* reference to figs. 1.50b and 1.50c: even a glance at the latter will give assistance which is better dispensed with unless it is found to be essential. The student who is in trouble will, however, be able to check his efforts by scaling these figures.

The essential result from this work is that the ram D has in the configuration shown an acceleration of od' or $+63$ m/s² (i.e. to the right), and a velocity to the left.

$$\therefore \text{ Force required to decelerate ram} = 70 \times 63$$
$$= +4\,410 \text{ N (i.e. to right)}$$
$$\text{but Force on ram due to cutting} = +270 \times 9{\cdot}81$$
$$= 2\,649 \text{ N (i.e. to right)}$$
$$\therefore \text{ Horizontal Force which } CD$$
$$\text{must exert on ram} = +1761 \text{ N (i.e. to right)}$$

Superimposed on fig. 1.50a are shown the directions of the forces on *QC*, and in figs. 1.50d and 1.50e the triangles for the forces *on* the ram and *on* the member *QC* respectively are shown. Note that in the former we have an *unbalanced* force available to provide the

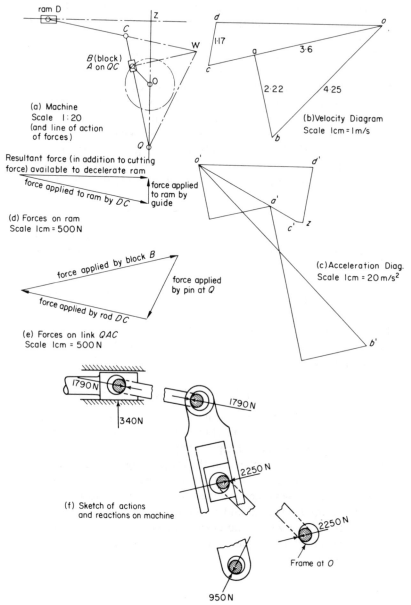

Fig. 1.50 Forces on machine tool with quick-return mechanism

horizontal force of 1761 N we have just calculated, while in the latter (since we are neglecting all other inertias) we must have equilibrium between the three applied forces: one at C due to the tension in CD, one at the block, which (neglecting friction) must be perpendicular to CQ, and the equilibrant at Q, which must be concurrent with these two and therefore pass through the point marked W in fig. 1.50a.

The directions of all the forces involved are made clear in fig. 1.50f by showing enlarged-clearance diagrams so that the position taken up by the various pins, etc., under the action of these forces can be shown unambiguously. This method of showing the direction of the forces acting is recommended to students as a valuable aid to the initial understanding of problems of this kind.

The torque which the crank OB is exerting on the machine is equal to the force acting through the block multiplied by the perpendicular distance from O to the line of action (BW in fig. 1.50a) of this force.

$$\text{Torque} = 2\,250 \times 0{\cdot}116 = 262 \text{ Nm}$$

From fig. 1.50e, and with even less danger of error from fig. 1.50f, it is seen that the crank OB is applying a clockwise torque to the mechanism. Since it is rotating anti-clockwise, the corresponding power input $\left[\dfrac{300}{60} \times 2\pi \times 262 \text{ W} \simeq 8{\cdot}2 \text{ kW}\right]$ is negative. This reflects the fact that the ram D is slowing down and imparting its kinetic energy to the crankshaft. At the instant shown this energy is in excess of that being absorbed in the cutting process, so that the crankshaft and driving motor will be speeding up.

§ 1.11 Flywheels

A flywheel is a device for reducing to acceptable proportions the cyclic variation in speed of a machine. Even if it were not a matter of everyday experience that a reciprocating engine requires to be provided with a flywheel, a glance at the torque diagrams illustrated in figs. 1.47–1.49 would show that these engines at least could not continue to run unless they included some means of storing energy when the torque was high and giving it up again when the torque was negative. Clearly the requirement varies with the type of engine, the number of cylinders, and so on: we may take these examples as illustrations of this.

Before we can attack the problem of designing the flywheel we must have full information about what is happening to the energy which the engine is producing. In very many cases the output required will be a constant torque, and for simplicity we shall assume that this is so; but it is as well to point out that if another condition obtains, for instance if the engine were called upon to drive a reciprocating pump, we should have to carry out an operation precisely similar to that described in § 1.10 in order to find the variation of torque *required* with crankshaft angle, and subtract it from the torque *supplied,* to find the surplus torque at any instant which requires to be catered for.

To return to the case of the single-cylinder engine for which some data were given in § 1.10 and whose torque diagram was illustrated in fig. 1.47, and the assumption that the torque output is required to be constant, we must first calculate the value of this output torque; and this can be done either from the original indicator diagram in fig. 1.46 or from fig. 1.47. Thus, the net area of the indicator diagram gives the energy produced in two revolutions as $A\int P\,dl$ (where A is the area of the piston, P the gas pressure at any instant, and l the stroke); the net area of the torque/angle diagram gives the same energy as $\int T\,d\theta$. The areas of these diagrams can, of course, be determined by any of the usual methods; by using a planimeter if one is available, by Simpson's rule, or even by counting squares. Numerically, from the original indicator diagram

$$\int P\,dl = \text{Area of diagram (of course to scale)}$$
$$= 3\!\cdot\!0 \text{ cm}^2 \times 8 \text{ bar/cm} \times 20 \text{ mm/cm} = 480 \text{ bar mm}$$
$$= 0\!\cdot\!48 \text{ bar m} = 48\,000 \text{ N/m}$$

Area of piston (from data in § 1.10) $= \dfrac{\pi}{4} \times 76\!\cdot\!2^2 = 4\,570 \text{ mm}^2$

$\therefore$ Work done/rev $= 48\,000 \text{ N/m} \times 0\!\cdot\!00457 \text{ m}^2 = 220 \text{ Nm or J.}$
Or from the torque/angle diagram,

Net positive area $= 0\!\cdot\!63 \text{ cm}^2 \times 214 \text{ Nm/cm} \times 1\!\cdot\!65 \text{ rad/cm}$
$= 222 \text{ Nm } (\simeq 220 \text{ Nm})$

The mean output torque is then this energy divided by the total angle, 4π radians, through which the crankshaft turns, and this value of $17\!\cdot\!7$ Nm is indicated in fig. 1.47.

Whenever the torque on the crankshaft exceeds the output torque the excess will cause the flywheel to accelerate. A glance at fig. 1.47 shows that such a condition obtains during nearly the whole of the firing stroke (in fact from $\theta \simeq 1°$ to $\theta \simeq 170°$) and that in each cycle of two revolutions the flywheel will reach its minimum and maximum speeds, say ω_2 and ω_1, at these two instants respectively. The excess energy, say E, which the flywheel must store is therefore represented by the area shaded in fig. 1.47. We can now write down

$$E = \tfrac{1}{2}I(\omega_1^2 - \omega_2^2)$$

where I is the moment of inertia of the rotating parts of the engine; for practical purposes we can often ignore everything except the flywheel itself, but if not it is an easy matter to subtract the moment of inertia of the other parts from the required total. If the fluctuation of speed is not large it may be convenient to express ω_1 and ω_2 as $(\omega + \delta\omega)$ and $(\omega - \delta\omega)$ respectively, where ω is approximately the mean running speed and $\delta\omega$ is the fluctuation above and below this mean. We then obtain immediately

$$E = 2I\omega \, . \, \delta\omega$$

Thus, if the flywheel inertia is known we can calculate the fluctuation in speed, or if the permissible speed fluctuation is specified we can determine the inertia of the flywheel required.

In the engine to which fig. 1.47 refers the flywheel is a steel disk 450 mm diameter, 68 mm thick. Its moment of inertia I is therefore

$$I = \tfrac{1}{2}MR^2 = \tfrac{1}{2}(7\,800\pi \times 0.225^2 \times 0.068)\,0.225^2 = 2.13 \text{ kg m}^2$$

We can measure the shaded area as 0·98 cm²; the scales are 214 Nm/cm, and 1·65 rad/cm, so we can calculate E as 346 Nm. ω we have stated to be 1 500 rev/min or 157 rad/s. It follows that $\delta\omega = 0.52$ rad/s, and the engine speed will vary from 156·5 to 157·5 rad/s.

It will be noted that in this particular case we have no need to go through the long and elaborate procedure described merely to design a flywheel. In an engine such as this we shall make only a trivial error if we assume that the minimum and maximum speeds of the crankshaft occur at the beginning and end of the working stroke. Between these instants all the energy given to accelerate the reciprocating parts has, of course, been restored, so this term does not enter the calculations; we are concerned only with the gas pressures. Let the mean effective pressure in the four strokes (working, exhaust, suction, compression) be numerically p_1, p_2, p_3, p_4, respectively. Under normal conditions the three latter obviously have signs implying that work is being done on the gas by the piston. We can therefore say that the mean effective pressure in the whole cycle is $p_m = \frac{1}{4}(p_1 - p_2 - p_3 - p_4)$ and therefore that the 'excess' work to be stored in the flywheel is $l \cdot A(p_1 - p_m)$, where l and A are the stroke and area of the piston, and equate it to the change in kinetic energy of the flywheel.

If the torque diagram has a number of peaks and troughs above and below the mean torque line the points in the cycle at which minimum and maximum speed will occur may or may not be apparent. There is, however, no difficulty in determining them. Thus, for instance, in fig. 1.48 it is clear that one cycle is repeated four times and that during each of these repetitions conditions are identical, but the question is (to consider the fourth during which $3\pi < \theta < 4\pi$): will the minimum speed occur at A or C, and the maximum at B or D? By inspection (or if necessary by measurement) we can see that more energy is given to the flywheel during the period CGD than was lost by it during the period BFC, hence D will indicate the point at which maximum speed, and A that at which minimum speed, is obtained. In more complicated cases each area above and below the mean torque line can be measured and added successively (algebraically, of course) to find the points of maximum and minimum speed.

Reference Books

1. F. REULEAUX. *The Kinematics of Machinery*, trans. by A. B. W. Kennedy. Macmillan, 1876.
2. A. F. C POLLARD. *The Kinematic Design of Couplings in Instrument Mechanisms*. New edn. Hilger, 1951.
3. P. GRODZINSKI. *A Practical Theory of Mechanisms*. Emmot, 1947.
4. K. H. HUNT. *Mechanisms and Motion*. English Universities Press, 1959.
5. A. S. HALL. *Kinematics and Linkage Design*. Prentice Hall, 1961.
6. R. BEYER. *Kinematic Synthesis of Mechanisms*, trans. by H. Kuenzel. Chapman & Hall, 1963.
7. B. B. LOW. *Mathematics*, 2nd edn. Longmans, 1948.
8. W. ABBOTT. *Practical Geometry and Engineering Graphics*, 6th edn. Blackie, 1957.

Examples

NOTE. A part of a mechanism can be given straight-line motion by providing for it a straight guide and sliding pair, but this would frequently involve difficulties in relation to unwanted mass and friction. Several mechanisms have been devised which use only turning pairs and which provide either accurate or approximate straight-line motion; others make use of linear motion at one part of the mechanism where it is easily obtained to provide it elsewhere. A few of the most interesting or important of these mechanisms are presented in examples 1.1 to 1.3.

1.1 The Watt straight-line mechanism is shown diagrammatically in two forms in figs. 1.51a and 1.51b. Originally devised by James Watt because of the difficulties in his days of providing a straight guide for the end of the piston rod of his steam engines, this mechanism gives only approximately straight-line motion. Its merit is its simplicity: so great is this merit that it is the only straight-line motion of technological importance today. An approximation to it is used in many different types of machines, for instance in motorcar suspensions, of which a typical example is shown in fig. 1.51c to illustrate again the relationship between the appearance in real life of the mechanisms we discuss and their diagrammatic form. When seen in plan view, each of the members attached to the chassis is of triangular form, so that the pivots are well separated in the fore-and-aft direction. The object of using the Watt type of mechanism in this case is to avoid unnecessary lateral motion between tyre and road surface as the springs operate, and so to save tyre wear.

(*a*) Compare the mechanisms in figs. 1.51b and 1.51c. Are the proportions in fig. 1.51c such as to give straight-line vertical motion to the part of the tyre in contact with the ground? If not, can you suggest a reason for the compromise adopted?

(*b*) In fig. 1.51b, for the general case when the ratio AC/BC is not determined, calculate the horizontal displacement of C when θ changes from zero to a *small* value. Hence (using the approximation $1 - \cos\theta \simeq \frac{1}{2}\sin^2\theta$) show that the ratio chosen by Watt is correct for small displacements.

(*c*) Draw the mechanism (*b*) with (say) dimensions $OA = 50$ mm, $PB = 75$ mm, $AC = 75$ mm, $BC = 50$ mm, in several configurations in the range $-30° < \theta < 30°$ to satisfy yourself that the motion of C is approximately straight, and perpendicular to the direction of OA when $\theta = 0$. Note that for large displacements a slightly *shorter* length of AC (say 65 mm) does in fact give a better overall approximation to straight-line motion than the standard length.

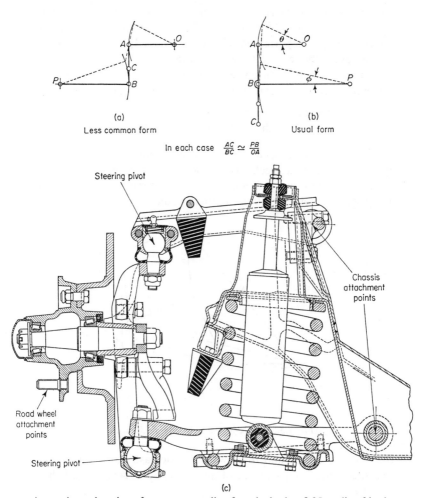

(a)
Less common form

(b)
Usual form

In each case $\frac{AC}{BC} \simeq \frac{PB}{OA}$

Steering pivot

Chassis attachment points

Road wheel attachment points

Steering pivot

(c)

Approximate location of tyre contact: dia of road wheel $= 2.35 \times$ dia of brake disk; centre line midway between wheel location surface and brake disk

Fig. 1.51 Watt straight line mechanism, and practical modification
(By permission of Jaguar Cars Ltd.)

(*d*) For an intermediate position (say when $\theta = 15°$) check by instantaneous centre and by velocity diagram that the velocity of *C* is approximately vertical, but that the point having true vertical motion lies between *B* and *C*.

(*e*) In fig. 1.51c the whole method of locating the front wheel of a car relative to the chassis is indicated. Consider this design in the light of: (*a*) the possible motion of the wheel relative to the chassis which must (1) be permitted, (2) be constrained; and (*b*) the forces which have to be resisted (1) when cornering, and (2) when braking.

In particular, do you consider the choice of (*a*) ball joints, (*b*) opposed taper-roller bearings is sound? What are some alternatives?

N.B. No answers are given to (*e*): these questions are for individual consideration in the light of the principles outlined in § 1.2.

1.2 The 'Peaucellier Cell', illustrated diagrammatically in fig. 1.52, is an example of a mechanism which gives theoretically accurate straight-line motion, perpendicular to the fixed base *OP*, of the point *A*. In practice, accurate motion would be obtained only if all

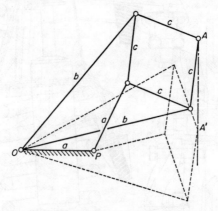

Fig. 1.52 Peaucellier straight-line mechanism

the links were rigid and there were no backlash at any of the pin-joints. It is too complex to be of much use, but it serves as a helpful exercise.

(*a*) Choosing, say, the dimensions $a = 40$ mm, $b = 95$ mm, $c = 38$ mm, draw a Peaucellier cell in two possible configurations chosen at random and satisfy yourself that the motion is correct.

(*b*) For *one* of these configurations check both by drawing the velocity diagram *and* by finding the necessary instantaneous centres that the velocity of *A* is perpendicular to *OP*.

1.3 In engines which run at a modest speed it is possible to estimate the work being done on the piston by using a device known as a 'pencil indicator'. The cylinder pressure is communicated to a small cylinder fitted with a spring-loaded piston whose displacement (provided that the motion is small, so that inertia is not important*) is proportional to this pressure. To make measurement easy this motion has to be amplified and communicated to a pencil which is pressed lightly against a paper-covered drum. This drum is caused to oscillate rotationally so that its angular displacement is proportional to the motion of the piston of the engine. With such a device one can draw on the paper an 'indicator diagram' such as the one reproduced in fig. 1.46. The area of this diagram is a measure of the work done on the piston. We are concerned here only with the mechanism to amplify the motion of the indicator piston. Clearly this is a case where the end of the piston rod can easily be guided in a straight line; the pencil must also have straight-line motion. A pantograph would suit the case, except that there are rather too many pivots in which lost motion would be detrimental. There are many practical approximate solutions; two good examples are illustrated in figs. 1.53 and 1.54:

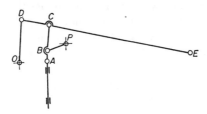

Fig. 1.53 'Crosby' indicator mechanism

(a) The 'Crosby' indicator mechanism, fig. 1.53. If the piston A has in the configuration shown a velocity of 3 m/s and an acceleration of 600 m/s² upwards, find the velocity and acceleration of the pencil E.

DATA (with lengths in mm): Co-ordinates of O (0, 0), P (26, 9), A (15, 0) $AB = 5.5$, $AC = 19$, $BP = 11$, $OD = 23$, $DC = 16$, $DE = 96$.

(b) The 'Dobbie–McInnes' indicator mechanism, fig. 1.54. Again if the piston A has in the configuration shown a velocity of 3 m/s and an acceleration of 600 m/s² upwards find the velocity and acceleration of the pencil E.

* Cf. § 5.4.2.

DATA (with lengths in mm): Co-ordinates of O (0, 0), P (46, 44), A (38, 25), $AB = 18$, $PB = 9$, $PC = 19$, $OD = 45$, $DC = 31\cdot5$, $DE = 89$, DCE collinear.

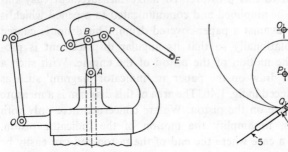

Fig. 1.54 'Dobbie–McInnes' indicator mechanism

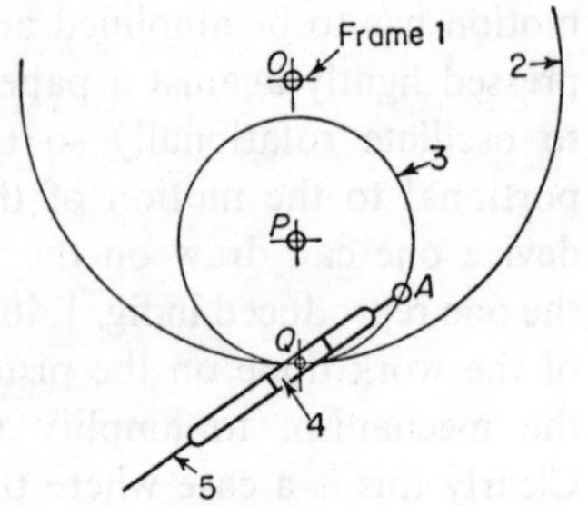

Fig. 1.55 Gear-tooth profile analogue

1.4 A mechanism, illustrated diagrammatically in fig. 1.55, consists of a frame (1) carrying three fixed pivots O, P, and Q. O and P are the centres of two gear-wheels, one (2) with internal teeth, the other (3) with external teeth, meshing together under the pivot Q, which does not interfere with them, since they can pass freely under it. On the pivot Q is a block (4) over which a rod (5) can slide freely. One end of this rod is pivoted to the rim of wheel (3) at the point A.

Find all the relative instantaneous centres of these five elements, and in particular centre I_{25}.

NOTES:

(*a*) The fact that gear-wheels have been mentioned is unimportant: (2) and (3) can be regarded simply as two circles which roll together without slipping.

(*b*) This example is a mechanical simulator of a situation which will be discussed fully in § 2.11.3. Solving it now will be of help when this situation is being considered.

1.5 Fig. 1.56 shows the retractable undercarriage mechanism of an aircraft. Points O, P, and Q are fixed points and APB is a bell crank.

Oil under pressure is fed into the space under the piston to retract the undercarriage. If the velocity of the piston relative to the cylinder is v m/s in the position shown, find the angular velocity of the arm QW.

DATA (with lengths in mm): $QC = 360$, $QF = 960$, $DF = 660$, $OD = 1\,380$, $OE = 300$, $BE = 480$, $PB = 360$, $AB = 450$, $AP = 240$.

1.6 Fig. 1.57 shows the operating mechanism for an element X in a machine. The crank OA rotates at 1 500 rev/min clockwise, and the crank PB at 3 000 rev/min anti-clockwise. O, P, and Q are fixed in the frame of the machine.

Find, for the given configuration, the angular velocity of X.

DATA (with lengths in mm): $OA = 26$, $PB = 13$, $AC = BD = 96$, $CE = ED = 53$, $QE = 90$.

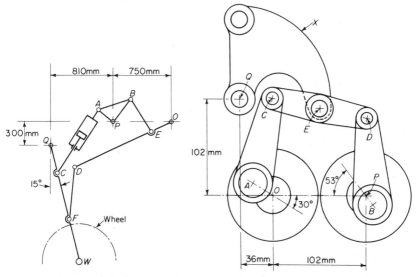

Fig. 1.56 Mechanism for undercarriage retraction

Fig. 1.57 Mechanism for combining
harmonic motion

1.7 Fig. 1.58 shows a mechanism used to provide a cyclically varying angular velocity. The crank OA rotating at a uniform speed of 20 rad/s drives through the connecting-rod AB a bell-crank lever BPC, which drives through a second connecting-rod CD the output crank OD, which is mounted on a sleeve concentric with O. Find for the configuration shown the angular velocity of the crank OD and the angular acceleration of the crank PB.

DATA (with lengths in mm): $OA = PC = 200$, $AB = CD = 150$, $PB = DO = 169$, $OP = 75$; $BPC = 135°$.

1.8 Fig. 1.59 shows a mechanism for operating an earth shovel. O and P are fixed centres. $OA = 480$, $PB = 390$, $AB = 600$, $AC = 1\,800$ mm. In the configuration shown the driving crank OA is rotating at 1 rad/s clockwise. Find:

    (*a*) by drawing a velocity diagram
    (*b*) by the method of instantaneous centres

the velocity (in magnitude and direction) of the bucket edge C.

1.9 With the data in question 1.8, and the additional information that the rotational acceleration of OA is $\frac{1}{2}$ rad/s² clockwise, find the linear acceleration of C.

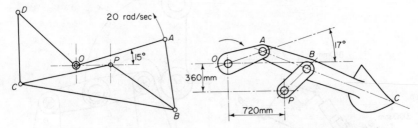

Fig. 1.58 Mechanism for introducing cyclical variation into rotational speed   *Fig. 1.59* Mechanism of earth shovel

1.10 In the machine shown in fig. 1.59 the member ABC has a total mass of 700 kg. The centre of gravity is on the line BC, 600 mm from B, and the radius of gyration about the centre of gravity is 900 mm.

For the conditions stated in questions 1.8 and 1.9, determine in magnitude, direction, and position the force required to give ABC its acceleration and (assuming the pivot P to be free) the torque which the crank OA must provide on this account *alone*.

1.11 Fig. 1.60 shows a toggle mechanism used for experiments on the rapid compression of gases. In the position shown, the pressure on the piston, which is 63·5 mm in diameter and has a mass of 2·5 kg, is 7 bar. The mass of all the other elements may be neglected. The angular velocity of the crank OA is 360 rev/min. Find the torque which it must be exerting in the configuration shown.

DATA (with lengths in mm): $OA = 125$, $AB = 500$, $PB = 375$, $BC = 375$.

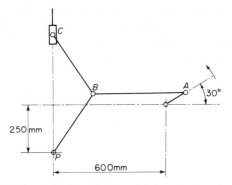

Fig. 1.60 Toggle mechanism for rapid compression of gases

1.12 Fig. 1.61 shows part of a vane-type compressor. The rotor holding the sliding vanes runs at 2 000 rev/min, and the tips of the vanes remain in contact with the stationary outer member. Find the magnitude and direction of the acceleration of the centre of the vane shown, at the instant depicted.

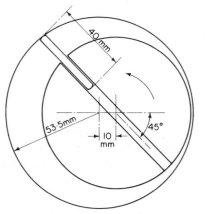

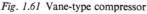

Fig. 1.61 Vane-type compressor

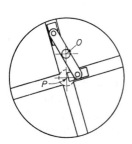

Fig. 1.62 'Burn' transmission

1.13 Fig. 1.62 shows a coupling for connecting two parallel shafts *O* and *P*, whose centres are 150 mm apart. *O* carries an arm 300 mm long between centres and terminating in two pins which carry blocks sliding in rectangular slots in a disk attached to *P*.

At the instant shown the arm has travelled 30° past the line of centres *OP* and is rotating at 1 000 rev/min anti-clockwise. Find the angular speed and acceleration of the disk.

1.14 Fig. 1.63 shows a 'Geneva mechanism' for obtaining inter-mittent rotation from continuous rotation. The pin A is carried in a constant-speed shaft, which is rotating at 100 rev/min, and it engages with the slotted member as sketched. $OP = 128$ mm, and $OA = 90$ mm. At the instant shown $\angle POA$ is $20°$; find the instantaneous angular velocity and acceleration of the slotted member.

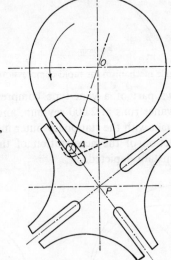

Fig. 1.63 'Geneva' mechanism

1.15 Fig. 1.64 shows part of a mechanism consisting of a crank PA oscillating about a centre P and carrying at its extremity a block which slides in the member OB. When in the configuration shown the angular velocity of PA is $+30$ rad/s (i.e. anti-clockwise) and the angular acceleration -120 rad/s². Find the instantaneous accelera-tion of the pin B.

DATA (with lengths in mm): $OP = 200$, $PA = 100$, $OB = 250$; $\angle OPA = 30°$.

1.16 Fig. 1.65 shows an unusual type of quick-return mechanism. The driving crank OA is 57 mm in length, and it rotates at 400 rev/min. For the configuration shown find the velocity and acceleration of B.

NOTE: The methods of § 1.4.3 have to be extended to the acceleration diagram.

1.17 The connecting-rod for a petrol engine is 267 mm long between centres and weighs 1·6 kgf. Its centre of gravity is 89 mm from the centre of the big end, and its radius of gyration about the centre of gravity in the plane of motion is 94 mm. Find in magnitude, direction and position the force required to accelerate the rod, and the effect of this force (in magnitude and sense) on the turning moment of the crankshaft, when the crank has turned through 255° past T.D.C.

The stroke of the engine, which runs at 3 000 rev/min, is 146 mm.

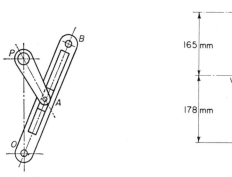

Fig. 1.64 Oscillating mechanism   *Fig. 1.65* Quick return mechanism

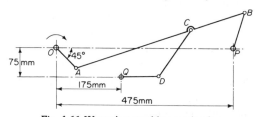

Fig. 1.66 Wrapping machine mechanism

1.18 Fig. 1.66 shows part of the mechanism of a wrapping machine. It is driven by the crank *OA*, which rotates at 1 000 rev/min. The only important mass may be considered as one of 20 kg concentrated at *D*. At the instant illustrated, find in magnitude and direction the reactions *on the frame* at the fixed pivots *O*, *P*, and *Q* required to give this mass its acceleration.

DATA (with lengths in mm): $OA = 75, AB = 475, AC = 325, PB = 100, CD = 150, QD = 100$.

1.19 Calculate from first principles the percentage fluctuation of speed in a four-stroke cycle engine from the following data: cylinder diameter 200 mm, stroke 400 mm, speed 250 rev/min, mass of rotating parts 1 200 kg, radius of gyration of rotating parts 600 mm, mean pressure in cylinder during expansion stroke 7·5 bar, and during compression stroke 1·7 bar above atmosphere. The pressure during suction and exhaust may be neglected and the resistance at the crankshaft assumed constant.

1.20 A single-cylinder, four-stroke petrol engine has a bore and stroke of 75 mm and 100 mm respectively. The mean effective pressures during the firing, exhaust, suction, and compression strokes are 10, 0·3, −0·1 and 1 bar when the engine is running at a mean speed of 3 000 rev/min and delivering power at a constant rate. Under these conditions the speed fluctuates by ±0·1 per cent. What is the moment of inertia of the rotating parts?

If the mean speed is reduced to 500 rev/min, but conditions are otherwise unchanged, to what extent will the cyclical speed fluctuate?

If a twin-cylinder engine (firing each revolution) were to be made utilising the same flywheel but providing twice the torque, what would be the speed fluctuation at 500 rev/min?

1.21 During the outward stroke of the piston of a double-acting steam engine the turning moment has a maximum value of 18 000 Nm when the crank makes an angle of 60° with the top-dead-centre position. During the inward stroke the maximum turning moment is 14 000 Nm, when the crank makes an angle of 280° with the top dead centre. The turning moment diagrams on a crank-angle base for both strokes are triangular, and the torque supplied by the engine is constant. Find the crank angles at which the speed has its maximum and minimum values.

If the crankshaft speed is 80 rev/min, the diameter of the rim-type flywheel is 3 m, and the speed is to be kept within 0·5 per cent of the mean speed, what weight of flywheel will be required?

1.22 The armature of an electric motor has a moment of inertia of 0·1 kg m^2. The motor runs at a mean speed of 1 500 rev/min and drives, through a reduction gear of 30:1 and a suitable linkage, a mass of 100 kg which oscillates through a total distance of 300 mm with simple harmonic motion. During its travel in one direction only the mass moves against a constant resistance of 50 kgf.

If all other masses and resisting forces are negligible, estimate the cyclical speed fluctuation of the motor.

2. Transmission of Rotational Motion

§ 2.1 Introduction

The transmission of rotational motion from one element or shaft in a machine to another is an almost universal requirement. The satisfaction of this requirement may at times introduce no problems worth discussion; at other times the problems may be very difficult indeed. If, for instance, the two shafts to be coupled are permanently and accurately in line with each other, the ends abutting, a rigid coupling will suffice, and the problems that arise are in the field of strength of materials. If the alignment, even if only because of distortions elsewhere, is seriously imperfect a rigid coupling may be inadmissible. The easiest imperfection to deal with is an angular error, and *provided* that the axes of the shafts intersect at the coupling, the insertion of a flexible member or its equivalent in the form of one of the proprietary couplings available may be a sufficient remedy. A greater angular error will call for a slightly more elaborate device, such as a Hooke joint (§ 2.3.1). Unfortunately this type of coupling introduces a cyclical variation in the angular velocity ratio between the two shafts, and if such a variation cannot be tolerated it becomes necessary to use two such joints in series, correctly arranged (§ 2.3.2), or a more elaborate joint (§§ 2.3.3 and 2.3.4). The choice is often determined by the disposition of the elements. A more awkward imperfection is one in which the axes of the shafts do not intersect at the joint. If the ends abut and cannot be separated an Oldham coupling (§ 2.2) or a proprietary counterpart involving a flexible element or an elementary link motion may provide a remedy, but only minor imperfections can normally be dealt with in this way. If the ends of the shafts can be separated axially the insertion of an otherwise unsupported intermediate shaft coupled at each end by a Hooke joint, which provides lateral constraint and so ensures that the axes intersect, may give a neat solution.

If the shafts to be coupled are not even approximately collinear an entirely new situation arises. The devices which can be used, apart from elementary mechanisms such as were discussed in chapter 1, include belts, chains, and gears, and with these devices it is easy to cause the driven member to rotate at a speed different from that of the driving member; indeed, this may well be the main reason for using such a device. Occasionally, for special purposes, a cyclical speed variation may be deliberately introduced by the use of an asymmetrical linkage (see, e.g. fig. 1.58) or elliptical gearing, but in the great majority of cases it is important to avoid cyclical variations: this is the prime consideration in the design of toothed gearing. The choice of the device to be used depends, of course, on the conditions which have to be met, and there is no point in attempting an exhaustive treatment of such a topic. On the other hand, there is something to be said for devoting a paragraph or two to it before considering individually the most important of these devices.

If the shafts to be coupled are parallel but a considerable distance apart the situation can be quite awkward, in that an apparently trivial problem may have no really neat solution. The early locomotive builders met this problem when, in order to obtain a sufficiently large tractive effort, they had to couple several of the axles to the one which was acting also as the engine crankshaft. Their solution was to add to the already existing frame, axles and wheels,

pins and a coupling rod linking the wheels on one side of the locomotive (so providing a four-bar chain) with a similar arrangement
set at 90° to the first on the opposite side. This is really a very elegant
solution for the particular conditions which have to be met, and it is
worth while spending a little thought on the possible alternatives: it
will probably be agreed that the choice was a wise one. Why, then,
is this solution so seldom adopted today for the transmission of
rotational motion?

If it is unnecessary to maintain an accurate phase relationship
between two shafts a belt can frequently be used. Belts will be discussed in some detail in § 2.4; for the moment it need only be
pointed out that the power which a belt can transmit is limited by the
facts that, having to be flexible, a belt cannot easily be made of
extremely strong materials, and that the power has to be transferred
through frictional forces. A belt of V form is much more effective
than a flat belt, and the use of the latter, for anything other than
light or 'casual' purposes, has diminished rapidly. Belts have many
advantages, however, such as cheapness, silence, and flexibility in
use, and we shall have to consider how to design a belt drive to
transmit a given amount of power without excessive slipping. We
might also note in passing that belts are sometimes made with
internal 'teeth' which engage with teeth on the pulleys, thus providing, within the limits of the force they can transmit, a positive drive.

If an accurate phase relationship has to be maintained the belt
can be replaced by a chain, which, of course, gives a positive drive.
A chain can be made of strong materials and does not rely on
friction: a chain drive is therefore much more compact for a given
power than a belt drive. However important technologically, a
chain need not detain us, since the problems involved are again
mainly in the field of strength of materials.

By far the most important device used to transmit rotational
motion at the present day is the toothed gear, and we shall devote
most of our attention to gearing. Even so, we shall have space to
deal only with the major principles involved in the most straightforward form of gearing, straight-tooth or spur gearing, and add only
brief notes on one or two other main types. It is at this point worth
noting that one of the chief virtues of gearing, namely its compact-
ness, can in specific instances be a disadvantage: for instance, if we
were to try to couple the locomotive axles referred to above by
means of gears we should probably be faced with having to provide

a whole train of gears between each axle, or having to drive through bevel or similar gears to a longitudinal shaft and back again through a right angle to each of the other axles – a somewhat clumsy alternative.

In fact, it is not difficult to see that although there are often a number of alternative possibilities, the conditions of the problem – the disposition of the shafts, the requirements for phase accuracy, the requirements for power, the absolute speed involved, the ease of lubricating the system or of excluding lubricants from the system, and so on – will frequently determine qualitatively for us the type of drive which we shall have to use.

§ 2.2 Transmission of Rotational Motion between Parallel but Misaligned Shafts: the Oldham Coupling

Each shaft in this coupling (fig. 2.1) is provided with a flange in which a single diametral groove is cut: between the two flanges is inserted

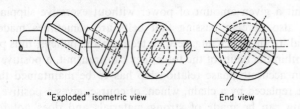

"Exploded" isometric view   End view

Fig. 2.1 Oldham coupling

a disk, on each face of which is a tongue to fit the groove in the flange; these tongues are at right angles to each other. Since there can be no relative rotation between the tongue and its groove, the angular speed of each member must at all times be the same, so no cyclical variation in angular velocity is introduced. It can, however, be seen from the figure that the centre of the disk will move in a circular path whose diameter is equal to the distance between the shaft axes: the maximum velocity of sliding in each groove is the product of the angular velocity of the joint and the radius of this circle. The disk is unbalanced and the sliding surfaces are awkward to lubricate. These features are unattractive for high-speed or high-torque applications, and the usefulness of the coupling is in consequence very restricted.

§ 2.3 Transmission of Rotational Motion between Shafts which are not Collinear but whose Axes Intersect

2.3.1 Hooke joint

This is the device most commonly used for this purpose. It is illustrated diagrammatically in fig. 2.2, and consists of a member in the form of a cross *AABB* linked by turning pairs to forked members attached to the ends of the shafts which are shown in a horizontal plane. That this device will transmit rotation is obvious enough.

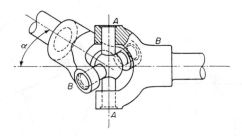

Fig. 2.2 Hooke joint

Unfortunately the velocity ratio is not constant; in other words, if one shaft rotates at constant speed the other is driven at a speed which varies cyclically, and this may have serious consequences in many applications, particularly if the inertias associated with the shafts are large. We must therefore determine the magnitude of the variation.

As a preliminary we may consider the geometrical theorem that if two lines at right angles to each other are projected on a plane parallel to one of them the projections will be at right angles to each other. This is illustrated in fig. 2.3, and the proposition so illustrated hardly requires formal proof.

Now consider fig. 2.4, which is a view looking along the right-hand shaft in fig. 2.2. When the arm *AA* of the cross is vertical it coincides with *YY*, and *BB* with *XX*. When *AA* rotates through an angle θ it is seen at *A'A'*. The ends of *BB* *appear* to move in an ellipse with major and minor axes *BB* and *BB* cos α respectively, and at this instant will be seen at *B'B'*. From the theorem just considered *B'B'* will be perpendicular to *A'A'*, but in reality the ends of *BB* move in a circular path, *B'* indicating the *true* vertical position of *B*, so that *OB''* shows the position of *OB* as seen when looking along the left-hand shaft. It follows that the true angle ϕ through which BB has rotated is $<B''OX$, and from the known properties of an ellipse we see that

$$\frac{EB'}{EB''} = \cos \alpha$$

We can therefore write

$$\tan \phi = \frac{DB''}{OD} = \frac{CB'}{EB''} = \frac{CB'}{EB'} \cos \alpha = \tan \theta \cos \alpha \qquad (2.1)$$

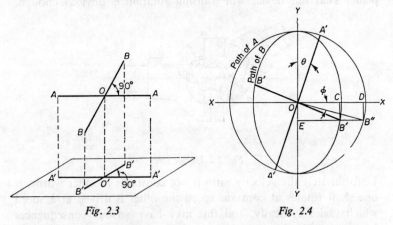

Fig. 2.3   Fig. 2.4

To obtain the velocity ratio we differentiate both sides of this equation with respect to time (α being, of course, constant)

$$\sec^2 \phi \cdot \dot{\phi} = \cos \alpha \sec^2 \theta \cdot \dot{\theta}$$

So the velocity ratio

$$\dot{\phi}/\dot{\theta} = \cos \alpha \sec^2 \theta / \sec^2 \phi$$

and if we wish to express this in terms of α and θ (say) only we can eliminate ϕ by writing

$$\sec^2 \phi = 1 + \tan^2 \phi$$

(and from equation (2.1)) $\qquad = 1 + \tan^2 \theta \cos^2 \alpha$

$$\therefore \frac{\dot{\phi}}{\dot{\theta}} = \frac{\cos \alpha}{\cos^2 \theta (1 + \tan^2 \theta \cos^2 \alpha)} = \frac{\cos \alpha}{1 - \sin^2 \theta \sin^2 \alpha} \qquad (2.2)$$

By inspection, the minimum value of this ratio occurs when $\theta = 0$ or π and is $\cos \alpha$; the maximum occurs when $\theta = \pi/2$ or $3\pi/2$, and is $1/\cos \alpha$. This gives a clear indication that the major cyclical speed variation is at a frequency twice that of the shaft speed: it can be shown* that

$$(\theta - \phi) = z \sin 2\phi + \tfrac{1}{2}z^2 \sin 4\phi + \tfrac{1}{3}z^3 \sin 6\phi + \ldots$$

where z has the value

$$(1 - \cos \alpha)/(1 + \cos \alpha)$$

If θ is assumed to be constant we can find the acceleration of the other shaft by differentiation, thus

$$\ddot{\phi} = \frac{d\dot{\phi}}{dt} = \frac{d\dot{\phi}}{d\theta} \cdot \frac{d\theta}{dt} = \frac{\cos \alpha \sin^2 \alpha \sin 2\theta}{(1 - \sin^2 \theta \sin^2 \alpha)^2} \dot{\theta}^2$$

To find the condition for the maximum value of this acceleration we should, strictly speaking, differentiate again and equate the result to zero. The result hardly justifies the labour involved, particularly since the assumption that θ is constant is quite unrealistic and implies that an infinite moment of inertia is fixed to one of the shafts. It is more sensible to say that (from considerations of symmetry) the maximum acceleration will occur when

$$\theta \simeq \pi/4$$

and that its value will therefore be roughly

$$\ddot{\phi}_{max} = \frac{\cos \alpha \sin^2 \alpha}{(1 - \tfrac{1}{2}\sin^2 \alpha)^2} \dot{\theta}^2$$

The error involved is trivial. In fact, if α is not very large this reduces to

$$\ddot{\phi}_{max} \simeq \dot{\theta}^2 \sin \alpha \tan \alpha$$

If the moment of inertia of the masses attached to the driving and driven shafts is similar this will in fact give a closer approximation to the total relative acceleration between the masses, and this acceleration can then be inversely apportioned to each mass in accordance with the principle of conservation of angular momentum.

If the masses, the angular velocity, and the angle α between the shafts are all small, and if the assumption that the shaft axes intersect at the centre of the cross-member is justified (for instance, by allowing one of the shafts lateral freedom so that its axis is positioned by the joint itself), a single Hooke joint may provide a satisfactory drive. It is relatively seldom that all these conditions are met, and in the great majority of cases a more satisfactory result

* See E. J. Nestorides, *Handbook of Torsional Vibration*, Cambridge University Press, 1958.

can be achieved by using a double Hooke joint, as in fig. 2.5, or a more sophisticated form of joint which will be discussed later.

2.3.2 Double Hooke joint

The double Hooke joint needs little further discussion. In fig. 2.5a the driving and driven shafts A and C are coupled by a shaft B arranged to be equally inclined to each. In fig. 2.5b the shafts A and C are parallel, so the inclination of each to B is automatically the same: this is the arrangement suggested in § 2.1 as preferable to an Oldham coupling. From consideration of symmetry it can be seen that if B were the driver A and C would receive the same angular

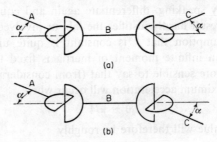

Fig. 2.5 Double Hooke joints

velocity at any instant. It follows that if A rotates at constant speed so does C, and only the member B, which can be made of negligible inertia, has cyclically varying velocity. Equally important is the fact that the problem of ensuring that the shaft axes intersect correctly is minimised or eliminated, since the joints can usually be arranged to support the shaft B without the need for additional, and kinematically redundant, bearings.

Cases, however, arise in which drives have to be taken abruptly through considerable and varying angles, and in which cyclical variations in velocity cannot be tolerated. A notable example is a car with front-wheel drive, since the front wheels must be steerable. Various ingenious solutions have been devised, of which most fall into the category of modified double Hooke joints. One such device and one of another type merit attention.

2.3.3 The Birfield joint

The Birfield joint is illustrated in a slightly simplified form in fig. 2.6. In view (*a*) the shafts A and B to be coupled are collinear; in view (*b*) at an extreme angle of 40° to each other. This joint is a development of the Rzeppa joint, which in turn can be regarded as a development of the double Hooke joint shown in fig. 2.5a, the intermediate shaft having in effect been reduced to zero length and replaced by a number of balls (usually six) in a 'cage' or part-spherical shell *C*, which constrains them to lie in one plane.

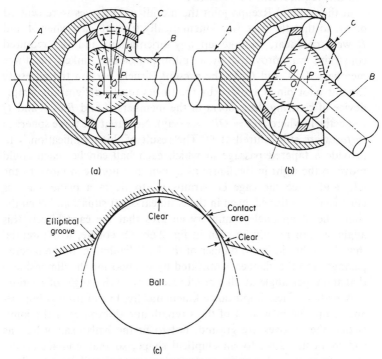

(a)   (b)

(c)

Fig. 2.6 Birfield universal joint
(By permission of Birfield Engineering Ltd)

The shell *C* has external and internal radii r_1 and r_2, and the shafts *A* and *B* terminate in internal and external spherical surfaces of radius r_1 and r_2 respectively, so without the balls these three members form a double spherical joint centred at *O*. To transmit the torque 'meridianal' grooves are cut in these spherical shaft ends: by meridianal is meant that if the sphere is regarded as having north and south poles on the axis of the shaft the grooves have the same situation as meridians of longitude on the surface of the earth. If therefore we place one or more balls in the grooves the joint will be capable of transmitting torque: one ball is kinematically sufficient; six give a more reasonable practical solution.

But, as we have seen in the case of the double Hooke joint, in order to give a constant velocity ratio the intermediate member (i.e. in effect the normal to the plane containing the centres of the balls) must be equally inclined to the axes of the shafts A and B. How can this requirement be satisfied?

In the original Rzeppa joint the meridianal grooves were centred on O, so that the cage was kinematically free, at least when A and B were collinear, to take up any orientation, and the required constraint was provided by a 'pilot pin' which linked the three members A, B, and C. In the present Birfield version the necessity for this pilot pin is avoided by the (in its result) delightfully simple device of centring the *grooves* in the members A and B at P and Q respectively where $OQ = OP = x$ (say). Note that the three spherical *surfaces* are still centred at O. The result of this modification is to provide a tapered passage in which each ball can lie; each could move to the right in the figure as drawn, but none can move to the left, and since the cage constrains all to lie in a plane, each is effectively positioned, even in the collinear-shaft situation. When the shafts lie at an angle symmetry ensures that the cage bisects this angle, as can readily be seen in fig. 2.6b. Of course, it is essential that there should be no danger of the balls 'jamming' in the tapered passage, and this danger is obviated by so choosing the dimension x that the taper angle at all times exceeds twice the angle of friction.

A matter of less importance kinematically, but of no less importance from the viewpoint of the strength and durability of the joint, is that the grooves are ground, not to fit the balls exactly but, as shown in fig. 2.6c, to an elliptical shape, so that contact occurs away from the edge as in an 'angular contact' ball bearing. It is chiefly in this respect that the views (*a*) and (*b*) are simplified, since to take exact account of this modification,* some rather difficult three-dimensional geometry is required which tends to obscure the basic issues.

2.3.4 The Chobham joint

The Chobham joint achieves the same result – that of providing constant velocity ratio – by the use of the entirely different and very simple *principle* illustrated in fig. 2.7.

* For a full description, see British Patent 810, 289, 1956.

Let AOA', BOB' be the axes of two shafts intersecting at an angle
α in O, and let aa', bb' be arms or rigid 'wires' symmetrically attached
to the shafts, i.e. so that $Oa = Ob$ and $\angle Aaa' = \angle Bbb'$, and
touching each other in the position shown. It is then clear from
symmetry alone that if the two shafts are rotated through the same

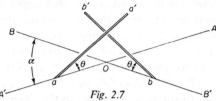

Fig. 2.7

angle the wires will continue to touch each other (why should one
move faster than the other in the plane of symmetry?), so that the
arrangement forms an embryonic and unidirectional constant
velocity joint, but it is also fairly obvious that the wires will rotate
relative to each other and that the point of contact will move along
them. Kinematically there is no difficulty in providing, say, four
such pairs of wires, of which two, preferably diagonally opposite
each other, would transmit torque in one direction, and the other
two, conveniently at right angles to the first pair, would transmit
torque in the other direction. (Such an arrangement additionally
provides a lateral or 'self-centring' constraint for the axis of one

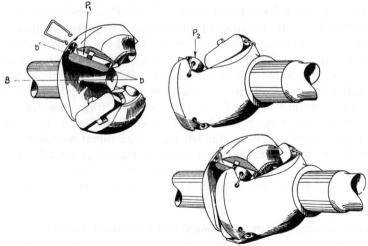

Fig. 2.8 Chobham universal joint
(Crown Copyright Reserved)

shaft relative to the other, but no axial or 'end-wise' constraint.) But of course this arrangement, although 'kinematically correct', is utterly impractical, and the 'wires' have to be replaced by opposed surfaces free to accommodate the rotating and sliding motions described above. The illustrations of the Chobham joint in fig. 2.8 show clearly how this is carried out. In the upper figure the two halves of the joint are separated: the left-hand member corresponds to the shaft BO in fig. 2.7, and the upper 'pad' P_1 to the 'wire' bb'; the pad P_2 on the right-hand member corresponds to the 'wire' aa'. To avoid interference these pads have to be carried by the forked members shown. The flat faces of P_1 and P_2 slide on each other and the pads are free to rotate in the forks to the degree required, but are restrained from axial movement by the collars shown. The spring clip serves to retain the pad in the fork until assembly is completed as shown in the lower illustration.

§ 2.4 Belt Drives

Belts are most frequently used to transmit rotational motion between parallel shafts, though they can, of course, if necessary, by the use of appropriate intermediate guiding pulleys, be used in other applications. Only the parallel-shaft case need be considered here, since the variants introduce no new principles of importance. Again, 'belts' can be of a number of types; of 'flat', 'V', or 'circular' cross-section. Flat belts are usually made of leather or rubberised fabric, but can, for instance, be made of steel. V-belts usually contain longitudinal textile strands to carry the load, impregnated and covered with 'rubber' to provide a high coefficient of friction, but the construction varies considerably. Circular 'belts' can be made of leather, but are usually fabric ropes or wire ropes: the former normally run in contact with the sides of V-grooved pulleys and can be treated as V-belts, the latter (i.e. wire ropes) normally run on the bottom of the grooves and can be treated as flat belts. We need therefore consider in principle only flat and V-belts, and we shall see that these two cases can be dealt with in an almost identical fashion.

2.4.1 Tension required to make a belt pass round a pulley

In order that a belt may be used to transmit rotational motion between two shafts, it must be made to pass round pulleys on the shafts, and it will not do so unless there is in it a tension, say T_0, which we shall now determine. In order that it may transmit power, additional tensions must exist, and, although the calculations could be completed in one step, we shall for convenience and clarity determine them subsequently and separately.

Fig. 2.9 shows diagrammatically a belt in which there is a tension T_0 approaching at a speed v a pulley which is free to rotate, travelling round part of its circumference, and leaving again at the same speed and (since the pulley is free) with the same tension. During its passage round the pulley any element of the belt has to be accelerated radially (cf. § 0.11), and the only force available to provide this acceleration is T_0. If T_0 is larger than necessary the belt will be pressed against the pulley surface: we consider the limiting case when this pressure is zero.

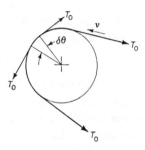

Let the *mass* of *unit length* of the belt be m, and let the pulley radius be r. Consider an element (fig. 2.9) of the belt which subtends at the centre of the pulley an angle of $\delta\theta$. Its length will be $r\delta\theta$, and its mass $mr\delta\theta$; this element has an acceleration of v^2/r towards the centre of

Fig. 2.9 Tension required to make a belt go round a pulley

the pulley, and must therefore be subjected to an inward radial force $= mr\delta\theta \cdot v^2/r$ or $mv^2\delta\theta$. But the only forces acting on the element are the two tensions T_0 acting at its ends tangentially to the pulley. We can therefore resolve these forces in the radial direction and write

$$2\,T_0 \sin \delta\theta/2 = mv^2\delta\theta$$

But since $\delta\theta$ is small we can put $\sin \delta\theta/2$ equal to $\delta\theta/2$, and so

$$T_0 = mv^2 \qquad\qquad (2.3)$$

(If information about the belt had been provided in the form that the cross-sectional area was a and the density was ρ we should have calculated m as $(\rho \cdot 1 \cdot a)$ so the tension T_0 would have been given by $\rho a v^2$. We could then have calculated the mean tensile *stress* in the belt as $T_0/a = \rho v^2$. This is the usual expression for the stress caused by rotation in the rim of a flywheel.)

2.4.2 Power transmitted by a belt

Now let the *additional* tensions in the belt to transmit power be T_1 in the tight side and T_2 in the slack side, fig. 2.10. The power transmitted to the pulley from the tight side of the belt is (Force × Velo-

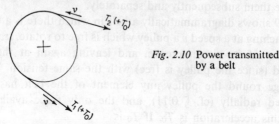

Fig. 2.10 Power transmitted by a belt

city) $= T_1 v$, and from the pulley to the slack side is $T_2 v$, so the net power transmitted is given by:

$$\text{Power} = (T_1 - T_2)v \qquad (2.4)$$

2.4.3 Maximum tension in a belt

If an attempt is made to transmit too much power the belt will either stretch excessively, break, or slip. Clearly to avoid the former contingencies the

$$\textit{Maximum total tension} = T_{\max} = (T_1 + T_0) \qquad (2.5)$$

must be limited to a value which experience or test has shown to be suitable: the determination of this value lies in the realm of strength of materials, but a note on this topic will be found in § 2.4.8. To determine the conditions necessary to avoid slipping we proceed as follows.

2.4.4 Condition for avoidance of slip

Let us consider first the force available to stop an *element* of a belt slipping round a pulley. If such an element is subject to an effective radial force N pressing it on to the surface of the pulley, then in the case of a flat belt (fig. 2.11a) the reaction of the pulley will, of course, be equal and opposite to N, so the frictional force available will be simply $F = \mu N$, where μ is the coefficient of friction. In the case of a V-belt (or rope) running in a V-pulley (fig. 2.11b) the reaction will be in the form of two component forces n, n, which we may assume to be normal to the sides of the V-groove. (This assumption is not strictly justifiable unless slip actually occurs after the belt is lodged in the groove, but in view of unavoidable uncertainties as to the coefficient of friction, it can be regarded as sufficiently accurate.) The frictional force available to stop slipping will then be $F' = \mu(2n)$, and by resolving the forces radially we see that

$$2 \cdot n \sin \alpha/2 = N$$

where α is the included angle of the groove, so

$$F' = \mu \cdot 2n = \mu N \operatorname{cosec} \alpha/2$$

We can, of course, regard the flat belt as running in a 'groove' of included angle 180°, when $\operatorname{cosec} \alpha/2 = \operatorname{cosec} 90° = 1$, and the two

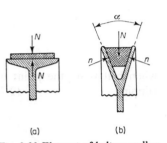

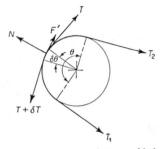

(a)   (b)

Fig. 2.11 Element of belt on pulley   *Fig. 2.12* Forces on element of belt

expressions agree. Thus we can deal with both cases simultaneously by writing

$$F' = \mu'N, \quad \text{where } \mu' = \mu \operatorname{cosec} \alpha/2 \qquad (2.6)$$

Now let us consider the whole angle of lap of the belt. In fig. 2.12 we have again the situation depicted in fig. 2.10, but we need give no further consideration to the tension T_0 required to cause the belt to pass round the pulley, just making contact. We are concerned only with the additional tension which varies from T_2 to T_1 as the belt goes round the arc of contact θ and which causes the belt to press on the pulley with a finite force. Let the belt be on the point of slipping.

Consider the element of the belt which subtends the angle $\delta\theta$ at the centre of the pulley. Let the tensions at the ends of this element be T and $(T + \delta T)$, and the effective normal and frictional forces exerted on the element by the pulley be N and F'. These forces must be in equilibrium. Resolving radially and tangentially we have

$$N = T \sin \frac{\delta\theta}{2} + (T + \delta T) \sin \frac{\delta\theta}{2}$$

$$F' = \delta T \cos \frac{\delta\theta}{2}$$

and since the belt is on the point of slipping,

$$F' = \mu'N$$

As $\delta\theta$ tends to zero these reduce to:

$$N = T\delta\theta$$
$$F' = \delta T$$
$$F' = \mu'N$$

whence

$$\delta T = \mu'N = \mu'T\delta\theta$$
$$\therefore \frac{\delta T}{T} = \mu'\delta\theta$$

and, integrating,

$$[lnT]_{T_2}^{T_1} = [\mu'\theta]_0^\theta$$

so

$$ln\frac{T_1}{T_2} = \mu'\theta \quad \begin{array}{l}\text{when belt is on the point of}\\ \text{slipping}\end{array} \quad (2.7)$$

or

$$\frac{T_1}{T_2} = e^{\mu'\theta}$$

A note on the values of μ to be expected in practice will be found in § 2.4.8.

2.4.5 Basic equations

Summarising these results we have

$$T_0 = mv^2 \quad \text{(tension to make belt go round pulley)} \quad (2.3)$$
$$ln\frac{T_1}{T_2} \not> \mu'\theta \quad \text{(condition for avoidance of slip)} \quad (2.7)$$
$$\mu' = \mu \operatorname{cosec} \alpha/2 \quad (2.6)$$
$$T_1 + T_0 \not> T_{max} \quad (2.5)$$
$$\text{Power transmitted} = (T_1 - T_2)v \quad (2.4)$$

Needless to say, when we get down to arithmetic we must be careful over our units (see § 0.9).

2.4.6 Equation for power transmitted

Trivial manipulation of these equations allows us to express the power transmitted by a belt on the point of slipping in the form

$$\begin{aligned} \text{Power} &= (T_1 - T_2)v \\ &= T_1(1 - e^{-\mu\theta \operatorname{cosec} \alpha/2})v \\ &= (T_{max} - mv^2)(1 - e^{-\mu\theta \operatorname{cosec} \alpha/2})v \end{aligned} \quad (2.8)$$

but the use of any such equation as a formula to be memorised is to be deplored, since it obscures the relatively simple steps in the argument, and offers the maximum opportunities for arithmetical errors to be committed without detection.

2.4.7 Theoretical condition for maximum power

The equation just developed is in a form which allows us to deduce the speed at which a given belt drive will transmit maximum power. If the maximum tension, the mass per unit length of belt, the coefficient of friction, the angle of lap, and the groove angle are constant we can write

$$\text{Power} = (T_{\max} - mv^2) \text{ const. } v$$
$$= \text{const. } (v \cdot T_{\max} - mv^3)$$

and differentiating

$$\frac{d}{dv}(\text{power}) = \text{const. } (T_{\max} - 3mv^2)$$

$$\frac{d^2}{dv^2}(\text{power}) = \text{const. } (-6mv) \text{ which is negative}$$

so the condition for maximum power is given by

$$T_{\max} = 3mv^2 = 3T_0$$
$$\text{or } T_0 = \tfrac{1}{3}T_{\max} \tag{2.9}$$

This result should be used only with the greatest caution. It is valid only for the conditions which have been postulated, and these include the assumption that the maximum actual tension $(T_1 + T_0)$ in the belt is in fact adjusted to the maximum tension which the belt can withstand: such a condition can seldom be achieved or maintained. The result is really useful only in providing a guide to the speed which could be regarded as a top limit for maximum power seldom approached in practice. There are exceptional occasions on which an extremely high output speed is required at very low power, and for such purposes the limiting condition is, of course, $T_0 = T_{\max} = mv^2$.

2.4.8 Note on values of $T_{\max}$ and μ in practical applications

To replace a design handbook is not one of the objectives of this book. Nevertheless, there are a few important general principles in relation to belts which can be most easily understood if a few factual numbers are introduced into the discussion. Let us compare a flat leather belt with a rubberised fabric V-belt.

'Leather' is, of course, a somewhat variable material, but good-quality leather will have a density of about 800–1 100 kg/m³* and an ultimate tensile strength of between 300 and 450 bar. A stress approaching this value, applied repeatedly, would very soon lead to failure, and the maximum allowable tension is more likely to be in the region of 40–50 bar. Even this stress will give a short life, and experience suggests that it is more economical to limit the stress to, say, 15–20 bar; i.e. about *one-twentieth* of the ultimate stress, or less. Again, the coefficient of friction between clean dry leather and, say, a cast-iron pulley may be of the order 0·5 or even more, but under practical conditions it would be unwise to rely on a figure greater that one-third of this. For the sake of argument, let us calculate the maximum power which can be transmitted by a belt, 1 cm² in cross-section, of density 1 000 kg/m³, in which the stress is to be limited to 18 bar, on the assumption that $\mu = 0\cdot167$.

If the pulleys differ in size the angle of lap on one will be less than 180°: suppose we guess at 3 rad, or 162°.

For maximum power we must choose the speed so that $T_0 = \frac{1}{3}T_{max}$. This gives

1 000 $\times a \times v^2 = \frac{1}{3}(18 \times 10^5 \times a)$, where a is the cross-sectional area of belt,

whence $v = \sqrt{600} = 24\cdot5$ m/s

$$T_1 = T_{max} - T_0 = \frac{2}{3}T_{max} = \frac{2}{3} \times 18 \times 10^5 \times 10^{-4} = 120 \text{ N}$$

$$\frac{T_1}{T_2} = e^{0\cdot167 \times 3} = e^{0\cdot5} = 1\cdot65 \quad \therefore T_2 = 73 \text{ N}$$

$$\therefore \text{Power} = (120 - 73) \, 24\cdot5 = 1 \, 150 \text{ Nm/s} = 1\cdot15 \text{ kW}$$

In the case of the V-belt we start with a material of somewhat higher density – about 1 100 to 1 400 kg/m³ – but capable of withstanding a higher working stress, of the order 40 bar at low speeds, but preferably less – perhaps 30 bar – at high speeds, and in all probability providing a somewhat higher coefficient of friction, say about 0·2 under reasonable working conditions. The standard belt† has an included angle of 40°. When we bend this round a pulley of reasonable size (which, of course, adds to the maximum stress to a degree depending on the form of the construction: we must not use too small a pulley) the outer layers stretch and the inner layers com-

---

* i.e. a relative density of 0·8–1·1
† B.S. 1440 : 1962.

press, and due to the 'Poisson's ratio' effect this angle decreases. The pulleys have usually a V-angle of 38° or so. If we now carry out a similar calculation for the maximum power which can be transmitted by a V-belt of 1 cm² cross-section, taking T_{max} as 30 bar, ρ as 1 250 kg/m³, μ as 0·2, and α as 38°, we get

$$T_0 = \tfrac{1}{3} \times 30 \times 10^5 \times a = 1\,250 \times a \times v^2$$

whence $\quad v = 28\cdot3$ m/s

$$\therefore\ T_1 = \tfrac{2}{3} \times 30 \times 10^5 \times 10^{-4} = 200\ \text{N}$$

$$\frac{T_1}{T_2} = e^{0\cdot2 \times 3\cdot07 \times 3} = 6\cdot3 \qquad \therefore\ T_2 = 32\ \text{N}$$

$$\therefore\ \text{Power} = (200 - 32)\ 28\cdot3 = 4\,760\ \text{Nm/s} = 4\cdot76\ \text{kW}$$

Thus there is a factor of 4 between the two drives in favour of the V-belt. Moreover, the width taken up by the V-belt will be considerably smaller. It is not surprising that the V-belt has largely ousted the flat belt.

Altogether it will be appreciated that at least two of the factors involved in these calculations are, in the general case, little better than informed guesses. The figure chosen for the allowable stress is a compromise which can be varied over fairly wide limits, depending on the relative importance of such factors as first cost, space available, and trouble-free life required. The value assigned to the coefficient of friction will depend less on the materials than on the degree of cleanliness which can be maintained under operating conditions. Designers can, however, obtain an answer, which no doubt represents as good a compromise as can be achieved, by consulting the British Standard publication B.S. 1440 : 1962, which deals with V-belts for industrial drives.

§ 2.5 Types of Gearing

Gears are used for transmitting rotary motion between shafts at a suitable distance from each other. If the shafts are parallel, *spur* or *helical* gears are used: in the former type the teeth also are parallel to the shafts, in the latter they are, as the name suggests, cut along a helix. If the shaft axes intersect, *bevel* or *spiral bevel* gears are used, the bevel gear being the counterpart to the spur, while the spiral bevel is the counterpart of the helical gear. If the axes are neither parallel nor intersecting, *skew* (or 'spiral', or 'crossed-helical') or worm gears are used; these are essentially similar in principle, the name worm gear being normally reserved for the case when one of the wheels is relatively small and has a small number of teeth. Several other types of gears are used in particular cases: for instance, hypoid (or offset helical bevel) gears are commonly used in car back-axles,

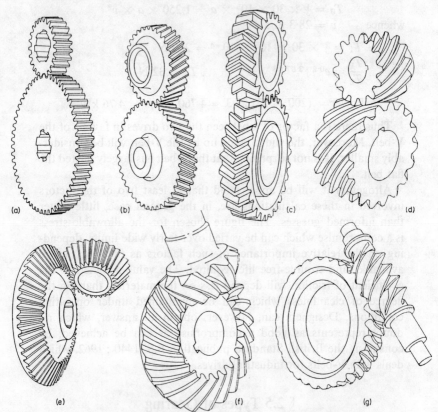

Fig. 2.13 Types of gears: *a*. spur, *b*. helical, *c*. double helical, *d*. skew or crossed-helical, *e*. bevel, *f*. spiral bevel, *g*. worm

(By permission of the David Brown Corporation (Sales) Ltd.)

but these are quite outside our scope. In fact, we have space to deal only with the basic principles of spur gearing, and to add brief notes on helical gears of conventional and unconventional types, and on worm gears. A few of these types of gear are illustrated diagrammatically in fig. 2.13.

§ 2.6 Spur Gears

A friction drive could be devised between two parallel shafts by
fixing to each a disk so arranged that their edges pressed against
each other and rolled together without slipping. The ratio of the
angular velocities of the shafts would be constant and inversely
proportional to the radii of the disks on them. Such an arrangement
would, of course, transmit very little power, since slipping would
occur at a very low torque, but if intermeshing teeth are provided
to prevent this the disks are converted into 'gears' (fig. 2.13a). In
order to illustrate some of the more important terms used in gearing,
fig. 2.14 shows a sketch of part of a spur gear. The (hypothetical)

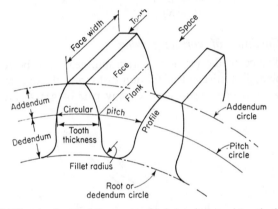

Fig. 2.14 Sketch of part of Spur gear-wheel showing usual terminology

disk is of fundamental importance, and is represented by the *pitch
circle*. The part of the tooth above the pitch circle is called the
addendum, and its bearing surface the *face*; the part below the pitch
circle the *dedendum*, and its surface the *flank*. The curve forming face
and flank is called the *profile*. The difference between the dedendum
and the addendum of the mating gear is called the *root clearance*.

Two terms not indicated on this diagram may be introduced at
this stage, *module* and *diametral pitch*. The *circular pitch* p_c is, as
shown on the diagram, the distance measured along the pitch circle
between corresponding points on adjacent teeth, and is therefore
equal to $\pi D/N$, where D is the pitch circle diameter and N the
number of teeth. If, as is frequently the case, the pitch circle diameter

is chosen as a convenient number the circular pitch is inevitably an incommensurable number (or non-terminating decimal) and to that extent awkward to use; and vice versa. The module m is defined as p_c/π and is therefore equal to D/N: to such a term there is no objection. Unfortunately, however, the term 'diametral pitch' P, defined as the 'number of teeth per *inch* of pitch-circle diameter', is likely to be used for some years to come and cannot be ignored. There are many objections to its use, of which the most important are that it is not a pitch at all but the reciprocal of such a quantity, and that it introduces entirely unnecessarily a specific unit.

For purposes of reference, but only if the units are coherent

$$p_c = \pi D/N$$
$$m = D/N$$
$$P = N/D \quad \therefore P = \frac{\pi}{p_c} = \frac{1}{m} \qquad (2.10)$$

2.6.1 Condition of 'correct gearing'

As explained in the previous paragraph, gears can be regarded as the kinematic equivalent of friction disks, and the first requirement is that the velocity ratio between any pair of gears as each tooth

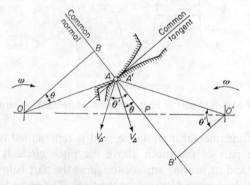

Fig. 2.15 Condition for 'correct' gearing

passes through its various phases of engagement *must* be constant. Any significant departure from this condition would clearly have disastrous consequences if the gears were coupled to considerable inertias. We must therefore investigate this requirement.

Fig. 2.15 illustrates the situation where a gear-wheel with centre O rotating at a speed ω clockwise drives another with centre O' rotating at a speed ω' anti-clockwise. At the instant shown the point A

on wheel O is in contact with the point A' on wheel O'. There will, of course, be a common tangent to the tooth profiles at the point of contact, and, perpendicular to this, a common normal: these are indicated on the diagram. The velocity V_A of point A will be $\omega.OA$ in a direction perpendicular to OA, and this can be resolved into components $\omega.OA \cos\theta$ and $\omega.OA \sin\theta$ along the common normal and common tangent respectively, where θ is the angle between V_A and the common normal: similar expressions $\omega'.O'A'.\cos\theta'$ and $\omega'.O'A'.\sin\theta'$ can be written down for the point A'. If contact is to be continuous between the teeth these component velocities along the common normal must be identical. We have, therefore,

$$\omega.OA.\cos\theta = \omega'.O'A'.\cos\theta'$$

If we drop perpendiculars OB and $O'B'$ from O and O' to the common normal, then since $V_A \perp OA$, we see that $\angle AOB = \theta$ (and similarly $\angle A'O'B' = \theta'$). Therefore

$$\omega.OB = \omega'.O'B'$$

Let P be the point in which the common normal cuts OO'. As $\triangle$s OBP and $O'B'P$ are similar, we can now write

$$\frac{OP}{O'P} = \frac{BP}{B'P} = \frac{OB}{O'B'} = \frac{\omega'}{\omega} \qquad (2.11)$$

But we started with the requirement that ω'/ω should be constant: we therefore now see that $OP/O'P$ must be constant. Moreover, we have already noted that the friction disks to which these gear-wheels are to be kinematically equivalent must have radii inversely proportional to the angular speeds; in other words, P must be the point where these 'disks' (i.e. the pitch circles of the gears) touch; this point is called the *pitch point*, and we can express our requirement for correct gearing by saying that *the common normal at the point of contact between two teeth must always pass through the pitch point.*

2.6.2 *Velocity of sliding between gear-teeth*

We shall return in the next paragraph to the meaning and implications of the result we have just proved, but it is convenient in passing to evaluate the velocity of sliding between the teeth shown in fig. 2.15. We saw that the velocity of point A along the common tangent was $\omega.OA \sin \theta$ and of point A' was $\omega'.O'A'.\sin \theta'$. The sliding velocity is therefore the difference between these quantities, i.e.

sliding velocity between teeth

$$= \omega'.O'A'.\sin \theta' - \omega.OA.\sin \theta$$
$$= \omega'.AB' - \omega.AB$$
$$= \omega'(AP + PB') - \omega(PB - AP)$$
$$= (\omega' + \omega)AP + (\omega'.PB' - \omega.PB)$$

(and from equation (2.11)) $= (\omega' + \omega)AP$ \qquad (2.12)

Thus we see that sliding between gear-teeth occurs at all stages except the instant when they touch at the pitch point, when the direction of sliding is reversed. This is important in relation to lubrication and wear.

2.6.3 Types of spur gearing

Although the condition for correct gearing established in § 2.6.1 is not theoretically restrictive, in that we can select any reasonable

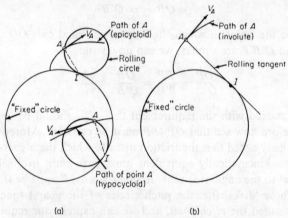

Fig. 2.16 Epicycloid, hypocycloid and involute curves

profile for one tooth and derive graphically or otherwise the form of a mating or conjugate profile to satisfy the condition, this fact is of little general interest. One type of profile, the 'involute', is used almost exclusively in modern engineering practice, but the 'cycloidal' profile, though of limited practical value, is of sufficient academic interest to justify mention. We shall now define these terms.

If a circle rolls on a straight line any point on its circumference traces out or 'generates' a curve which is called a *cycloid*. If it rolls on the outside of another circle the curve is called an *epicycloid*; if on the inside, a *hypocycloid*. If a straight line rolls on a circle any point on the line traces out a curve which is an *involute*. (Alternatively, we can think of an involute as the curve traced out by a point on a string being unwrapped from a cylinder.) An involute could be regarded as an epicycloid traced out by a rolling circle of infinite radius, but since we cannot have a circle of infinite radius rolling inside a finite circle, we clearly cannot have a corresponding hypocycloid. These various curves are sketched in figs. 2.16a and 2.16b. In each case, at the instant depicted, the point of contact is *I*, which is, of course, the instantaneous centre for the rolling member. The direction of motion of the generating point *A* is therefore perpendicular to *IA*: in other words, *the normal to such a curve at any point passes through the point of contact I.* We can now show that tooth profiles formed from such curves may satisfy the condition for correct gearing.

2.6.4 Cycloidal gear-teeth

Let *AA, BB* in fig. 2.17 be the pitch circles of two gears with centres at *O* and *O′* respectively, and let a circle of radius *r* roll (*a*) on the outside of *AA*, and (*b*) on the inside of *BB*, so generating the epi- and hypo-cycloids shown, which can be regarded as partial tooth

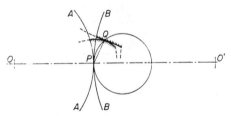

Fig. 2.17 Cycloidal teeth

profiles in contact at *Q*. (This rolling circle is shown touching both pitch circles at the pitch point *P*, and the generating point *Q* is taken at random. The choice of any other point *Q′* would, of course, give identical curves at a different stage in the progression of contact.) By the theorem just proved, the normal to both cycloids passes through *P*, so the condition for correct gearing between these parts of the tooth profiles (that is, the addendum for wheel *O* and the dedendum for *O′*) is satisfied.

Similarly, by taking another generating circle (which need not be the same size) rolling inside AA and outside BB we can construct the dedendum curve for O and the addendum curve for O': these parts are shown in dotted lines.

Note particularly that we have assumed that the pitch circles of the two wheels touch at P. If this condition is not satisfied (for instance, if the gears are mounted at an inaccurate distance OO' apart) the condition for correct gearing will be violated. Note, too, that the whole of the tooth profile is constructed from the correct geometrical curves, and that there is therefore no fundamental limit (apart, of course, from such obvious practical limitations as the size of the rolling circle) to the height of each tooth which in the diagram has been cut off arbitrarily, again by dotted lines. We shall return to these points when we have considered 'involute' gears. Note, thirdly, that the point of contact between the teeth is Q, and since Q was *any* point on the arc PQ, this arc (and a similar arc on the other side of OO') is the locus or path of the point of contact.

2.6.5 *Involute gear-teeth*

It is helpful to tackle the consideration of the involute type of gear-tooth from a slightly different viewpoint. Again in fig. 2.18 we have gear centres O and O', but this time let us draw not the two pitch

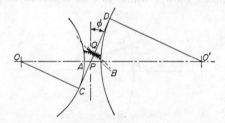

Fig. 2.18 Involute teeth

circles required but two somewhat smaller circles, which we shall call base circles, cutting OO' in A and B respectively, and the common internal tangent CD to these base circles cutting OO' in P. Again choose *any* point Q on CD: let the tangent roll on each base circle in turn: as it does so Q will trace out two involute curves. Alternatively, we can think of this common tangent as a piece of string wound round each of the base circles and held taut by applying small torques of opposite sign about O and O'. Take Q on CD and consider the string being unwound from one base circle on to the other while still being kept taut; the point Q will then trace out the involute curves relative to the two base circles. These two involute curves can be regarded as tooth profiles, which are in contact at Q. Again the fact that these curves satisfy the condition for correct gearing is self-evident, since the common tangent to the base circles is (by the theorem proved in § 2.6.3) the normal to each involute curve. Again the fact that Q is *any* point on CD shows that *this line is the path of contact*; moreover, we see that (neglecting friction) the force exerted by one tooth on the other will act along this line, and consequently the angle ϕ between CD and the normal to OO' is usually called the *pressure angle*.

If we were designing gears we should probably be given the distance OO' and the required velocity ratio ω/ω'; we should select the pitch point P so that $\omega.OP = \omega'.O'P$, *choose* a pressure angle and draw CD, and then draw the two base circles to be tangent to CD. There is no fundamental reason for the choice of one pressure angle rather than another, but there are considerable advantages in adopting a standard value, both from the production point of view and also if we propose to make a range of gears which will engage or mesh with each other. The first standard generally adopted was $14\frac{1}{2}°$ (for the somewhat inadequate reason that, $\sin^{-1} 14\frac{1}{2}°$ being approximately $\frac{1}{4}$, the angle was convenient for both the draughtsman and the craftsman), and although this angle is now outmoded, we shall learn much from a consideration of the reasons for the change.

2.6.6 Length of path of contact and number of teeth in contact
So far we have given no thought to the question of the height of each tooth. We have already noted that with cycloidal teeth there is, within reason, no limitation on the height of the tooth, as the profile above and below the pitch circle can be constructed from 'correct' curves. With involute teeth the situation is radically different because an involute curve must be wholly outside the base circle. We might at first sight conclude that the complete height of the tooth could not exceed the distance between the two base circles. If we look at fig. 2.19, however, and remember that the path of contact is CD we can see that even a tooth as long as that shown on the right-hand gear in the position EC would make *contact* with the profile CF of its mating tooth only outside the base circle CA, i.e. where this profile CF has the 'correct' involute shape, so the condition for correct gearing will not be violated. Clearly, however, we shall have

to cut away the space between the teeth on the left-hand wheel to at least a depth *CG* below the base circle to make room for the tooth *EC* when this tooth is along the line of centres *OO′*, and the shape which this clearance space must have is by no means obvious. This is a matter to which we shall give further consideration in § 2.6.7. For the moment we note that it *may* be possible to use a tooth as

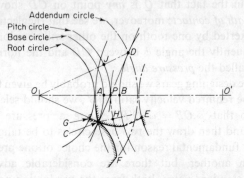

Fig. 2.19 Involute tooth height

long as this, i.e. an outer radius on the right-hand wheel = *O′C*, or an addendum height of *HC* (and similarly an addendum on the left wheel = *JD*), but that any greater size may lead to trouble: again we shall enquire more closely into this possibility in § 2.6.7.

From this argument the maximum addendum on one wheel is determined by the size of the base circle on the *other* wheel. Thus, to provide this addendum on a wheel which must be capable of meshing with a series of wheels of different sizes is impossible, and even to provide it for a wheel which is to mesh with a given size of mating wheel poses very awkward manufacturing problems. For this reason various ranges of 'standard proportions' have been established. It must be emphasised that these proportions have no fundamental significance: they are, like the pressure angle, compromises which may be reasonably satisfactory for the general run of gears for which they are used, but there are examples where non-standard proportions have very considerable advantages. It is, however, instructive to look first at gears which conform to these standards.

One set of standard proportions is that the addendum should be equal to the module *m* (or the reciprocal of the diametral pitch) and the dedendum (to give radial or tip clearance between the wheels)

greater than this by an amount of one-twentieth of the circular pitch
or 0·157 m, while the thickness of the teeth on each wheel at the
pitch circle is to be equal.

Given the standard proportions of the teeth of a pair of gears in
engagement with one another, we can calculate the length of the
path of contact and hence determine the number of pairs of teeth in
contact at any instant. We shall see in § 2.11 that this information is
of considerable importance when we come to consider the strength
of gear-teeth. Fig. 2.20 shows the base circles, addendum circles,

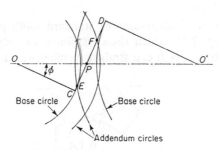

Fig. 2.20 Length of path of contact

and the common tangent for the pair of gears. With the addendum
sizes shown, contact clearly cannot occur outside the points E and
F: EF is therefore the path of contact. Moreover, if we regard the
mating involutes as being produced by a point on a piece of string
being unwound from the base circle, centre O, and wound on the
base circle, centre O', then the length of string unwound from the
beginning to end of contact between the two mating teeth is EF.
The angle of rotation of the gear centre O during this period will be
$\dfrac{EF}{OC}$ rad, and the arc subtended by this angle at the pitch circle will be

$$\frac{EF}{OC} \times OP = \frac{EF}{\cos\phi}$$

where ϕ is the pressure angle. Consequently, the number of pairs of
teeth in contact will be

$$n = \frac{EF}{\cos\phi} \times \frac{1}{\text{circular pitch}} = \frac{EF}{\cos\phi} \times \frac{1}{\pi m} \qquad (2.13)$$

where m is the module.

The 'number of pairs of teeth in contact' usually lies between one and two (for instance it might be 1·73 or some other odd number), which implies that the choice of this conventional phrase was not a particularly happy one, but of course means that during part of the period of engagement of one pair of teeth a second pair is simultaneously in engagement, and during the rest of the period only the one pair of teeth is in contact. Clearly the 'number of pairs in contact' must be made to exceed unity, otherwise one pair of teeth would become disengaged before the next pair was in engagement.

2.6.7 Interference

One of the great disadvantages of standard tooth proportions is indicated in fig. 2.21, which shows a pinion of only 12 teeth in engagment with a *rack* (i.e. a straight member with teeth cut in its

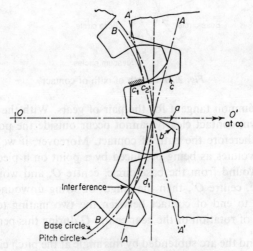

Fig. 2.21 $14\frac{1}{2}°$ rack and 12-tooth pinion

surface). In the involute tooth system the rack is of great importance as we shall see in § 2.6.10. It can be regarded as a section of the rim of a gear-wheel of infinite diameter, and as the point of tangency of the pressure line and its base circle is an infinite distance away, the small part of the involute which is in contact with the pinion teeth is a 'straight' line, which will be normal to the pressure line. Consequently, the sides of the rack teeth are inclined at an angle ϕ to the line joining the centres O and O'.

In drawing fig. 2.21, then, there is no difficulty in constructing the rack profiles, of which one lettered *a* is shown passing through the pitch point. The addendum, dedendum, and tooth width can be calculated from the standard proportions, and the tooth shape completed. Similarly, the pressure line or path of contact *AA* can be drawn at $(90° − 14\frac{1}{2}°)$ to the line of centres *OO'* and the base circle *BB* for the pinion drawn tangent to it. The addendum and dedendum circles can be drawn and the profile of a pinion tooth *above* the base circle, such as *b*, can be constructed easily enough. (The usual method adopted is shown in fig. 2.22. Selecting a starting point such as *P*, we can draw a tangent *PA* and mark out with a pair of dividers a series of small steps back to the base circle at *A* and round it a short distance to *B*, and then the same number of steps forward again on a new tangent at *B* to a new point *C* on the profile.) Having constructed the profile outside the base circle, what shape do we adopt below the base circle? We know it cannot be an involute.

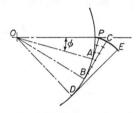

Fig. 2.22 Construction of an involute curve

Reference to a handbook may well provide the statement that this part of the profile is a radial line, and the profiles have been so drawn in fig. 2.21. However, the student may and should be unhappy when confronted with an arbitrary rule of this sort: let us see if there is any justification for it.

If the profiles *a* and *b* alone are considered the result may seem satisfactory, but if more teeth are drawn, as in fig. 2.21, the trouble becomes manifest. For instance, the opposite flank *c* of the same rack tooth happens in this configuration to be almost exactly tangent to the next pinion tooth along the whole of the length c_1 to c_2. Now if the rack moves upwards the tip c_1 will inevitably jam against the radial flank of the pinion tooth. The pinion tooth will have to be *'undercut'* to avoid this, and so will be weakened. Again the figure shows that there is in fact insufficient room for the tip d_1 of the next rack tooth *even when it has moved up the pinion tooth as far as the base circle*, and this is a much more serious matter, since if material is removed from the pinion at *this* point to make room for it we shall have to cut away part of the involute profile, which will reduce the effective length of the path of contact and the 'number of pairs of teeth' in contact.

This effect is called '*interference*', and as we shall see we may be able to ensure in the design stage that interference will not occur between properly cut mating gears, but unless care is taken, it may still arise in the manufacturing process. Fig. 2.21 affords a useful illustration, since one of the processes for producing gears is to use a rack cutter of the form shown, reciprocating normal to the paper, while the cutter and the 'gear blank' (i.e. a solid disk having an outer diameter equal to the addendum circle diameter) are moved with the same pitch-line velocity. Obviously in the case shown both undercutting of the 'radial' flank and interference with the involute profile of the pinion teeth would occur.

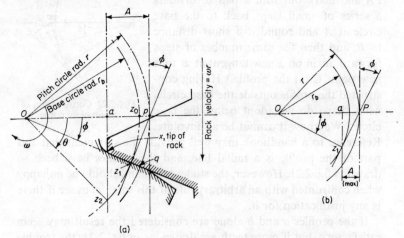

Fig. 2.23 Condition for interference between rack and pinion

In order to establish the conditions for avoiding interference between a rack and pinion consider the general case depicted in fig. 2.23a, which has been drawn in such a configuration that the prolongation of a rack profile passes through the centre O of the pinion. The teeth have a pressure angle ϕ and the rack an addendum A. The tip of the rack tooth is x, and z denotes the base of the involute profile on the pinion tooth, which is at z_1 in this configuration. The rest of the notation will be clear from the figure. By definition the pinion and rack must be moving with the same pitchline velocity ωr. If z reaches the point z_2 before the tip of the rack tooth x, then the pinion will be clear before interference with the involute part of its profile can occur; if not, interference is inevitable.

Now the time for z to travel from z_1 to z_2 is $\dfrac{\theta - \phi}{\omega}$.

During this time x travels a distance $\omega r \dfrac{(\theta - \phi)}{\omega} = r(\theta - \phi)$.

But the distance $xz_2 = az_2 - ax$

$$= r_b \sin \theta - (r - A) \tan \phi \quad \text{and} \quad r_b = r \cos \phi$$

$$\therefore \ xz_2 = r \left\{ \cos \phi \sin \theta - \left(1 - \frac{A}{r} \right) \tan \phi \right\}$$

So interference will occur if

$$r(\theta - \phi) > r \left\{ \cos \phi \sin \theta - \left(1 - \frac{A}{r} \right) \tan \phi \right\}$$

Now $az_0 = r_b(1 - \cos \theta) = r \cos \phi \, (1 - \cos \theta)$.

Also $az_0 = A - z_0 p = A - z_1 q = A - r(1 - \cos \phi)$.

$$\therefore \ r \cos \phi - r \cos \phi \cos \theta = A - r + r \cos \phi$$

$$\therefore \ \left(1 - \frac{A}{r} \right) = \cos \phi \cos \theta$$

Consequently, interference will occur if

$$(\theta - \phi) > \{ \cos \phi \sin \theta - \cos \phi \cos \theta \tan \phi \}$$

or $\qquad\qquad\qquad \{ \cos \phi \sin \theta - \cos \theta \sin \phi \}$

or $\qquad\qquad (\theta - \phi) > \sin (\theta - \phi)$

But as an angle is always greater than its sine interference is inevitable if $\theta > \phi$, i.e. if the addendum circle for the rack cuts the base circle for the pinion beyond z_1, the point of tangency between the base circle and the common normal. The limiting condition is shown, free from unnecessary additional lines, in fig. 2.23b, to which we shall refer again later.

Although we have taken the case of a rack and pinion it is not difficult to see that the limiting condition applies equally to the case of two wheels of finite size. If in the configuration illustrated in fig. 2.23a the rack tooth is replaced by a wheel tooth, it will be seen that contact still occurs at z_1, and that *in the region around z_1* the two teeth are effectively identical in shape, since the gradient of the profile is the same and over a short distance the curvature of the wheel tooth will make no appreciable difference. Thus any part of the wheel tooth extending beyond z_1 will still interfere with the base of the pinion involute, even though the total damage done by the wheel tooth may be less than that done by the rack tooth.

Fig. 2.24 shows the general case of two mating gears of different sizes. The limiting addendums according to this condition are BD and AC, and a glance is sufficient to show that $BD > AC$, so if we keep to the 'standard proportions' mentioned in § 2.6.6, whereby

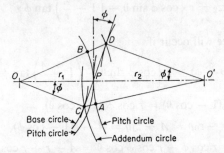

Fig. 2.24 Limiting addendum to avoid interference

the addendum on each wheel is the same, interference will first occur between the tip of the larger wheel and the flank of the smaller. The addendum must therefore not exceed AC. Let the pitch circle radii OP and $O'P$ be r_1 and r_2. Now from geometry

$$O'C^2 = O'P^2 + PC^2 - 2.O'P.PC \cos(\pi/2 + \phi)$$
$$= r_2^2 + (r_1 \sin \phi)^2 - 2r_2.r_1 \sin \phi(-\sin \phi)$$
$$= r_2^2 \left\{ 1 + \left(\frac{r_1^2}{r_2^2} + 2\frac{r_1}{r_2} \right) \sin^2 \phi \right\}$$

But addendum $\not> AC$
$$\not> O'C - r_2,$$
$$\not> r_2 \sqrt{\left\{ 1 + \left(\frac{r_1^2}{r_2^2} + 2\frac{r_1}{r_2} \right) \sin^2 \phi \right\}} - r_2$$

Since the number of teeth on each wheel is proportional to its radius we can write $\frac{N_1}{N_2}$ for $\frac{r_1}{r_2}$. Moreover, if we accept the value of the addendum as equal to the module m, or $1/P$, we know that

$$m = \frac{D}{N} = \frac{2r_2}{N_2}$$

so
$$m > \frac{mN_2}{2}\left|\sqrt{\left\{1 + \left(\frac{N_1^2}{N_2^2} + 2\frac{N_1}{N_2}\right)\sin^2\phi\right\}} - 1\right|$$

or
$$\sqrt{\left\{1 + \left(\frac{N_1^2}{N_2^2} + 2\frac{N_1}{N_2}\right)\sin^2\phi\right\}} - 1 < \frac{2}{N_2} \qquad (2.14)$$

If $N_2 \gg N_1$ the second term under the root sign will be small so we can expand the root as a binominal: moreover $\left(\frac{N_1}{N_2}\right)^2$ will be very small so

$$1 + \tfrac{1}{2}\left(2\frac{N_1}{N_2}\right)\sin^2\phi - 1 < \frac{2}{N_2}$$

or
$$N_1 < \frac{2}{\sin^2\phi} \qquad (2.15)$$

a result which in the extreme case of the rack and pinion, where $N_2 = \infty$, can be deduced much more expeditiously from fig. 2.23b, thus

$$A > aP$$

but $aP = z_1 P \sin\phi = OP \sin^2\phi = r_1 \sin^2\phi = \frac{mN_1}{2}\sin^2\phi$

and if
$$A = m$$
$$m > \tfrac{1}{2}mN_1 \sin^2\phi$$

or
$$N_1 < \frac{2}{\sin^2\phi} \qquad \text{(as before)} \qquad (2.15)$$

And if $\phi = 14\tfrac{1}{2}°$,
$$N_1 < \frac{2}{(\tfrac{1}{4})^2} \text{ or } 32$$

If there is to be no interference in this case, then the minimum number of teeth on the pinion must be 32. To have a minimum number of teeth of 32 may be inconvenient, and since this number has been dictated by the arbitrary choice of a pressure angle of $14\tfrac{1}{2}°$, it is not surprising that other standards have been introduced. A number of angles have been used, but in 1940 a committee of the British Standards Institution in B.S. 436 'unanimously agreed to recommend a tooth form having a pressure angle of 20°, a working depth of twice the module and a substantially semicircular clearance curve at the bottom of the tooth space', and stated that 'the standard tooth form represents a well-balanced compromise between strength, resistance to wear, and quietness of running . . .'. We may therefore accept this value of 20° for further discussion.

The minimum number of teeth on a pinion to avoid interference when working with a rack is now

$$N \not< \frac{2}{\sin^2 20°} \text{ or } 17$$

if the addendum is still 1 module. Moreover, it is immediately obvious that another effect of increasing the pressure angle will be

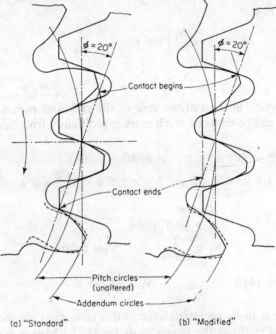

Fig. 2.25 20° rack driving 18-tooth pinion

to provide stronger teeth, since the base will be widened. Fig. 2.25a shows a section of a standard 18-tooth pinion meshing with a 20° rack: comparison with fig. 2.21 shows the improvement due to the two simultaneous changes (of angle and number of teeth) and for emphasis the outline of one $14\frac{1}{2}°$ tooth, undercut to provide clearance for the rack, is superimposed on one 20° tooth.

2.6.8 Teeth with addendum not equal to the module

If we are producing a standard range of gears, any one of which must be capable of meshing with any other at a standard centre distance to the specification laid down in B.S. 436, no further modifications are permissible, but if we are relieved of this requirement for interchangeability we can produce tooth forms appreciably better for particular purposes. A glance at these rack–pinion combinations is sufficient to show that the rack tooth is very much stronger as a cantilever* than the pinion tooth, and that a better balance could be achieved either by departing from the principle that the width as measured along the pitch circles should be equal, and thickening up the pinion tooth at the expense of the rack tooth, or by departing from the principle that the addendum on each tooth should be the same, and increasing the pinion addendum and reducing the rack addendum. If now we make both these modifications simultaneously and in the appropriate proportions as in fig. 2.25b we find that we have redrawn the *same* rack at a greater distance from the pinion centre and that we have achieved a markedly stronger pinion tooth: again the difference is emphasised by superimposing the outline of two teeth in fig. 2.25b. We have not affected at all the condition for 'correct' gearing though we have slightly reduced the length of the path of contact, and displaced the positions along the path of contact at which contact between the teeth begins and ends, and hence altered to a minor extent the sliding velocity between the teeth. When we consider in § 2.6.10 how gears are manufactured we shall return to this subject, and we shall see that this degree of 'flexibility' in design does not involve any extra problems in production.

In passing, we may note that another type of modification which is possible is merely to reduce the addendum and dedendum below the values which have been taken as standard. Such teeth are usually described by the adjective 'stub', and the modification may for special purposes be useful, in that the teeth are stronger as cantilevers and interference may be avoided, but the improvement is obtained at the expense of a much greater reduction in the length of the path of contact.

2.6.9 Comparison between cycloidal and involute gearing

At this stage it may be worth while to recapitulate briefly the major points we have been considering in relation to spur gears. Correct gearing (that is, constant velocity ratio) can be obtained by choosing a reasonable but arbitrary profile for the teeth on one wheel and

* See § 2.11.

designing the other to match it, but this is rarely of practical importance. More generally it can be obtained by the use of cycloidal or involute teeth, and we have seen that in the former case the whole profile can be constructed from geometrically correct curves, whereas in the latter case only the profile above the base circle is used, and the 'profile' of the tooth below the base circle is in fact the outline of a space required to give clearance to the tip of the mating tooth; moreover, in the latter case there is a fairly severe limitation, depending mainly on the pressure angle chosen and to a minor extent on the size of the mating wheel, on the minimum number of teeth in any gear.

Again, the shape of a cycloidal tooth is such that both from the

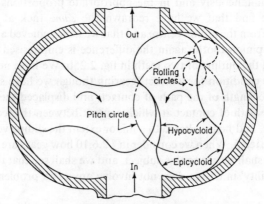

Fig. 2.26 'Root Blower'

point of view of strength as a cantilever and of resistance to wear – matters which we shall discuss more fully in § 2.11 – it is to be preferred to a comparable involute tooth.

It might at first sight appear that the balance of advantage lay with the cycloidal tooth, and for particular applications involving gears with a small number of teeth this is in fact the case. For instance, in clocks and watches it may be desirable to obtain a very large gear ratio in one pair of gears, and this can most easily be achieved by using a pinion with a very small number of teeth. Perhaps the extreme case of the rotors in the Root blower, shown in fig. 2.26, is worth noting. The rotors are examples of a pair of wheels, each of which may have only two teeth. Such a combination would, of course, be useless for transmitting power, since the length of the path of contact is much less than the circular pitch; another pair of conventional gears has to be used to drive these rotors at equal and opposite speeds. Nevertheless, the rotors are cycloidal 'gears' in which the profiles are wholly generated by a rolling circle whose diameter is $\frac{1}{4}$ that of the pitch circle. They are enclosed in a casing as indicated, and it can be seen that when they rotate any fluid in the system is trapped in the space between the rotor and casing and transferred from the inlet to the outlet passage.

In less extreme cases the advantages of cycloidal gears diminish, and they suffer from the serious restriction that the centre distance between two gears has, as pointed out in § 2.6.4, to be correct; otherwise the condition for constant velocity ratio is violated. *No such requirement exists for involute gears.* This follows directly from the considerations in § 2.6.5. The shape of the involute curve depends only on the diameter of the base circle, so if the two centres O and O' in fig. 2.18 are moved farther apart the internal tangent CD will be at a smaller inclination to OO' (i.e. the effective pressure angle will be increased), but it will still pass through the pitch point, so the constancy of the velocity ratio will be unaffected. If therefore the centre distance between two gears of finite diameter is increased above its nominal value the effective pressure angle and the backlash will be increased and the length of the path of contact will be reduced, but correct gearing will be maintained. If one of the wheels is a rack the pressure angle will not change, but the backlash will still be increased.

A second, even more important, advantage which the involute-gear system has over the cycloidal system is the simplicity of the basic rack, which as we have seen is bounded by plane surfaces, and is therefore much easier to produce than a cycloidal tooth of any type, each profile of which is essentially bounded by epicycloidal and hypocycloidal surfaces which must be accurately located relative to each other. The importance of this consideration will be more fully appreciated when we have discussed in outline the methods of manufacture of gears. For general engineering purposes the involute system has completely ousted the cycloidal system.

2.6.10 *Manufacture of spur gears*

Although manufacture as such is not the concern of this book, the methods adopted to make gears are of considerable theoretical interest and have a bearing on many points raised in this chapter.

Given the shape of a gear-wheel which has to be produced – for instance, the one shown meshing with the rack in fig. 2.25 – it is obviously possible, though by no means easy, provided that undercutting does not complicate the problem, to produce a form cutter which approximates to the shape of the gap between the teeth, and to use this cutter as shown in fig. 2.27 to machine one gap after another in a gear blank whose outer diameter is that of the addendum circle. The gear blank has to be rotated through the angle subtended by the circular pitch after each gap has been machined. Though this

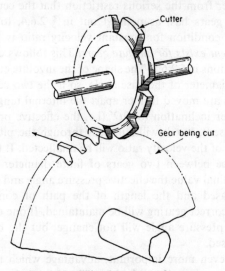

Fig. 2.27 Milling form-cutter for gears

method is still used extensively, particularly for roughing out the gap, it requires, even for teeth of standard proportions and pressure angle, a different cutter, not only for each pitch, but ideally for gears with differing numbers of teeth. For general work this demands a large range of cutters which is in practice impossible, and each cutter is used for a range of tooth numbers, with the result that the tooth profile is inaccurate. Of more recent times the method, in a modified form, has been used for mass production, the gap being finished by a grinding wheel accurately shaped for the particular gear being produced.

We can approach the problem of producing involute teeth more elegantly by making use of the fact that the required profile is a

simple geometrical curve. We have seen (fig. 2.16) that an involute
curve is traced out by any point on a tangent which rolls on the
base circle, and it would not be difficult to produce a mechanical
model, with a cylinder to replace the base circle and a straight edge
as the tangent, with attached to it a pencil to draw such curves on
paper attached to the cylinder. We should have to take precautions
to prevent slipping between the cylinder and straight edge, such as

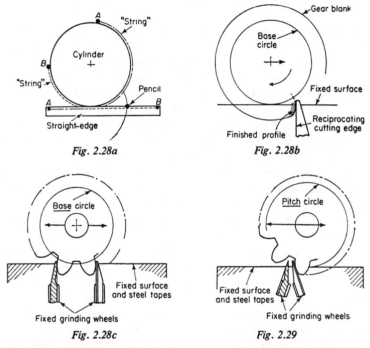

Fig. 2.28a   Fig. 2.28b

Fig. 2.28c   Fig. 2.29

Figs. 2.28 and 2.29 True *and* approximate generation of involute tooth profiles

the 'strings' or 'tapes' *AA* and *BB* in the sketch of the arrangement
shown in fig. 2.28a. We have tacitly assumed that the cylinder and
the paper on which the curve is being drawn are 'fixed' in space
because this is the natural approach to such a problem. But of
course there is no special virtue in this: we can as easily assume that
the straight edge is fixed in space and that the cylinder, with the
paper attached to it, is rolling on the straight edge; this will not
affect the relative motion or the shape of the curve. Now if we take
the further step of making the whole mechanism three-dimensional,

replacing the 'pencil' by a cutting point which reciprocates in the plane perpendicular to the 'paper', and replace the paper by a solid disk or gear blank, we have an embryonic device which could be used to 'generate' the involute profile of a tooth. Note that here the word 'generate' is used in the strict sense of originating or bringing into existence a curve without the use of a formed cutter. However, a 'point' tool is a geometrical abstraction, and we should need to take an infinite number of cuts with such a tool to produce a smooth surface. There is no need for such pedantry: we know that the involute curve is always normal to the straight edge, so there is nothing to keep us from using a tool with a straight cutting edge perpendicular to the straight edge, as shown in fig. 2.28b: we shall then achieve a good approximation to the involute with a finite number of cuts. We should note that one point on the tool produces the final involute curve; the rest of the tool is removing some of the excessive material and consequently reducing tool wear. Again, particularly if we are concerned with finishing roughly cut, and perhaps hardened, teeth by grinding we can adapt this principle, as shown in fig. 2.28c, and, using only two approximately flat grinding surfaces, produce simultaneously two profiles ground as accurately as our machine flexibilities and grit sizes will allow.

At this stage it may occur to us that we could achieve a very similar result in a rather different manner. Let us look back to a sketch of a gear engaging with a rack, for instance in fig. 2.25. We know that as the rack is translated, the gear of course rotating simultaneously, the rack tooth sweeps out the space between the teeth of the gear. If then we replace the rack by a cutter of rack form reciprocating in a direction perpendicular to the plane of the drawing, and combine this reciprocation with a slow translation of the rack and appropriate rotation of the gear blank, we shall clearly cut out the involute profiles as before. Does this mean that we are again *generating* these profiles? Strictly, the answer is 'no'. We are using a *form* cutter in the sense that the *shape* of the cutter affects the shape of the tooth we are cutting. After all, we know that the point of contact between the teeth moves along each tooth, which implies that each point on the gear-tooth profile is cut by a different element of the cutter-rack tooth, so the process is intermediate between form cutting and generation. This, however, is seldom appreciated, probably because the form of the involute rack is so simple, and the process is usually called 'generation'. However, the advantages of the process as compared with form cutting are obvious: the accuracy of the final profile depends essentially on the shape of a straight-sided rack cutter which can be accurately made without difficulty.

We have referred to 'simultaneous' and 'appropriate' rotation of the gear blank, and a moment's thought is enough to show that this implies that the pitch circle of the gear must roll without slipping along the pitch line of the rack. We can achieve this result in any one of several ways. For instance, we can use a device exactly analogous to the mechanism we used for the generation of a profile, as in fig. 2.28, but the steel tapes must now be arranged to pass round a cylinder of diameter equal to the pitch circle, not the base circle. Alternatively, if we have already available a master gear and rack of sufficient accuracy we can use them to control the rotation of the gear blank. This takes us one further theoretical step away from the concept of pure generation of the tooth form and therefore introduces additional sources of error: the method might now be better described as a copying process. In all these cases we shall, of course, have to provide a means of 'indexing' the blank forward from one tooth space to the next.

The basic principle having been established, it is relatively easy to devise modifications which will have particular advantages in special cases. For instance, we have talked of a 'cutter rack', but for final grinding of hardened gear-wheels we can replace this by two large grinding wheels whose side-cutting faces occupy the position of the rack profiles as depicted in fig. 2.29. Again we have discussed a *reciprocating* cutter rack, but if we were to mount a series of cutter racks in succession round an axis we should have an equivalent rotary cutter which could be traversed across the width of the blank; we can do even better by arranging such a series of racks helically round an axis so that they resemble in appearance an interrupted or 'gashed' worm wheel. Such a tool, called a *hob*, is shown in fig. 2.30. The axis has to be set at an angle to the plane of the blank, and the cutting edges must therefore have the form of a helical gear. This method allows a gear blank to be cut *continuously* as the hob is fed across the width, the blank being rotated at the appropriate speed to accommodate the apparent axial progression of the 'racks'.

Another method, at first sight rather different but which in reality has much in common with the reciprocating-rack method, is that adopted in the gear shaper depicted in fig. 2.31. The cutter is in the form of a pinion having an involute profile. It is reciprocated along

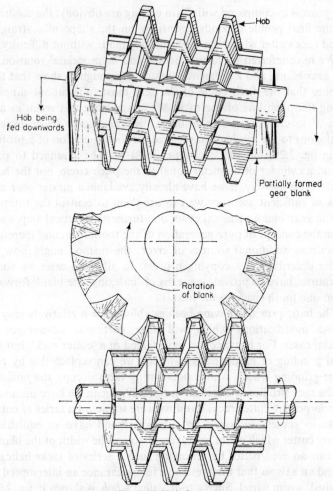

Fig. 2.30 Hob for gears

its axis, and the cutter and gear blank are simultaneously rotated at the same pitch-line velocity. This method has the advantage of being able to produce internal gears and, like the hob method, of obviating the requirement for indexing the gear blank, since the cutter is effectively of infinite length. Clearly it is again, academically speaking, a compromise between a form-cutting and a generating process. To the extent that the shape of the cutter tooth influences the shape of the tooth cut, the process is form-cutting, and only

more obviously so than the rack method, because the cutter profiles
are no longer straight. To the extent that the shape of the tooth is
dependent on the relative motion of cutter and blank, it is a generating method, and so for teeth of standard proportions and pressure
angle only a single cutter is needed for a given pitch to machine gears
with differing numbers of teeth. The only limitation is that imposed
by interference which can occur with either external or internal
gears, and the drawing of the gear issued to the workshop should
have a note on it informing the machinist of the permissible number
of teeth on the cutter to be used. Further details and discussion of
such topics are more appropriate, however, to a book dealing with
production engineering.

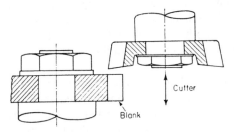

Fig. 2.31 Shaping a gear, using circular cutter

We can now return to the peculiar virtue of the involute system
which we noted in § 2.6.8 above. Here we saw that if we wished to
increase the strength of a pinion tooth as a cantilever in relation to
the strength of a mating tooth of a rack, and to eliminate interference, we could do so by increasing its addendum and reducing its
dedendum, and that this pinion would mate correctly with the
standard rack positioned at a greater distance from the pinion centre.
It follows from the discussion above that we can cut such a pinion
with the standard cutter by the simple expedient of using a larger
blank and cutting it only to the standard tooth depth. Moreover,
since the pinion meshes correctly with the standard rack, it will also
mesh correctly with any standard wheel, though at an increased
pressure angle and with an increased centre distance: the path of
contact and the rubbing velocity between the gears will also be
modified. Alternatively, we can cut a wheel (larger than the pinion)
to mesh with it with a correspondingly decreased addendum and
increased dedendum. This will weaken the wheel teeth as cantilevers,
but (cf. fig. 2.25) since they were originally inherently stronger, we

can gain on balance: we now restore the original centre distance and pressure angles to the combination. The latter is the more commonly applied alternative, and it will be noted that in the quotation from the British Standard in § 2.6.7, no mention is made of addendum and dedendum separately, but that it is recommended that the working depth should be twice the module. Later in the same British Standard recommendations are made as to these values, which depend on the number of teeth in both pinion and wheel; such teeth are usually described as 'corrected'. Provided that we have taken care to avoid the danger of interference, these methods of cutting gear-teeth relieve us of anxiety as to the shape of the space below the base circle. The cutter will automatically remove enough material to allow the mating teeth clearance. It is a simple matter to add a little to the addendum of the cutter to provide root clearance for the mating tooth. The tip of the cutter is 'eased' or slightly rounded to give a smooth fillet radius at the bottom of the gear-teeth.

When a pair of teeth are coming into engagement there will be a certain degree of misalignment and consequently impact, as the preceding pair of teeth will be deflected by the forces being transmitted between the gears. To minimise this the tips of the teeth are frequently 'eased' by removing a small and arbitrary amount of material below the true involute profile. This again can be achieved by modifying the flank of the basic cutter.

§ 2.7 Helical and Skew Gears

As stated in § 2.5, the name helical gear implies that the teeth are cut not parallel to the axis of the shaft but along a helix. The sketch, fig. 2.13, will make this clear and show that two helical gears on parallel shafts which are to mesh together must have the same helix angle but be of opposite hand. If any section perpendicular to the axes of the shafts be considered, it will be seen that the contact progresses as in an ordinary spur gear, but if a section parallel to the axes be considered it is also evident that contact will start at one side of the wheels and progress simultaneously across their width: thus contact starts at one *point* (at the tip of the leading edge of the driven tooth) and spreads along a lengthening inclined *line* which again diminishes to a point before contact between these two teeth ceases. Noise arises from suddenly applied force, so it is not surprising that such gears run much more quietly (for a given quality of cutting) than spur gears. It is obvious that such gears will produce on the shafts and their bearings axial forces which may be a nuisance. These forces can be eliminated by the use of 'double helical' gears, in which each wheel consists of two otherwise identical gears of opposite hand; such gears are inevitably more expensive.

Helical gears are usually cut with standard cutters for spur gears, the direction of the cutting action being inclined to the axis of the shaft. In the case of a helical rack it is easy to see that the result will be to produce a tooth whose profile normal to the direction of cutting has the standard shape and, of course, the standard pressure angle. It follows that the effective pressure angle is greater than the standard one. This, though less obvious, is equally true for all such helical gears. More important, it follows that all such gears, spur and helical, will mesh together if mounted on shafts at the appropriate angle. Thus, such gears can be used to couple non-parallel shafts, when they are called *skew* or *spiral* gears.

§ 2.8 Unconventional Helical Gears: the Wildhaber–Novikov System of 'Circular Arc' Tooth Profiles

Regarded from the point of view we have so far adopted, there would seem to be little additional matter of fundamental importance to be considered in relation to helical gears as opposed to spur gears, but if we can free our minds from such 'traditional' thinking, and consider the problem of helical gearing entirely afresh and not as a derivative of spur gearing, an interesting new possibility begins to appear. We have seen that (geometrically speaking, that is on the assumption of truly rigid material) contact between helical gears occurs along an inclined *line* which progresses across the face of the tooth and we know that (practically speaking, that is taking into account the elasticity of real materials) this 'line' is bound to have finite if very small width, so that at all times a finite area is available to transmit the driving force from one tooth to another. In § 2.11.2 we shall be discussing the problem of contact stresses a little more fully, but for the time being we can take it as obvious that the greater the difference in curvature between two surfaces, the greater will be the distortional stresses in the material to provide the finite area; thus, two flat surfaces would not need to distort at all, two large cylinders slightly, and two small cylinders would have to distort greatly. Now involute teeth are inevitably convex in profile, so from this point of view they are a poor choice. Cycloidal teeth might be better, but they, too, inevitably have a difference in curvature, and so are not ideal. What happens if we go the whole hog and decide to design our teeth primarily to provide the maximum degree of 'conformity of surface', for instance, by making one tooth convex and the other concave, and of nearly the same radius of curvature? Well, of course, we violate the principle on which we have been basing all our consideration of gear-teeth, that of providing uniform velocity ratio, and for spur gears that would be unthinkable; but are the results necessarily so catastrophic for helical gears?

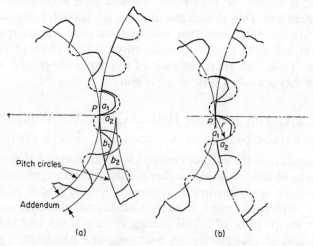

Fig. 2.32 Novikov gearing

Careful thought shows that the answer is 'no', and an alternative solution is shown in fig. 2.32. Here we picture stages in the meshing of a particular cross-section of two helical gears, of which one, say the driven wheel, has no addendum and the driver no dedendum, but only root clearance. Since the pitch-line velocities are constant and equal, the working surface of the driver a_1 is travelling the faster: at the pitch point (fig. 2.32a) most of it is well clear of the mating surface a_2, but at a later stage, which we can choose, makes up on it and contact occurs as in fig. 2.32b. If we now design our teeth so that when in this configuration the two profiles a_1, a_2 are both arcs of a circle struck from the pitch point, then we have arranged that at this instant every common normal to the whole contacting profile passes through the pitch point. It follows immediately from the argument in § 2.6.1 that each pair of contact points has the same velocity along the common normal, i.e. that the surfaces have come together without impact; it is not difficult to see that this implies that they will begin to separate again, as shown, for instance, by the teeth b_1, b_2 in fig. 2.32a. Geometrically speaking, then, contact at any instant between 'rigid' gears of this type takes place over a line in a transverse plane, but between gears of real material under load (that is when allowance is made for elasticity) it may well spread over the area required to support the load with less distortion and hence lower stresses than occur between involute teeth.

Summing up, we can see that if two *spur* gears were made on this principle, and were *made* to run together at constant velocity ratio, they would be in contact only for a very brief period when in the phasing shown in fig. 2.32b. Two helical gears (in which the face width must be such that the next tooth engages before the one in which we are interested loses contact) will transmit the load through an area of contact which, in a cross-sectional view such as that given, is fixed in space, but moves along the face-width of the wheels as rotation proceeds.

This type of gearing was originally proposed by Wildhaber in 1926, revived by Novikov in 1955, and is claimed to be particularly suitable for the transmission of very high tooth loads.

§ 2.9 Worm Gears

Worm gears, of which an example is sketched in fig. 2.13, can be regarded as an extreme form of skew gears in that they are used to couple skew shafts, usually though not necessarily at right angles to each other. The smaller member of the combination, the worm, has a small number of teeth – from one to, say, seven – so that it resembles a single or multi-start thread, and its tooth or 'thread' may be made of arbitrary form, for instance, 'straight sided' (so that an axial section has the same form as an involute rack) or preferably of the same form as that of a helical gear. This form is produced by generating the worm thread by means of a straight-sided involute rack cutter operating in the direction of the thread helix at the pitch circle: the thread so obtained has convex profiles when viewed in axial section. The teeth of the worm wheel can be generated by a cutter of the same shape as the worm. When the worm rotates, the apparent motion of the thread is, of course, a translation in the axial direction, so the apparent relative motion between worm and wheel is similar to the motion between a rack and pinion, the worm-wheel representing the pinion.

§ 2.10 Efficiency of Gearing

The power loss in properly lubricated spur or helical or similar types of gearing is usually very low, that due to tooth friction being of the order of only 1 per cent or less of the power transmitted at full load. To this loss have to be added those due to oil churning and bearing friction. In such gears, as has been pointed out, there is inevitably sliding at all points in the path of contact except at the pitch point, and it can be deduced that the 'coefficient of friction' is low and that lubrication must therefore be effective in spite of the extremely high contact pressures which exist. Some attention will be given to this in chapter 7.

In the case of skew, and more particularly worm, gearing, sliding occurs not only as in spur gears, but, much more importantly, in a direction at right angles to this. In fact, we can obtain a sufficiently close approximation to the situation in a worm gear by ignoring the pressure angle of the thread and thinking of this thread as perpendicular to the axis. We can then regard the thread as an inclined 'plane' which moves relative to the surface of the worm-wheel; and the analogous situation of a block being pushed up an inclined plane by a horizontal force is one with which we are familiar in mechanics, so we can write down the expression for the efficiency η of such an operation

$$\eta = \frac{\tan \alpha}{\tan(\alpha + \lambda)} \qquad (2.16)$$

where α is the inclination of the plane, or in this case the pitch angle of the worm, and λ the 'angle of friction'. For the case of the worm-wheel driving the worm we have the corresponding expression:

$$\eta' = \frac{\tan(\alpha - \lambda)}{\tan \alpha} \qquad (2.17)$$

Now in the case of a 'single-start' (or single-tooth) worm α may be only a few degrees, and if the surfaces are dry or poorly lubricated λ may well exceed α: in this case η will be less than 0·5 and η' will be 'negative': in other words, the drive will be irreversible. Such a gear has its uses, but would be unthinkable for power transmission. For a multi-start worm, however, α can be made of the order 45°, and if the gears are well lubricated λ under running conditions, particularly at high speeds, may well be of the order 1° or less. The efficiency then is of the order 0·97–0·98, i.e. of the same order as that for spur gears. As far as power *loss* is concerned, the difference is probably negligible, but it should be noted that the losses have to be dissipated as heat, and since the amount of heat which has to be got rid of is almost directly proportional to the effective coefficient of friction, it is vital to ensure that the best possible lubrication is maintained, and in the case of highly loaded gears that sufficient cooling is provided.

§ 2.11 The Strength of Gear-teeth and their Resistance to Wear

The strength and resistance to wear of gear-teeth are so bound up with the subjects we have been discussing as to make their brief consideration here wholly appropriate. Teeth which 'fail' usually do so either by breaking off at the root – clearly a major disaster – or, much more frequently, because the contacting surfaces become seriously worn or otherwise damaged. These two types of failure can be dealt with separately in that order.

2.11.1 Strength

We can approach the question of tooth breakage in a fairly naïve way. The load responsible is, of course, the force, say F, by means of which torque is transmitted, and (since as we have seen the coefficient of friction is small) we shall make no serious error if we assume that it acts along the path of contact or pressure line, so its magnitude is easily calculated by dividing the torque on one of the wheels by its base circle radius. But as we have seen (§ 2.6.6), the length of the path of contact must exceed the circular pitch p_c, and usually does so by a considerable margin, so that for much of the time this force is shared by two teeth, while for part of the time one tooth alone must carry it. Fig. 2.33a shows a typical case of two gears in mesh at the instant at which the second tooth is entering into contact at the point A, and it is fairly obvious that this will give the condition for the greatest bending moment at the root of the tooth M, since at all earlier times the arm of the force is smaller, and at all later instants two teeth are sharing the load while the arm will never be nearly doubled. (The greatest bending moment on the meshing teeth can be determined similarly.) Following the methods outlined in § 2.6.6, we can in any given case easily determine this position: A is the point where the addendum circle cuts the pressure line, $AB = p_c \cos \phi$, and $BC = \frac{1}{4} \cdot AB$; so $\angle POC = \theta$ can be found.

In fig. 2.33b tooth M is again shown, with the force F acting on it.

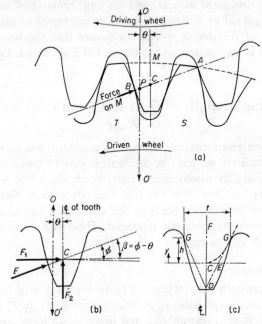

Fig. 2.33 Strength of gear teeth

F can be resolved into tangential and radial components F_1 and F_2, and it is conventional to treat these separately and *assume* that the former causes at the tooth root stresses exactly the same as those due to a pure bending moment remotely applied, and the latter a uniformly distributed compressive stress (in other words, that the tooth is long enough to allow de St Venant's equivalence principle to be applied) and that the shear stress is unimportant. The errors arising from these assumptions are probably small compared with the uncertainties we shall encounter later. To determine the weakest section of the tooth in bending we could easily proceed by trial and error: as we take sections farther down the tooth the bending moment increases, but as we reach the fillet radii the section modulus increases. Alternatively, we can use the graphical construction shown in fig. 2.33c, where a parabola GCG is fitted to the tooth form so that the vertex is at C and the sides are tangent at G.* GG indicates the weakest section. With the notation shown in fig. 2.33 and taking the width of the gear as w, we can now write down from elementary bending theory

$$\text{Maximum bending stress} \left(= \frac{M}{z} \right) = \pm \frac{6F_1 h}{wt^2} = \pm \frac{6Fh \cos \beta}{wt^2}$$

while the direct compressive stress due to F_2 is

$$\frac{F_2}{wt} = -\frac{F \sin \beta}{wt} \ (- \text{ indicating compression})$$

Although the resultant compressive stress will be numerically larger than the tensile stress, it is the latter which is the more likely to cause failure in fatigue, which is what we have to fear. The compressive stress, usually taken as the criterion, is

$$\frac{F}{wt^2} (6h \cos \beta + t \sin \beta)$$

or in terms of the *equivalent* tangential load F_b *per unit width* acting at the pitch-circle radius the compressive bending stress = σ_b, where

$$\sigma_b = \frac{F_b}{t^2 \cos \phi} (6h \cos \beta + t \sin \beta)$$

or

$$\sigma_b = \frac{F_b}{Y'}$$

where

$$Y' = \frac{t^2 \cos \phi}{6h \cos \beta + t \sin \beta} \tag{2.18}$$

is called the 'strength factor' for the tooth, though it clearly depends not only on the tooth itself but also on the gear with which the one we are considering engages.

---

* To construct the parabola, fit a rule by trial and error to fig. 2.33c so that $DE = EG$, where D is on the centre-line of the tooth, $DCE = 90°$, and DEG is tangent to the profile at G.

To show that GG is weakest section, let the equation to the parabola GCG, with the origin at C and axes CE, CF, be $y = C_1 x^2$. Then BM on parabolic cantilever = $F_1 \cdot y$; modulus of section = $\frac{1}{6}w \cdot x^2$, where w is tooth width; so stress $= \frac{6F_1 y}{wx^2} = \frac{6F_1 C_1}{w} = $ constant. But tooth is stronger than cantilever everywhere *except* section GG.

To calculate the strength factor in any specific case is somewhat laborious, but not intolerably so. However, the necessity is obviated by the fact that tables have been worked out which cover all reasonable combinations: they are to be found in the British Standards publication B.S. 436,* calculated for a module m (or diametral pitch P) of one *inch*. From the form of the expression for Y' we can see that its value changes linearly with a linear dimension of the tooth, so that if Y is the value in the B.S. tables we can write

$$Y' = m \,.\, Y = Y/P$$

hence

$$\sigma_b = \frac{F_b}{Y \,.\, m} = \frac{F_b \,.\, P}{Y}$$

We have so far assumed that the equivalent tangential tooth load F_b at the pitch circle is calculable simply by dividing the transmitted torque by the pitch-circle radius. This makes no allowance for impact, which as we have seen at the end of § 2.6.10 is inevitable, or for the dynamic stresses due to profile imperfections, both of which will increase with speed. That such effects are in the general case incalculable is obvious, and in B.S. 436 they are 'allowed for' by a frankly empirical factor X_b, which depends on the speed and also the number of hours per day for which the gears are expected to run. The latter allowance appears to be illogical, but until we have some better criterion we shall do well to accept this factor, when our 'calculated' stress becomes

$$S_b = \frac{F_b}{Y \,.\, m \,.\, X_b} = \frac{F_b P}{X_b Y} \qquad (2.19)$$

How far such a calculated value will inevitably deviate from the true stress in the material of the tooth can be appreciated by a glance at fig. 2.34, which shows the photoelastic stress pattern obtained by viewing a pair of 'gears' made of a suitable transparent material in a polariscope. Close crowding of the light and dark bands indicates a region in which the stress is changing rapidly: the 'stress concentration' in the contact zone is obvious and this will be discussed shortly. The stress concentration in the fillet of 'zero radius' is too intense to be shown up; but it may be taken that any such tooth would in practice fail prematurely. Even in the left-hand wheel, which has the 'substantially semi-circular curve at the bottom of the tooth space' recommended by B.S. 436, the stress concentration in the region of the fillet is very considerable, yet we have ignored it

---

* At the time of going to press B.S. 436 had not been converted to SI units.

Fig. 2.34 Photo elastic stress pattern for loaded gear teeth

From M. A. I. Jacobson 'The photoelastic investigation of bending stresses in spur gear teeth', *International Conference on Gearing 1958*, Inst. Mech. Eng.

Plate 3 Finite elastic stress pattern in a loaded resin tooth.

completely. Moreover, we have given no consideration whatsoever to a number of factors which could be very important, such as the very high 'residual stresses' which can easily be left by such processes as machining, grinding, heat treating, flame hardening, and carburising – some perhaps favourable, some certainly unfavourable. We are driven to the conclusion that the stress we have calculated is a purely 'nominal' one. We compensate for this by using a 'working stress' which bears an equally nominal relationship to the fatigue strength of the material under repeated stress,* but which has been found to give the right answers: in other words, we again use an empirically determined value for this working stress.

Of course, it would be academically more satisfactory if we could calculate accurately the true stress in any gear-tooth, determine the true fatigue strength for the material we propose to use, and hence find the true limiting torque (or minimum wheel width). This very brief discussion of the problems involved should have made clear how very far we are from being in this happy situation. Nevertheless, it would be wrong to deduce that the ultimate errors involved are necessarily grievous. The fund of knowledge built up over the years is such that designs based on these techniques and the data in the British Standard referred to are probably conservative. It is to be hoped that steady progress can be made towards refining the methods used. Meanwhile, the Standard has been drawn up in such a fashion that no deep knowledge or understanding of the problem is required. All the 'designer' is required to do is to look up in the various charts and tables the values of Y, X_b, and S_b appropriate to his design and insert them in the simple formula whose genesis we have been discussing.

2.11.2 Wear

So far we have considered only the bending strength of gear-teeth, but as has been hinted several times, the problem of tooth profile damage is even more intractable mathematically. If two parallel curved surfaces, such as the profiles of meshing spur gear-teeth, made of a truly rigid material, were pressed together they would make contact along a line, which implies that the area of contact would be zero, and the pressure infinite. No materials are rigid,

---

* In fact, the 'allowable basic stress' for steel gears will be found to be of the order of half the fatigue strength of the material in *reversed* stress.

however, so distortion (one hopes, entirely elastic) occurs, and a finite though small area carries the load. The case of two cylindrical surfaces of uniform radii r_1 and r_2 was solved by Hertz, who showed that the maximum compressive stress σ_c was given by

$$\sigma_c = \sqrt{\left[\frac{F_c(1/r_1 + 1/r_2)}{\pi(k_1 + k_2)} \right]} \qquad (2.20)$$

where F_c is the compressive load per unit length of the cylinders, and $k = (1 - \nu^2)/E$, ν_1 and ν_2 being Poisson's ratio and E_1 and E_2 Young's modulus for the two materials. If we take the case of two steel cylinders for which $\nu \simeq 0.286$ this reduces to

$$\sigma_c = 0.416 \sqrt{\{F_c E(1/r_1 + 1/r_2)\}}$$

and if we define the 'radius of relative curvature r_r' of the cylinders by $1/r_r = 1/r_1 + 1/r_2$ we obtain

$$\sigma_c = 0.416 \sqrt{(F_c E/r_r)}$$

It should be noted that this stress is one of three (triaxial) compressive stresses, and as such is unlikely to be an important factor in failure of the material. The maximum shear stress occurs at a small depth *inside* the material, and has a value of $0.3 \, \sigma_c$; *at* the surface the maximum shear stress is $0.25 \, \sigma_c$. These stresses are more likely to be responsible for failure. The important point for us at this stage is that each is proportional to σ_c, and therefore for any given material proportional to $\sqrt{(F_c/r_r)}$.

For several reasons, we cannot apply this result directly to gear-teeth. The analysis assumes two surfaces of constant radii of curvature, and elastic homogeneous isotropic stress-free material. First, a gear-tooth profile has a continuously varying radius of curvature, and the importance of this departure from the assumption may be emphasised by considering the case of an involute tooth where the profile starts at the base circle. The radius of curvature, say r_1, is at all times the length of the generating tangent, so at this point it is, mathematically speaking, zero; but it remains zero for no finite length of the involute curve, growing rapidly as we go up the tooth and having an unknown value within the base circle.* Clearly the Hertz analysis is completely inapplicable at this point; all we can say is that the stresses are likely to be extremely high, and we may

---

* If contact were to occur at this point the stress would not be infinite, as an infinitely small distortion would throw load on to the adjoining part of the involute profile, so that there would be a finite area of contact.

note with relief that if we provide a margin of safety against interference, contact will never occur at it. (Cf. figs. 2.25a and b.) In the
regions where contact between well-designed gear-teeth does occur
the rate of change of r_1 is much less rapid, and it is not unreasonable
to take a mean value for the short length in which we are interested
at any instant. Second, the assumption that the material is elastic
will certainly break down if the resulting shear stress exceeds the
shear yield strength of the material. The consequences are quite
beyond our ability to predict mathematically. We might manage the
calculations if one load application at one instant were all we had
to deal with; but the microscopic plastic flow which then occurred
would completely upset our calculations for contact at the next point
on the tooth profile, and so on. The situation when the original
contact re-occurred would be quite different; and we have to deal
with millions of cycles of loading as the gears revolve. All we can
say is that the repeated plastic flow is likely to lead to fatigue failure,
but that it will not necessarily do so, since the material may perhaps
build up a favourable system of residual stress, and will probably
'work-harden' to some extent. If such a process does go on we certainly no longer have a homogeneous isotropic stress-free material,
even if – which is more than doubtful – we had in the first place.
Third, gears which are transmitting more than nominal power must,
as a matter of common experience and for reasons which we shall
consider more fully in chapter 7, be lubricated. The introduction of
a film of lubricant between the surfaces might be expected to alter
the situation drastically. In actual fact, we shall see that modern
investigation leads to the astonishing conclusion that, at least under
a wide range of conditions, the contact stresses are hardly altered;
not because the lubricant is squeezed out, but because it assumes the
form of an extremely thin film of almost constant thickness.

In view of all these qualifying remarks, it is hardly to be expected
that one could design gear-teeth on the basis that, let us say, the
maximum shear stress $= 0.3\,\sigma_c = 0.104\sqrt{(F_cE/r_r)}$ was to be equated
to the shear strength of the material in fatigue. Nevertheless, the
Hertz analysis is of vital qualitative value in indicating the parameter F_c/r_r, which, for any given material, can be taken as a criterion
of the maximum stress, the actual value to be allowed being determined empirically.

In order to proceed, we have to determine the minimum value of
r_r when only a single pair of teeth (cf. fig. 2.33) is in contact. We

have seen that for involute teeth the radius of curvature is the length of the generating tangent, so, with reference to fig. 2.35, we can write down for contact at X

$$r_1 = AX$$
$$r_2 = BX$$
$$r_r = 1/(1/r_1 + 1/r_2) = \frac{r_1 r_2}{r_1 + r_2} = \frac{AX \cdot BX}{AX + BX}$$

r_r will therefore have a minimum value at either E or F, depending on which is nearer the adjoining base circle. In this case the critical

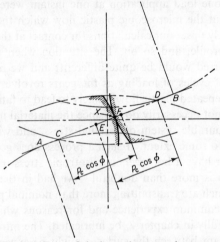

Fig. 2.35 Zone of single tooth contact between gears

point is E, and we can calculate r_r. If then we decide to call S_c the 'permissible surface stress factor' (to be determined empirically) we can write

$$F_c = S_c \cdot r_r$$

and the permissible tangential load at the pitch circle *on unit width of tooth* will be

$$F_c \cos \phi = S_c \, r_r \cos \phi$$

Again as in the case of bending stresses the need, for the individual designer, to work out laboriously each particular case is obviated by the provision in B.S. 436 of tables of a 'zone factor Z' which corresponds to ($r_r \cos \phi$) for meshing teeth of module 1; and again a 'speed factor X_c' is introduced to allow for impact and 'dynamic' stresses. In fact, these tables are based on a slightly *more* empirical approach, which suggested that $(r_r)^{0.8}$ gave better agreement with practice than $(r_r)^{1\cdot0}$, but the discrepancy could more logically* have been corrected by a modification in X_c. The Standard gives the final simple form of the critical factor S_c for surface wear as

$$S_c = \frac{KF_c}{X_cZ} \tag{2.21}$$

where K is not directly proportional to the diametral pitch P but is equal to $P^{0\cdot8}$ and S_c, as should by now be amply clear, is an empirical factor for which values are provided for commonly used materials.

Thus, by applying equations (2.19) and (2.21) to a pinion tooth and a meshing wheel tooth in turn we obtain four distinct values for the load which can be regarded as limiting; obviously the allowable load is the *smallest* of the four.

2.11.3 Strength and resistance to wear of cycloidal teeth

It is of interest to consider very briefly the strength and resistance to wear of cycloidal teeth, if only to correct some misleading impressions which are current. We need not spend much time on the strength problem, since the methods do not differ substantially: we need only notice that if the generating circle is less than half the

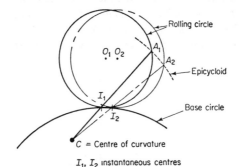

Fig. 2.36 Curvature of cycloid

diameter of the pitch circle the hypocycloid will give a stronger tooth than one with radial flanks (cf. § 1.6). Probably the most interesting point in relation to wear is the determination of the radii of curvature of the teeth. It is sometimes stated that, because the addendum of one tooth and the dedendum of the mating tooth are being generated by the same point on the rolling circle (cf. fig. 2.17),

* See H. E. Merritt, *Gears*, 3rd edn., Pitman, 1958.

the radii of curvature of the two profiles are the same. This is a fallacy, arising from the implicit assumption that the *instantaneous centre* of the rolling circle is the *centre of curvature* of the profile at that point, as indeed at first sight seems reasonable: after all, we have made the same 'obvious' assumption about the centre of curvature of an involute, when it happens to be true. That it cannot be true in the case of a cycloid becomes clear when we remember the case illustrated in fig. 1.28, and referred to above, in which for the particular case of a rolling circle half the size of the pitch circle the hypocycloid becomes a straight line, with a centre of curvature at infinity. The student who is puzzled by this may find fig. 2.36 a help. This shows simply enough why *I* is *not* the centre of curvature, and a similar figure drawn for an involute will show why in that case *I* and the centre of curvature do coincide. In order to find the centre of curvature we can construct the 'equivalent mechanism' which has been given as example 1.4 at the end of chapter 1. It will now be appreciated that this can be regarded as a pitch circle and rolling circle with fixed centres *O*, *R*, while a straight link *QB* attached to the generating point *Q* on the rolling circle passes through a guide pivoted at the pitch point *P*. By definition, *Q* will describe a cycloidal curve *relative to the pitch circle*. The instantaneous centre *X* for the member *QB relative to the pitch circle* should now be found. The radius of curvature of the hypocycloid at *Q* is then *XQ*. This

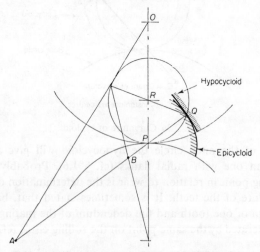

Fig. 2.37 Centres of curvature of cycloidal teeth

example should be worked out in detail: it is instructive. The solution is, however, provided for those who have difficulty.

The same argument applied to the epicycloid *and* hypocycloid shown meshing together in fig. 2.37 gives the two radii of curvature as BQ and AQ respectively, and since one of the curves is convex and the other concave, the radius of relative curvature is

$$\frac{AQ \cdot BQ}{BQ - AQ} = \frac{AQ \cdot BQ}{AB}$$

This is in general much greater than for comparable involute teeth in contact, when both curves are essentially convex, so we might expect the cycloidal teeth to withstand much heavier contact loads.

2.11.4 Note on modes of failure of gear-teeth

It may be worth while to end these brief remarks on the strength of gear-teeth with a short note on the ways in which they fail in practice. As has been indicated, they may fail by a major breakage usually at or near the root, which is almost always due to repeated 'bending' stresses and has the normal characteristics of a fatigue failure: an apparent absence of ductility, whether the material is itself ductile or, as frequently is the case, brittle; and sometimes a clear indication of a 'nucleus' or crack origin, and of the spreading of the crack as the high stresses are repeated. Alternatively, the failure may be confined to damage to the profile. We have discussed the 'contact' stresses, and their severity is well illustrated in fig. 2.34, to which we have already referred. The basic cause is unlikely to be the direct compressive stress, which is triaxial in nature, but may more reasonably, in many cases, be ascribed to the shear stresses which accompany the compressive stresses. Whether the shear stress which occurs at the surface, or the slightly greater shear stress which occurs inside the material, is likely to be responsible we simply do not know. Indeed, the failure often takes the form of 'pitting' (a word which describes the phenomenon perfectly), and there is ample evidence to suggest that this type of failure may be basically due to the effect on the solid material of the repeated application to its surface of very high fluid pressure, generated in the lubricant while it is in the contact zone.* Even if we were sure of this, however, there is no obvious remedy, since the presence of a lubricant is absolutely essential to minimise wear. If, as from the evidence sometimes

* S. Way, *Trans. Am. Soc. Mech. Eng.*, 1935, p. A49.

happens, the lubricant fails to keep the teeth apart under the immense contact pressures they will wear rapidly, often by local 'welding' on a microscopic scale, followed by the tearing apart of the welds.* The resulting damage is referred to as 'picking-up', 'scuffing', or 'scoring', and again these words are sufficiently descriptive of progressive states of this type of damage. We do not attempt specifically to 'design against' these latter forms of damage, other than by keeping the contact stresses to the empirically determined satisfactory values and using a lubricant* which again has been found in practice to be suitable.

§ 2.12 Gear Trains: Simple and Compound

A system of gear-wheels which is used to transmit motion is called a gear train. If, as is usually the case, the axes of the shafts on which the gears are mounted are fixed the determination of the ratio of shaft speeds, or 'gear ratio', is a matter of trivial difficulty. For any pair of gears, whatever the type, the angular speeds will be inversely proportional to the number of teeth. In the case of a pair of spur or similar type of gears on parallel shafts, the pitch-circle diameters are proportional to the number of teeth, so the angular speeds are also inversely proportional to the pitch-circle diameters, but this is not the case for skew gears (including, of course, worm gears), since for a given pitch and number of teeth the diameter depends on the spiral angle chosen, and this will not in general be the same for the two gears. Again in the case of gears on parallel shafts the direction of rotation will be reversed by each pair of gears, unless, of course, one of the pairs is an 'internal' gear, i.e. has its teeth facing inwards towards the shaft: if the shafts are not parallel the direction of rotation cannot be so easily defined by a sign, but can be shown very simply on a sketch of the gears, either by an arrow or by the usual convention of a dot on a tooth which is travelling upwards out of the paper and a cross on a tooth travelling downwards into the paper, these signs representing the point and tail of an arrow respectively.

A gear train in which each shaft carries one wheel only, as in fig. 2.38a, is called a simple train, and one in which two wheels are mounted on a common shaft, as in fig. 2.38b, is called a *compound* train. It is convenient to ascribe a letter to each wheel to denote at

* See chapter 7.

the same time the number of teeth in it; let us indicate its angular speed by ω with the corresponding subscript. For the simple train we can then write down

$$\frac{\omega_A}{\omega_B} = -\frac{B}{A}, \quad \frac{\omega_B}{\omega_C} = -\frac{C}{B}$$

and so on, and if we multiply these ratios together we find that the overall gear ratio is simply $\frac{\omega_A}{\omega_D} = -\frac{D}{A}$: the intermediate gears

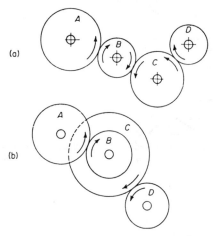

(a)

(b)

Fig. 2.38 Simple and compound gear trains

do not affect the numerical ratio and are usually called 'idlers'. Similarly, we can write down the ratio for the compound train

$$\frac{\omega_A}{\omega_D} = \left(-\frac{\omega_A}{\omega_B}\right)\left(\frac{\omega_B}{\omega_C} = 1\right)\left(-\frac{\omega_C}{\omega_D}\right) = \frac{B}{A}\frac{D}{C}$$

If the gears were bevel or skew we should have to state unambiguously the direction of rotation, but this offers no difficulty.

§ 2.13 Gear Trains: Epicyclic

An epicyclic gear train is one in which there is relative motion between the axes of the shafts. The rotations of the wheels are much less easy to visualise, and we must devote some attention to this case. Several methods of dealing with epicyclic gears are available, but we shall choose a simple one.

In fig. 2.39 is sketched an elementary epicyclic gear train, which we shall use to illustrate the method. Two meshing gears A and B, represented diagrammatically by their pitch circles, are coupled by the link X. Now if, for instance, the wheel A is 'fixed' in space and X rotates about its centre the wheel B will be caused to rotate: our problem might be to determine the amount of this rotation.

The solution is very easy if we tackle the problem in two imaginary stages. First, let us pretend that the link X is fixed while the wheel A

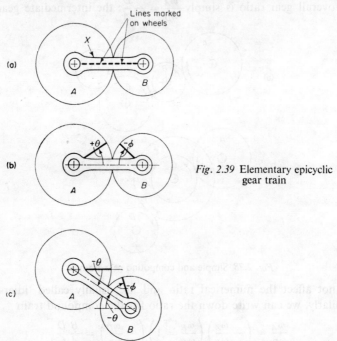

Fig. 2.39 Elementary epicyclic gear train

rotates through an angle $+\theta$ (so causing B to rotate through an angle $-\phi$, which we can determine without trouble); and second, that the train is locked so that no relative motion can occur while it is rotated as a whole about the centre of A through an angle $-\theta$. In fig. 2.39a,b, and c we show the system before motion occurs and after the two component movements; to show the rotations we have marked a line on each wheel.

The wheel A has returned to its starting-point, so the condition that it is 'fixed' has been satisfied; it is amply clear that the total rotation of B is $-(\theta + \phi)$. It was not really necessary, or rather should not be necessary again, to draw all these diagrams: we can express the situation much more concisely by setting out the rotations in a tabular form thus:

Table 2.1

	MEMBER		
CONDITION IMPOSED	X	A	B
X 'fixed' (A turned through $+\theta$)	0	$+\theta$	$-\phi$
$-\theta$ added to system (to 'fix' A)	$-\theta$	0	$-(\theta + \phi)$

This is so important that it is worth while making sure that we thoroughly understand the principle, so that we can safely apply it to more difficult problems. First we 'convert' the epicyclic train to a straightforward train by 'fixing' the axes of the shafts; in other (and more strict) words, we write down in the first line of the table a possible relative motion of the gears while their axes are fixed. (Clearly there is no restriction on the absolute motion. We can choose one number arbitrarily.) Second, we add a constant (which we can think of as corresponding to a rotation of the 'locked' system). The result will be a possible or 'compatible' rotation of each element in the system.*

Of course, we choose our numbers to satisfy any condition laid down, such as, in the above example, that A is to be 'fixed'. Had we been told that B was fixed, we should have added $+\phi$ to each number in the table. Had no member in the train been fixed, we should have had to exercise a little more ingenuity: thus, for instance, if we were told that X rotates through an angle $+\chi$ while A rotates through an angle $+\alpha$, we should proceed as follows:

Table 2.2

	MEMBER		
CONDITION IMPOSED	X	A	B
X 'fixed'	0	$+\theta$	$-\phi$
Multiplying by $(\alpha - \chi)/\theta$	0	$(\alpha - \chi)$	$-(\alpha - \chi)\phi/\theta$
Adding χ	χ	α	$-(\alpha - \chi)\phi/\theta + \chi$

* This statement is clearly a little too broad to cover the case of non-parallel axes, but the reservation is of trivial importance, and the modification easy to make.

The principle is that we obtain first in the *A* column a number ($\alpha - \chi$) to which we can afford to add χ.

Worked Examples

Although the principle described above is understandable enough, it has been discussed in relation to so elementary an example of an epicyclic gear train that it may be helpful to show its application to a few more complicated cases. This will have the additional merit of illustrating some of the peculiar advantages of epicyclic systems, in particular the way in which they can be used to combine or 'split' rotational motion, to provide different gear ratios from a unit without the need for bringing individual gear-wheels into and out of mesh, and to measure the torque being transmitted through a system.

Worked Example 2.1 Epicyclic 'differential' gear

This is a device which is almost universally used to allow torque to be transmitted from the engine to the two driving wheels of a road vehicle, not only while they are rotating at the same speed

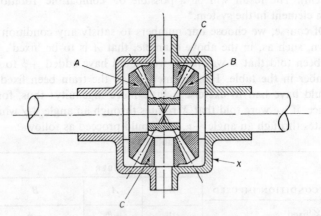

F g. 2.40 Differential gear

but also when, because the vehicle is turning a corner, they must to avoid skidding be allowed to turn at different speeds. In its simplest form it is sketched in fig. 2.40. The casing *X* is driven (through a bevel, spiral bevel, hypoid, worm, or similar gear) from the gearbox:

this casing carries a number (for symmetry, usually three or four) of bevel wheels such as C which mesh with wheels A, B fixed to the 'half-shafts' shown, to which in turn the road driving wheels are fixed. The whole mechanism is enclosed in a 'fixed' outer casing.

We can make out the table of possible motions as follows:

Table 2.3

		MEMBER		
CONDITION	X	A	C	B
Fix X	0	$+n$	$\sim nA/C$	$-n$
Add N	N	$N+n$	Not applicable	$N-n$

Note that as always we obtain the first line by 'fixing' the axes of the shafts: in this case it is convenient to assume that, say, A rotates through an arbitrary number n (of revolutions, radians, or any other unit: if desired in terms of velocity, we can assume in unit time) whereupon the corresponding values in columns C and B can immediately be inserted. The sign $\sim$ has been used to indicate that the direction of rotation of C cannot be specified by a simple algebraic sign. If we were interested in this we should have to use a vector notation, but all that really concerns us here is the fact that B is driven in the *reverse* direction as compared with A.

Again in this case it is convenient to add an arbitrary number (of revolutions in unit time, say) N to the system, which shows in general terms the possible relative speeds of the elements. It follows at once that the speeds of A and B are not individually defined, but only that their arithmetic mean must be that of the casing, and this is exactly the condition we want.

This device is here used for 'splitting' the rotational motion of the casing between the two elements A and B: if the action were reversed it would become a device for *combining* the rotational motion of these two elements.

This same device can in principle be used as a 'transmission dynamometer', i.e. as a means of measuring the torque which is being transmitted along a shaft. The shaft must be interrupted by the differential unit, so forming the two 'half-shafts' which have already been described by this conventional name. The 'casing' does not rotate, but is provided with a device for measuring the torsional reaction. The half-shafts therefore rotate at the same speed, but inevitably in opposite directions (cf. first line of table above). Now let us apply Newton's third law of motion to the system. At the input shaft we apply a torque which we call T to the system. The output shaft applies a torque $-T$ *to some other piece of mechanism*: the reaction on this system must therefore be $+T$. Thus, the two shafts together apply a torque of $+2T$, and since the casing is not being accelerated, it must be transmitting this torque to the measuring device.

This is an example of the sort of situation which is apt to cause a lot of trouble to the man who is careless about signs, partly because it is so apparently simple that he gives it too little attention. If it is thoroughly understood more complicated cases will give no trouble. There is something to be said for going through the mechanism with care, inserting the force on each bearing and on each gear-tooth, and making sure that the final result is in agreement with the one we have just proved. Perhaps it is even worth adding that if the mechanism were not to change the direction of rotation of the shaft the reaction torque would be zero: after all, the mechanism could then be eliminated in favour of a straight-through shaft (and would not in its original guise provide a differential action). It follows that in any gear train in which the direction of rotation is reversed the reaction torque on the gearbox is the numerical sum of the input and output torques; if the direction is not reversed, the numerical difference.

Worked Example 2.2 Simple change-speed gearbox

Fig. 2.41 shows diagrammatically the essential elements of one such device. The axis OO is in practice the centre-line of the gearbox,

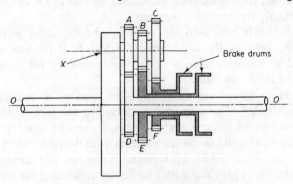

Fig. 2.41 Simple epicyclic change speed gearbox

and three or more systems of integral planet wheels A, B, C encircle it. The input is to the member X, which carries stub axles on which the planet wheels revolve freely: the output is taken from the shaft to which wheel D is fixed. Wheels E and F can be brought to rest by operation of the brakes shown (and frequently can be locked relative to each other by a brake not shown: the result is to prevent all relative rotation and provide a straight-through drive). The problem is to find the ratio between the speeds of X and D when either of the wheels E or F is stationary.

SOLUTION

The solution is obtained in table 2.4:

Table 2.4

CONDITION	MEMBER				
	X	ABC	E	F	D
Fix X	0	$+1$	$-B/E$	$-C/F$	$-D/A$
To fix E, add B/E	B/E		0		$B/E - D/A$
To fix F, add C/F	C/F			0	$C/F - D/A$

whence it is immediately apparent that the ratios required are

$$\text{with } E \text{ fixed, } \frac{\omega_X}{\omega_D} = \frac{B/E}{B/E - D/A} = \frac{BA}{BA - DE}$$

$$\text{with } F \text{ fixed, } \frac{\omega_X}{\omega_D} = \frac{C/F}{C/F - D/A} = \frac{CA}{CA - DF}$$

and that the direction of rotation will be reversed if the denominator is negative. Any number of ratios can be obtained by adding additional wheels and brakes, but the arrangement soon becomes too clumsy to be useful. Furthermore, although very large ratios can be obtained by making D and the fixed wheel nearly equal in size, ratios approximating to unity are in practice unobtainable.

Worked Example 2.3 Compound change-speed gearbox

Fig. 2.42 shows again purely diagrammatically a much more sophisticated four-speed and reverse gearbox* which need not be de-

* The 'Wilson' gearbox as used in a car transmission system, see *Proc. I.A.E.* Vol. 26, p. 216.

scribed in detail, since, although relatively complicated, its operation will become clear in the course of the calculation. It need only be said that top gear is obtained by engaging the cone clutch, while 3rd, 2nd, 1st, and reverse gears are obtained by applying brakes to the drums as indicated, one at a time.

Although in principle we could, as in example 2.2, write out the gear ratios in algebraic terms, the expressions we obtained would become excessively awkward. We shall therefore in this instance adopt the following particular sizes of gears, and work in arithmetical terms:

Wheel	E	F_1	F_2	F_3	G	H	J_1	J_2
No. of teeth	93	92	76	41	93	22	19	28

Note that we do not need to be told the numbers of teeth in the other wheels, but that they are calculable from those given. Thus, for instance, the (pitch-circle) radius of wheel E is the sum of the radius of F_3 and the diameter of D. The pitch or module of these teeth must, of course, be the same: it follows immediately that $D = \frac{1}{2}(E - F_3) = 26$; and so on.

PARTIAL SOLUTION

Table 2.5 giving the results follows: then explanatory notes.

Table 2.5

LINE		F	J	K	G	H	E	ω_j/ω_k
1	Fix K	$+1$	$-92/28$	0				
2	$\times 28$	$+28$	-92	0				
3	Fix F	0	-120	-28				$+4\cdot29$ (1st)
4	Fix F	0	-120	-28	$+24\cdot5$			
5	Fix G		$-144\cdot5$	$-52\cdot5$	0			$+2\cdot75$ (2nd)
6	Fix G							
7	Fix H							$+1\cdot67$ (3rd)
2	(K fixed)							
8	Fix E							$-6\cdot45$ (reverse)

In the first place, we *must* see that the 'basic' train is that giving first gear, consisting of $F_1(C)J_2$ and the planet carrier K. We therefore adopt the standard method of fixing K and giving (say) F one turn, which causes C to rotate through $+F_1/C$ turns and therefore

J_2 to rotate through $-C/J_2 \times F_1/C$ or $-F_1/J_2$ or $-92/28$ turns. This gives the first line: the second is inserted only to eliminate awkward fractions. The third line follows again the standard method: the member F is 'fixed' by adding -28 to line 2.

If we now look at the second gear train, consisting of $G(B)J_1$ and the planet carrier F, we see that to analyse its motion our first step will have to be to 'fix' F. But this is what we have just done in line 3: we already know the corresponding speeds of J and K, and J

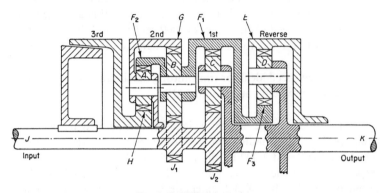

Fig. 2.42 'Wilson' gearbox

is coupled through the 'simple' train J_1BG to G, so we can immediately calculate the speed of G as $-J_1/G$ times that of J_1 and insert this value $(-\frac{19}{93}) \times (-120)$ or $+24\cdot5$ in the G column. To avoid confusion line 4 shows line 3 repeated with this addition. We can now 'fix' G by adding $-24\cdot5$ to line 4, giving line 5 and the second gear ratio, and so on.

The remainder of this problem should be tackled by the student without assistance except that suggested by the form of the table. The full table will be found at the end of the book, at the head of the answers to the problems at the end of this chapter. It is hardly necessary to say that when the cone clutch is engaged members J and H are required to rotate together, and this, as will be seen when line 7 of the table is completed, is a condition which cannot be obtained if relative motion occurs between the gears of the trains: the whole assembly is therefore locked, and a straight-through drive is provided from J to K.

Book for Reference and Further Reading

H. E. MERRITT. *Gears*, 3rd edn., Pitman, 1958.

Examples

2.1 A 100×5 mm section leather belt is used to connect two 750 mm pulleys and the *mean* tension adjusted to 100 kgf. One pulley is held fixed, and it is found that the belt slips when a torque of 20 kgf m is applied to the other. What power will this arrangement transmit if the speed of the pulleys is 800 rev/min and the *maximum* belt tension is limited to 125 kgf? The density of leather is 1 100 kg/m³.

Is the pulley speed ideal for transmitting maximum power? If not, what alteration would you recommend?

2.2 For a modern type of light-weight flat belting, driving between equal pulleys, the manufacturer's power rating figures at linear speeds of 1 000 and 3 000 m/min are 1·5 and 3·4 kW per cm width respectively. On the assumption that the coefficient of friction between belt and pulley is 0·2, calculate the value of the maximum recommended tension per cm width and the weight per square metre of the belt material.

2.3 A V-belt of included angle 30° is required to transmit 2 kW from a 150 mm-diameter pulley round which it has a 170° angle of lap. The pulley speed is 1 500 rev/min. The belt material weighs 2 500 kgf/m³ and has an effective safe tensile strength of 20 bar. The coefficient of friction between belt and pulley is 0·2. What minimum cross-sectional area of belt will be required?

If the pulley speed were increased, would the power which could be transmitted increase?

2.4 A V-belt has to transmit 4 kW from a 150 mm pulley running at 2 000 rev/min to a 300 mm pulley. The centres of the two pulleys are 450 mm apart. The included angle of the V is 30°, the belt weighs 0·45 kgf/m and the coefficient of friction between belt and pulley may be taken as 0·2. What is the least value of the outward force which must be applied between the shafts to avoid slipping of the belt?

2.5 A flat belt is 80 mm wide, 6 mm thick, and is made of material of density 1 400 kg/m³ and in which the stress must not exceed 28 bar. It connects two pulleys, the lesser angle of lap being 165° and the coefficient of friction 0·25.

At what speed should this belt run in order to transmit maximum power? How much will it transmit? What must the mean tension in the belt be under these conditions?

Plot curves showing how the power which can be safely transmitted varies with belt speed (*a*) if the mean tension is kept at this value; (*b*) if the mean tension is adjusted to the optimum value.

2.6 A Root blower consists of two cycloidal 'gears', each with 3 teeth. The centre distance is 150 mm. What must be the diameter of the rolling circle to generate the tooth form, and the radius of the ends of the casing enclosing the rotors?

Draw full size the rotors in mesh.

2.7 A 15-tooth pinion is to mesh with a 50-tooth wheel. The pressure angle chosen is 20°, and the module 15 mm. Draw full size a pair of teeth in contact at the pitch point, taking the addendum on each as equal to the module, with the tooth widths at the pitch circle equal.

B.S. 436 recommends addenda of $1·28m$ for the pinion and $0·72m$ for the wheel. Draw also such teeth as cut with a standard rack. What advantages does this modification confer on the gear pair?

2.8 The following information is given for a pair of involute gears transmitting 200 kW:

Pinion	28 teeth
Wheel	40 teeth
Face width	50 mm
Pressure angle	20°
Module	15 mm
Addendum	0·8 × module
Material	steel, for which $E = 2·07$ Mbar
Pinion speed	2 000 rev/min

From first principles, determine:

(*a*) the length of the path of contact;
(*b*) the 'number of teeth in contact';

(c) the maximum value of the bending stress, neglecting stress concentrations, and assuming that the maximum stress occurs at the junction between the flank and the fillet radius at 0·8 module below the pitch circle (see fig. 2.14);

(d) the minimum value of the 'radius of relative curvature' of the tooth profiles when one pair of teeth is taking the full load;

(e) the maximum value of the 'Hertz contact stress'.

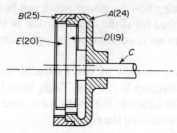

Fig. 2.43

2.9 An epicyclic gear train in a cyclometer, fig. 2.43, consists of:

An internally-toothed wheel *A* (fixed) of 24 teeth.

An internally-toothed wheel *B* of 25 teeth, revolving loosely on the axis of *A*, carrying the recording drum.

An arm *C*, also revolving on the axis of *A*, and driven by the cycle, carrying two wheels *D* and *E* of 19 and 20 teeth meshing with *A* and *B* respectively and fixed to each other.

Assuming that the arm *C* rotates once for each five turns of the road wheel, and that *B* should rotate through one turn per km, for what effective road-wheel diameter is this cyclometer suitable?

2.10 In an epicyclic wheel train two wheels, *A* and *B*, are keyed to the ends of two shafts, which are in line. *A* has 75 teeth and *B* has 74 teeth. *A* meshes with a wheel *C* having 25 teeth, and *B* with a wheel *D* having 26 teeth. If the arm carrying the compound wheels *C* and *D* revolves about the centre of the wheel *A* at 1 000 rev/min, and (*a*) *A* is fixed, (*b*) *A* revolves at 50 rev/min in the same direction, what will be the speed of the wheel *B*?

2.11 The sketch (fig. 2.44) shows the arrangement of a two-speed and reverse epicyclic gear drive: for simplicity the supporting bearings and operating members have been omitted, but it may be assumed that they are adequate and that the three conditions are obtained by locking, one at a time, the two brake drums shown with the clutch free, and by locking the clutch with both brakes free.

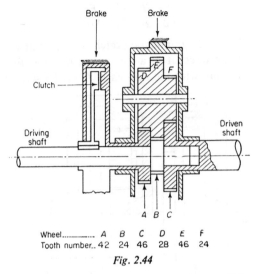

Wheel	A	B	C	D	E	F
Tooth number	42	24	46	28	46	24

Fig. 2.44

Find the available ratios, stating clearly the condition associated with each.

2.12 An epicyclic gear drive for a portable compressor is shown diagrammatically in fig. 2.45. The wheel A is driven by the engine, and the output to the compressor is taken from the arm D, which

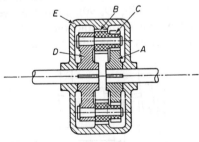

Fig. 2.45

carries shafts on which the compound pinions B, C are free to rotate. C is in mesh with A, and B with an internal wheel on the fixed casing E. The engine supplies 30 kW at 1 000 rev/min. Find, neglecting losses, the torque which the casing must provide and the torque on, and speed of, the output shaft. The wheels have the number of teeth shown in the table:

A	B	C	E
60	28	22	110

2.13 Fig. 2.46 is a diagrammatic sketch of a gear drive. The input shaft drives a sun-wheel *A* of a first epicyclic train; the annulus *B* of this train is solid with the sun-wheel *C* of the second train; the

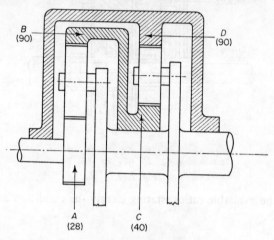

Fig. 2.46

annulus *D* of this second train is fixed. Both planet carriers are fixed to the output shaft. The number of teeth in each wheel is as shown (*A*, 28; *B*, 90; *C*, 40; *D*, 90). Find the overall gear ratio between input and output shafts, and the reaction torque carried by the fixed casing in terms of the input torque.

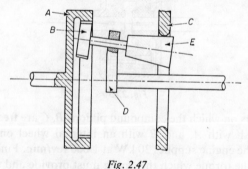

Fig. 2.47

2.14 Fig. 2.47 shows diagrammatically a proprietary variable-speed drive. The cone E is held by an arm D, which is driven at constant speed. There is a friction drive between the cone E and a rotationally fixed member C, and it may be assumed that there is no slip. This fixed member can be moved along the axis of the gear to change the gear ratio. If the fixed member C has an inner diameter of 150 mm and the diameter of the cone varies from 60 mm to 45 mm, find the range in output speed which can be obtained from an input speed of 5 000 rev/min. A has 90 teeth and B 30.

Find also the torque transmitted to member C when it is in the extreme right-hand position, when the input power is $\frac{1}{2}$ kW.

3. Balancing of Machines

§ 3.1 Introduction

Any mass, in particular any element of a machine, which is being accelerated must be being subjected to a force: to this force there must be an equal and opposite force – a reaction. So much we know from Newton's laws of motion. Thus the acceleration of a single mass in a 'fixed' machine implies inevitably that a reaction is applied to the foundations. We could equally well say that the foundation must supply the force needed to accelerate the mass.*

If, and only if, we can arrange that another mass or other masses† in our machine shall at all times have accelerations which in total require a force equal and opposite to that required by the original mass the foundations will be freed from these forces: under these conditions the machine is said to be 'balanced'.

It should be noted that we have made no mention of other forces which may arise in a machine and which may be greater by an

---

* It is not difficult to see that we are concerned here with varying and probably, in the widest sense, 'harmonic' forces, since if the mass were subjected to a constant acceleration, it would not long remain part of any finite machine. Thus, the reaction forces on the foundation will also be harmonic, and hence tend to give rise to distressing or even dangerous vibration. This will be discussed fully in chapter 5: for the moment it is enough to appreciate that such forces on foundations should if possible be eliminated.

† The frame of the machine itself may be one of these masses: see § 5.5.

order of magnitude than those we are discussing. Unless these forces give rise to accelerations of masses they *must* be self-balancing and so do not concern us – or the foundations. For instance, in a rolling-mill the enormous forces which may be required to deform the material being rolled are applied in opposite senses to the upper and lower rolls, transferred to the bearings and thence to the frame, where they cancel out. Again, in a single-cylinder piston engine large forces are generated when the charge is fired. If we *neglect* modifications arising as a result of the acceleration of the parts, the gas force on the piston, modified by the side thrust of the cylinder walls, is transferred to the connecting-rod and thence to the crankshaft. The vertical component of this force is transferred to the frame and counterbalances the force in the cylinder-head: the horizontal component together with the equal and opposite side thrust on the cylinder walls provides the torque reaction on the frame which alone makes it possible for the crankshaft to exert torque on some other system. This torque reaction is therefore inevitable if the engine is to do useful work on an external system and is the only reaction that the foundations must supply. But as we have seen in chapter 1, the forces applied to the frame through the piston and its mechanism are modified by the forces required to accelerate the piston and connecting-rod, so they are not equal and opposite to the force on the cylinder-head: the foundation has to supply the additional forces for equilibrium. In other words, we are back where we started, with the conclusion that the foundation has to supply the forces required to accelerate any parts of the engine which move.

§ 3.2 Balancing of Rotating Masses

3.2.1 Single out-of-balance mass

We start with the simplest possible case. Let the rigid shaft OO in fig. 3.1 carry at one transverse section a mass m_1 at a radius r_1 from the axis. Let the speed of rotation be ω. Neglect any gravitational effects. Then we know (§ 0.11) that the mass is being continuously accelerated towards the axis, the amount of this centripetal acceleration being $\omega^2 r_1$, and that a force of magnitude (mass × acceleration) $= m_1 \omega^2 r_1$, must be continuously exerted inwards on the mass by the shaft. The mass therefore exerts *on the shaft* (and in turn the shaft exerts *on the bearings*) an outward reaction equal and opposite to this, which can be represented by a vector rotating with the shaft as sketched in fig. 3.1, the so-called 'centrifugal' force. (We must never forget that 'centrifugal force' does *not* act on the mass itself: see Introduction.)

If we want to eliminate this rotating force from the bearings we need only mount diametrically opposite m_1 a 'balancing' mass, say M at radius R, as sketched in fig. 3.2. The force exerted by M will be $M\omega^2 R$, so for equilibrium, or balance, we need only ensure that

$$M\omega^2 R = m_1\omega^2 r_1$$

or
$$MR = m_1 r_1 \tag{3.1}$$

From this simple result three important conclusions of general validity can be deduced.

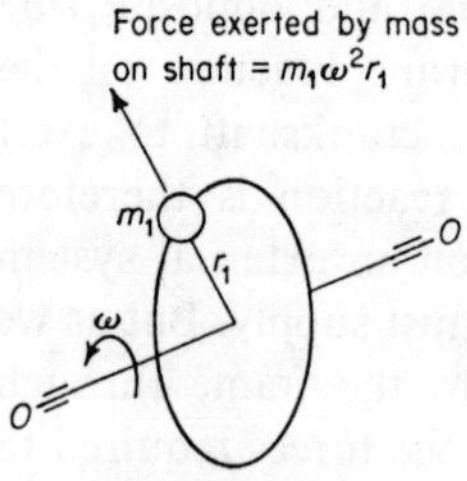

Fig. 3.1 Single unbalanced rotating mass 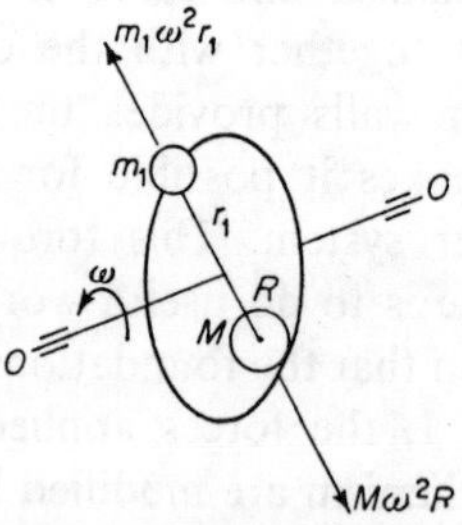   Fig. 3.2 Balancing single rotating mass 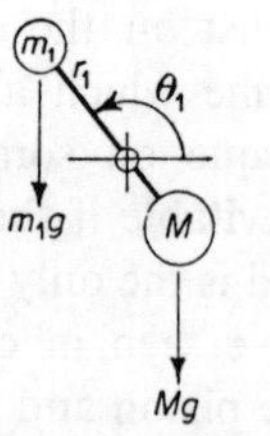   Fig. 3.3 Static balance of shaft in fig. 3.2

First, we are in problems of balancing concerned not with the absolute value of a mass or its distance from the axis of rotation separately, but with the product of these quantities, the 'mass-radius'.

Second, if a system is in balance dynamically it will also be in balance statically. (The converse, as we shall shortly see, is *not* necessarily true.) Thus if we look at an end view, fig. 3.3, of this same simple system, stationary in any arbitrary position and subjected to gravity, we see that the sum of the moments about the axis is

$$m_1 g r_1 \cos\theta_1 + MgR \cos(\theta_1 + \pi)$$

and from equation (3.1), this is identically equal to zero. The shaft would therefore have no tendency to rotate in its bearings. Indeed, the normal practical method of achieving 'static' balance is to mount the shaft on horizontal knife edges, which are effectively frictionless, and adjust the masses until the shaft will remain in equilibrium in any position.

Third, if a system is balanced for one speed it is balanced for every other speed, since ω^2 is automatically eliminated from the equation.

3.2.2 Several out-of-balance masses in one transverse plane

If we have a number of masses (m_1, m_2, ... at radii r_1, r_2, ... and angular dispositions θ_1, θ_2, . . . relative to any arbitrarily chosen datum) in one transverse plane, as shown in fig. 3.4, each will produce an outward radial force ($m_1\omega^2 r$, $m_2\omega^2 r_2$, ...) on the shaft which can be represented by a vector rotating with the shaft. Relative to the shaft, the force system will therefore be constant and can be

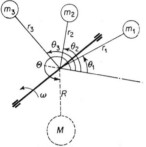

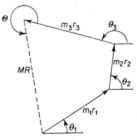

Fig. 3.4 Balancing of several rotating masses in one transverse plane

Fig. 3.5 Vectors for fig. 3.4

added by the usual vectorial methods. In order to balance the system we need only add a single mass M at radius R and angular disposition Θ, such that the vector force polygon closes (fig. 3.5). It helps to eliminate errors if we make a habit of assembling our data in tabular form as indicated. The force polygon, fig. 3.5, is drawn in terms of mass-radius (i.e. to the scale $\omega = 1$), since it is clear

Table 3.1

MASS	RADIUS	mr	θ
m_1	r_1	$m_1 r_1$	θ_1
m_2	r_2	$m_2 r_2$	θ_2
:	:	:	:
$\|M\|$	$\|R\|$	$\|MR\|$	$\|\Theta\|$

enough that the conclusions reached in the previous paragraph are still valid. When the length and orientation of the closing vector have been ascertained from fig. 3.5 the results can be inserted in the appropriate space left for them. Of course either M or R can then be chosen to suit the practical conditions, and the remaining quantity calculated.

3.2.3 Masses in different transverse planes: static v. dynamic balance

In § 3.2.1 it was pointed out that the conditions for dynamic balance satisfied the condition for static balance also, and it was stated that the converse was not true. The simplest illustration of this is the case

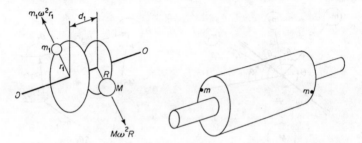

Fig. 3.6 Static balance: *Fig. 3.7* Static balance: dynamic
dynamic unbalance   unbalance

sketched in fig. 3.6, in which the two masses in fig. 3.2 are displaced along the shaft a distance d_1 relative to each other. That this system is in static balance is obvious enough: in fact, in end view it would be indistinguishable from fig. 3.3. When this system rotates, however, the centrifugal forces on the shaft form a *couple* of magnitude $m_1\omega^2 r_1 d_1$, which, like its component forces, rotates with the shaft: it is therefore in a state of static balance, but of dynamic unbalance. It is worth while to point out in passing that although such a situation is easy enough to appreciate when two discrete masses are involved, it is one which can in practice arise in a much less obvious form. For instance, let us suppose that a rotor is cast in the form of a drum, to some degree imperfect, so that in effect there is excess mass in the two regions indicated in fig. 3.7. It may easily be (or adjusted in the workshop to be) in perfect static balance as judged by the usual technique of mounting the shaft on horizontal knife edges, but give rise to a serious out-of-balance couple when running. Only a dynamic test will reveal this danger.

3.2.4 Representation of unbalanced couples and moments

Throughout this book, the importance of the correct representation of vectorial angular quantities has been stressed. It should not therefore be necessary to state that the out-of-balance couple arising in the system sketched in fig. 3.6 and reproduced in fig. 3.8 is correctly represented by a vector perpendicular to the plane of the forces, of magnitude $(m_1\omega^2r_1d_1)$, drawn according to the 'corkscrew' rule as

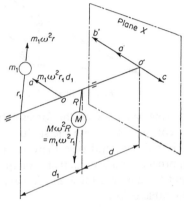

oa. A couple can be translated in its plane without alteration in its effect, so this vector can equally well be drawn in an arbitrary plane X as $o'a'$. It is also clear that in relation to the plane X we can represent the individual *moments* of the equal and opposite centrifugal forces $m_1\omega^2r_1$ and $M\omega^2R$ by vectors $o'b'$ and $o'c'$ respectively, where $o'b' = m_1\omega^2r_1 (d_1 + d)$ and $o'c' = M\omega^2R \cdot d$; and that their sum is the couple $m_1\omega^2r_1 \cdot d_1 \equiv o'a'$.

Now it will be seen that in effect we are drawing, in the

Fig. 3.8 Representation of unbalanced couples and moments

arbitrary plane X, vectors which are always rotated through $\pi/2$ relative to the force which is producing the moment, and *since we are going to deal with no other types of moment or couple,* we could without confusion and with considerable simplification achieve exactly the same answers by omitting this rotation, in other words, by drawing the vectors $o'a'$, $o'b'$, $o'c'$, *in the direction of the forces.* This is the convention universally adopted in dealing with balancing problems, and there is no objection to following it, *provided* that we are perfectly clear about our reasons and justification for doing so.

3.2.5 Masses in different transverse planes: general case

At any instant in time at which the shaft in fig. 3.9 occupies the position shown the 'centrifugal' forces exerted by the masses on the shaft act radially outwards and have magnitudes $m_1\omega^2r_1$, $m_2\omega^2r_2$, ... To find the forces exerted at this instant by the shaft on its bearings involves therefore only the methods of elementary statics and need not detain us long. The problem is exactly analogous to a loaded beam, except that such loads usually act in one plane. Using either calculation or graphical methods to suit the particular problem, we have in effect *first* to take moments about one bearing to find the force on the other, and *then*, *either* take moments about the other

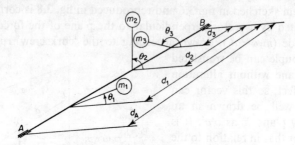

Fig. 3.9 Shaft with unbalanced rotating masses

bearing *or* equate all the forces acting to zero. The forces so found, of course, rotate with the shaft.

For the general case there is much to be said for the graphical approach, and a numerical example may be helpful. Suppose, then, that the data are those in ordinary type in the first four columns of the table below:

Table 3.2

COLUMN	1	2	3	4	5	6
PLANE	MASS, kg	RADIUS, mm	DISTANCE FROM R.H. BEARING, m	$\theta°$	'FORCE', mr kg m	'MOMENT', mrd kg m²
Left-hand bearing	—		4·5	229	11·4	51·3
1	40	250	3·5	20	10	35
2	25	500	2·0	80	12·5	25
3	10	400	1·0	150	4	4
Right-hand bearing	—	—	0	266	9·2	0

If the data are not presented in tabular form it is highly *desirable* that they should be so expressed as the first step in the solution; it is, of course, *essential*, since we propose to take moments about one of the bearings, that the distance *d* should be measured from it and so displayed in the table. We then calculate the corresponding values in the columns headed *mr* and *mrd* (taking the opportunity to eliminate the mixed units, but not bothering at this stage to multiply these numbers by ω^2 to obtain the correct numerical values of, and correct units for, the forces and moments). Because of our choice of reference plane we can now draw the moment polygon, fig. 3.10a, according to the convention described in § 3.2.4 and find the missing entry in line 1 of columns 6 and 4. Note that our entries (in italics) refer to the force exerted by the bearing *on* the shaft, i.e.

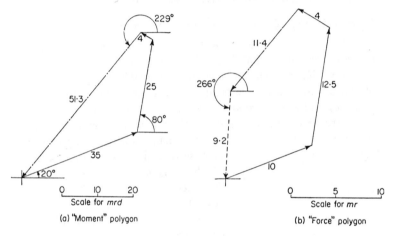

Fig. 3.10 Vector diagrams for balancing problems

to the equilibrant or closing line of the polygon. From column 6, using the value of *d* in column 3, we obtain the entry in the first line of column 5; this enables us to draw the force polygon, fig. 3.10b, and so to complete the table.

Now suppose this system is rotating at, say, 600 rev/min, or 62·8 rad/s: the forces exerted by the bearings on the shaft are

left-hand: force = 11·4 × 62·8² kg m × s⁻² = 45 000 N
right-hand: force = 9·2 × 62·8² kg m × s⁻² = 36 200 N

at the angles shown in the table: the forces exerted by the shaft on the bearings are equal and opposite: the whole system of forces rotates with the shaft.

Thus we find the out-of-balance forces in the system as given. If, on the other hand, we are required to find the masses which we should have to add to the system to balance it our procedure is identical except that we ignore the planes of the bearings (which no longer concern us, since the system is to produce no dynamic forces for them to support), but deal instead with the planes in which we are permitted to add the balancing masses. If for the sake of brevity we assume the system in fig. 3.9 to be unchanged, except that the bearings are located elsewhere, and the balancing masses are to be placed in their original planes, the whole table is unaltered and the first and fifth lines show the values and orientations of the balancing mass-radii which have to be added.

3.2.6 Balancing of rotating masses: general comments

It will be clear to anyone who follows through the example in the preceding paragraph that the methods outlined are perfectly general: it follows that provided that we have available *two* planes in each of which we can place the balancing mass we calculate to be necessary, *any system of rotating masses can be so balanced.*

It is true that some such systems, though not all, could be balanced by a *single* mass placed in a *particular* plane. It may be a good mental exercise to define the conditions under which this is possible, but the likelihood that such a solution will be practically possible is so remote, and the rewards so small, that the point is hardly worth pursuing.

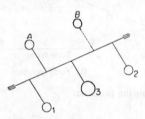

Fig. 3.11 Distribution of balancing masses

It is equally true, and of more practical importance, that the use of more than two balancing masses can be advantageous when lack of rigidity in the system is of importance. These points can perhaps be illustrated by the very simple example in fig. 3.11. Here we have a shaft with only two equal unbalanced masses A and B at the same orientation. We therefore do not need to go through the standard procedure: by looking at the problem we can see that masses 1 and 2, numerically equal to A and B but oppositely orientated in symmetrical planes, would produce balance. So would the single mass 3 $(= A + B)$ in the central plane. The shaft balanced in the first manner would when running bow out in the direction AB; the shaft balanced in the second manner would bow in the opposite direction. Clearly to minimise bowing (or loads on bearings in the neighbourhood of the central plane) it would pay to split the balance mass into all three planes. Again a crankshaft for a four-cylinder petrol engine resembles fig. 3.11 with masses 1, *A*, *B*, 2 attached, and usually with additional bearings. It may well pay to add (apparently quite unnecessary) 'self-balancing balancing masses' to this shaft to minimise distortion and the bearing loads.*

3.2.7 Note on dynamic balancing machine and the four vector method of balancing

Fig. 3.12 shows the principle of a machine on which it is possible to obtain the information required to balance dynamically a rotating system. The rotor to be balanced is mounted in bearings on a base-plate which is supported on springs, and driven either by a separate

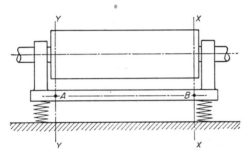

Fig. 3.12 Sketch of machine for balancing rotors

motor through a flexible drive or by a well-balanced motor mounted on the same base-plate. This base-plate can be pivoted about either of two axes which are chosen to be in the two reference planes in which balancing masses can be added.

If the speed of the unbalanced rotor is slowly raised while the base-plate is so pivoted the assembly will vibrate. The amplitude of this vibration† will be very dependent on the speed, as shown in fig. 3.13, being very large when the speed coincides with the natural frequency of oscillation, a condition known as 'resonance'. This particular condition gives maximum sensitivity in the detection of unbalance, but it is very difficult to control the speed accurately

---

* For an interesting discussion of the importance of such considerations in complex practical cases, see S. H. Grylls, 'The History of a Dimension', *Proc. I. Mech. E.*, A.D. L1/64.
† See chapter 5.

enough to obtain reliable measurements of the maximum amplitude. Normally the machine is operated at much higher speeds, where the vibration, though small, is relatively insensitive to variation in the speed and is directly proportional to the out-of-balance couple. The required sensitivity of measurement is then obtained by magnification of the amplitude by electronic or optical means.

Such a machine, if calibrated, could be used to find the magnitude of the unbalanced moment about each reference plane in turn. We also, however, need to know the angle relative to the rotor at which these moments are acting. From fig. 3.13 we see that the movement

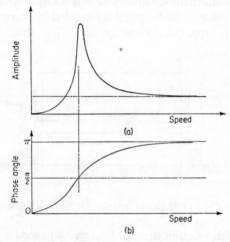

Fig. 3.13 Change of amplitude and phase with speed

of the base-plate does not indicate this directly. At low speeds this vibration is nearly in phase with the moment, but as we pass through resonance conditions change very rapidly, and at speeds well above resonance the vibration and moment are almost exactly in antiphase; which means, for instance, that at the instant when the unbalanced moment about the pivot is acting vertically *upwards* the base-plate is in its *lowest* position. One simple but very effective way of determining with reasonable accuracy the angular position of the moment is to run well above resonance speed and to adjust an electrical contact for triggering a stroboscope so that it is *just* made as the base-plate reaches this lowest position. Under the flashing light so produced the rotating shaft will be illuminated only when the moment is acting upwards: it will appear to be stationary in this position. A more sophisticated method is to couple a sine-wave generator to the rotor and compare the phase of its output with that of an electrical transducer used to measure the amplitude of the vibration.

It will be appreciated that it is easier to measure the amplitude of a vibration than its phase. We shall now show that by using the so-called 'four-vector' technique we can deduce the magnitude and phase of an unbalanced force or moment from four amplitude measurements. Consider, for example, a rotor being balanced in the machine shown in fig. 3.12, the pivots AA being in use. We want to

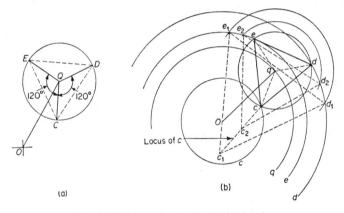

(a)   (b)

Fig. 3.14 'Four-vector' method of balancing

find the unbalanced moment about these pivots, but let us first assume that it is known to have a magnitude and phase represented by the vector OQ in fig. 3.14a. If now we add to the rotor at a known angle a known out-of-balance mass in the reference plane XX which gives rise to an additional out-of-balance moment $mrd = QC$, the total moment will be the vector sum of OQ and QC which is OC. If we remove this mass and replace it at 120° to its original position on the rotor, so that its moment about AA is QD, we shall have a total unbalanced moment OD; and similarly if we resite the mass at 240° to its original position we shall have a total moment OE. If now we run the rotor at a speed well above resonance (so that the amplitude is proportional to the out-of-balance moment and the phase 180°) in each of these four conditions, we know that we shall obtain four amplitude measurements proportional to the lengths of the vectors OQ, OC, OD, OE respectively.

If the original vector OQ is unknown all we have to work from is these four amplitude measurements, from which we have to re-construct fig. 3.14a. We can start by drawing four concentric circles from a centre O with radii proportional to the amplitudes, and labelling them q, c, d, and e respectively, as in fig. 3.14b. We now have to construct an equilateral triangle cde whose vertices lie on the appropriate circles and whose centroid lies on the q circle. This construction can be carried out by trial and error. Probably the easiest method is to select any arbitrary point q on the appropriate circle and draw a circle to cut (say) the d and e circles in d_1 and e_1. On the base $d_1 e_1$ an equilateral triangle $d_1 e_1 c_1$ will not in general meet the c circle in c_1. A second attempt $d_2 e_2 c_2$ gives an indication of the locus $c_1 c_2 \ldots$, and it is not difficult to arrive at the correct solution dec as indicated in fig. 3.14b. The whole figure can then be orientated so that the lines qc, qd, qe are parallel to the vectors QC, QD, QE, when the line oq must be parallel to the vector OQ which we set out to find. The scale is, of course, immediately ascertained, since we know the magnitude of the added moment $QC = mrd$ which is represented by qc.

§ 3.3 Balancing of Reciprocating Masses

3.3.1 Analysis of the usual motion of a reciprocating element in a machine

The balancing of reciprocating masses is rather more difficult, but of no less importance, since so many machines incorporate recip-rocating parts. Because the mechanism is almost universal, we shall

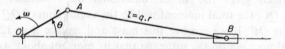

Fig. 3.15 Simple engine mechanism

first assume as a general condition that the part is driven by (or drives) a member rotating at constant speed through a crank-connecting-rod system, but the results can easily be modified for other conditions, as suggested in the next paragraph.

We have already, in § 1.3, analysed this 'simple engine mechanism', which is reproduced in fig. 3.15 with the same notation as in fig. 1.13, and found the acceleration of the reciprocating part, or 'piston' B.

Multiplying this by the mass m of the piston, we have the force F required to accelerate the piston.

$$F = m\ddot{x} = -m\omega^2 r \left(\cos\theta + \frac{1}{q}\cos 2\theta + \ldots \right)$$

the series being one in which the coefficients diminish rapidly, particularly after the second term. It is never in practice worth while to attempt deliberately to balance the forces represented by these later terms (though one can as an academic exercise discuss whether they do in fact automatically balance themselves or add up to something which could be of importance) and they are normally omitted from consideration altogether. We have, then, as a sufficient approximation

$$F \simeq -m\omega^2 r \left(\cos\theta + \frac{1}{q}\cos 2\theta \right) \tag{3.2}$$

So we see that the total force can be regarded as two separate forces, one, called the primary force, a simple harmonic force of frequency the same as the crankshaft (since $\theta = \omega t$), the other, called the secondary force, a simple harmonic force of smaller ($1/q$, or roughly $\frac{1}{4}$) magnitude but twice the frequency.

The qualitative cause of these separate component forces is very easy to understand, and such a picture is frequently helpful: we might as well pause for a moment to get it clear. If the connecting-rod were 'infinitely' long the motion of B would, of course, be identically the same as that of the projection of A on the line of stroke; in other words, simple harmonic (cf. § 0.14). This, then, accounts for the primary force. But the connecting-rod is of finite length, and during each *half* revolution of the crankshaft the end A is constrained to move, perpendicular to OB, through a distance r and back again. As a result, B is pulled inwards towards O through an *additional* displacement *and* pushed back again. In other words, it suffers an additional complete harmonic displacement during the half revolution of the crankshaft: hence the secondary force.

The forces we have been discussing are those required to accelerate the reciprocating parts. The reactions on the crankshaft, which will tend to be transmitted to the foundations, are, of course, equal and opposite, acting in the positive direction or outwards from O towards B when the crank is at 'top dead centre' (i.e. when $\theta = 0$).

They can most conveniently be pictured and discussed as indicated in the sketch fig. 3.16, i.e. as the projections on the line of

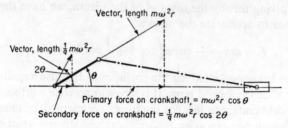

Vector, length $m\omega^2 r$

Vector, length $\frac{1}{q} m\omega^2 r$

2θ

θ

Primary force on crankshaft, $= m\omega^2 r \cos \theta$

Secondary force on crankshaft $= \frac{1}{q} m\omega^2 r \cos 2\theta$

Fig. 3.16 Component unbalanced forces on crankshaft due to reciprocating mass in simple engine mechanism

stroke of two separate force vectors,* the primary one rotating with the crank, the secondary one coinciding with the first at top dead centre and rotating at twice crankshaft speed.

3.3.2 The connecting-rod

In § 1.9 we dealt with the problem of determining accurately the forces required to accelerate a distributed mass such as a connecting-rod. When we are dealing with balancing problems we can seldom afford this degree of refinement, and the invariable procedure is to pretend that the connecting-rod can be replaced by two point masses, one located at the crankpin and therefore having its rotary motion, the other located at the gudgeon pin and therefore having the same reciprocating motion as the piston. The magnitude of these masses is calculated in accordance with equations (1.4a) and (1.4b), the third condition, (1.4c), namely

$$m_1 d_1{}^2 + m_2 d_2{}^2 = Mk^2$$

being ignored. In other words, the total mass of the rod is split up as the weight would be if the rod were placed horizontally on two weighing machines, one at each of the pin centres. It will be readily appreciated that since the quantity which is thus incorrectly substituted is the moment of inertia of the rod, the only error involved is in the magnitude of the harmonic torque required to give the rod its angular acceleration. Because the mass of the rod is largely concentrated at its ends, this error is usually quite small.

* Cf. § 0.14.

3.3.3 Less usual reciprocating motions

We may very briefly discuss two or three variants of the simple engine mechanism.

(a) There are machines in which it is vitally important that a reciprocating part shall have true *simple* harmonic motion. One way of achieving this is to use the mechanism sketched in fig. 3.17, which is variously known as a 'donkey-engine' mechanism or 'scotch yoke'. This is so inferior to an ordinary crank–connecting-rod system for load-carrying purposes that it would never otherwise

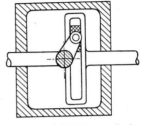

Fig. 3.17 'Donkey-engine' mechanism or 'scotch yoke'

be used. As far as balancing is concerned, it is, of course, simpler in that the secondary force is zero.

(b) In order to minimise the side thrust on the cylinder wall of an engine during the firing stroke the crankshaft may be offset from the line or axis of the piston stroke – the so-called 'désaxé' arrangement sketched in fig. 3.18. It is not difficult to apply the methods of

Fig. 3.18 'Désaxé' modification of simple engine mechanism

§ 1.3 to obtain an expression for the acceleration of the piston in this arrangement, and hardly more laborious. The student could well check for himself that the acceleration of the piston is given, to a similar degree of approximation, as

$$-\omega^2 r \left\{ \left(\cos\theta + \frac{e}{qr}\sin\theta \right) + \frac{1}{q}.\cos 2\theta \right\} \tag{3.3}$$

(c) In engines or other machines in which there are two or more 'banks' of cylinders, or in which the cylinders are disposed radially round a crankshaft, it may be decided to use a 'master' connecting-rod from one piston to the crankshaft, and to couple the other pistons through 'articulated' rods attached to 'knuckle' pins on the master rod, as indicated in fig. 3.19. Again there is no *basic* difficulty in writing down the displacement of any of the pistons in terms of the crank angle, and deriving therefrom an expression for its acceleration, but the task will be found to be extremely laborious in comparison with the simple connecting-rod–crank case. For purposes of reference the acceleration of an articulated rod piston is given as

$$-\omega^2 r \left\{ \cos(\theta - \phi) + \frac{1}{q} \cos 2(\theta - \phi) - \frac{2a}{ql_2} \cos(2\theta - \phi) \right.$$
$$\left. + \frac{a}{q^2 r}\left(1 + \frac{a}{l_2}\right) \cos 2\theta \right\} \quad (3.4)$$

where l_2 is the length of the articulated rod, q the ratio of master rod to crank radius r, θ the crank angle measured from the master cylinder, and ϕ the angle between the cylinders.

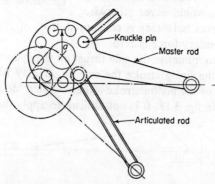

Fig. 3.19 Outline arrangement of radial engine

What is important is to remember that in such arrangements as the 'désaxé' and 'master-and-articulated rod' systems the simple expression in § 3.3.1 is quite inapplicable and must be replaced by the appropriate expression for the acceleration of *each* piston.

3.3.4 The problem of balancing a single reciprocating mass

In dealing with the balancing of rotating masses we saw that provided we could add to an existing shaft the masses we wanted, in at least two planes, the problem presented no basic difficulty. In dealing with reciprocating masses we have an entirely different situation.

Let us assume first of all that we have only one reciprocating mass m_1 – as, for instance, in a single-cylinder engine. We have worked out in § 3.3.1 the out-of-balance force. What do we do about it? If we had negative masses at our command . . . but this is clearly absurd. In their absence, we must try to give another mass an acceleration exactly equal and opposite to that of m_1 and in the same line of motion. But that probably means (because we cannot get another crank on the original shaft in the same plane as the engine crank) either an entirely separate additional shaft, or two additional cranks on the same shaft, and the associated connecting-rods, and some means of defining the line of motion of the balancing mass . . . it is all getting very complicated and expensive.

Can we do any good by adding a rotating balance mass to the crankshaft? Let us be very clear what we mean by this. In fig. 3.20

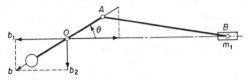

Fig. 3.20 Effect of mass opposite crank

the engine is sketched again. Associated with the piston is its crank OA, which, of course, has an effective mass-radius which we can calculate, and to which we have learnt (§ 3.3.2) to add an appropriate fraction of the mass of the connecting-rod as though it were concentrated at the centre of the crankpin. All this is a rotating mass which we can certainly balance completely by an appropriate mass, probably split into two equal masses attached to the crank webs as shown, for instance, in fig. 1.1: so much we take for granted. We are therefore talking now only about the reciprocating mass m_1 (including the piston, etc., and the remaining part of the connecting-rod mass).

Suppose we add a mass m_1 at crank radius r opposite the crank-pin. This will produce a force which we can represent by the rotating vector Ob, of magnitude $m_1\omega^2 r$, which we can split into the components $Ob_1 = m_1\omega^2 r \cos\theta$ along the line of stroke and $Ob_2 = m_1\omega^2 r \sin\theta$ perpendicular to it. But the *primary* out-of-balance force we are trying to eliminate has, as we have seen in fig. 3.16, a value $m_1\omega^2 r \cos\theta$ along the line of stroke *only*. We have succeeded in balancing this force, but only at the expense of introducing a force of identical amplitude in a plane at right angles to the line of stroke. The same is true of any similar compromise: we can eliminate a fraction of the original by introducing the same fraction into the plane at right angles; if we make the fraction $\frac{1}{2}$ the reciprocating forces, of magnitude $\frac{1}{2}m_1\omega^2 r \cos\theta$ in the line of stroke and $\frac{1}{2}m_1\omega^2 r \sin(\theta + 180°)$ perpendicular to it, add up in effect to a force of constant magnitude $\frac{1}{2}m_1\omega^2 r$ rotating in a direction opposite to that of the crank. The secondary force is, of course, wholly unaffected.

It may be true in specific instances that there is reason to hope or believe that a force so transferred will be less harmful than the original, but it can hardly be doubted that many 'balance weights' are attached to crankshafts not for any such reason but simply because it is customary to do so.

If we can afford to add *two* shafts and appropriate masses, rotating in opposite directions, to our machine it becomes very easy to produce any single reciprocating force we want. The principle* is illustrated in fig. 3.21, and hardly needs comment. Each rotating mass produces a rotating force on its shaft: the components in the direction *OO* cancel; those in the direction marked ₵ (which may be regarded as the line of stroke of a single-cylinder engine) add up. (If we had to deal with a secondary force as well we should have to provide two additional shafts running at twice crankshaft speed.) The principle of this 'Lanchester balancer' is well worth remembering: the device is used not only for balancing an unwanted force but equally often to produce an unbalanced harmonic force for the investigation of modes of vibration of structures.

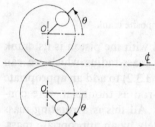

Fig. 3.21 'Lanchester' balancer

The more complicated methods briefly adumbrated at the beginning of this paragraph hardly need further discussion here; some possibilities will be indicated in § 3.4.

3.3.5 The problem of balancing several masses reciprocating in a plane containing the crankshaft

It will be appreciated that the practical difficulty of balancing a single reciprocating mass arises simply because there is no member with an opposed motion to which one can attach a balancing mass. If, however, we have a number of reciprocating masses, and some freedom in their disposition, we may well be able to arrange that

* F. W. Lanchester, *Proc. I.A.E.*, Vol. 8, p. 222.

they shall balance each other, completely or to some extent. To make the discussion less abstract, let us consider an engine with N cylinders in line as sketched in fig. 3.22. Due to piston No. 1, there will (with the usual notation) be on the frame in the plane 1 *at this instant* an unbalanced force $m_1\omega^2 r_1\left(\cos\theta_1 + \dfrac{1}{q}\cos 2\theta_1\right)$. Now let O be any plane normal to the axis of the crankshaft which we choose as a reference plane.* We can replace the force in plane 1 by an identical force in plane O together with the appropriate couple, so,

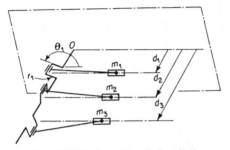

Fig. 3.22 Engine with cylinders in line

splitting our total force into its primary and secondary components, we can say that piston No. 1 produces, relative to plane O, at this instant

a primary force $= m_1\omega^2 r_1\cos\theta_1$

a secondary force $= \dfrac{1}{q}\cdot m_1\omega^2 r_1\cos 2\theta_1$

a primary moment $= m_1\omega^2 r_1 d_1\cos\theta_1$

a secondary moment $= \dfrac{1}{q}\cdot m_1\omega^2 r^2 d_1\cos 2\theta_1$

$\left.\vphantom{\begin{array}{c}1\\1\\1\\1\end{array}}\right\}$ all acting in the direction of the line of stroke

The same argument will hold for each piston, so for the whole engine we can write (assuming q to have the same value for each crank)

* For clarity in the diagram we have shown the reference plane O at one end of the crankshaft, but in the consideration of a practical engine it is more sensible to choose as a reference plane the one containing the centre of gravity of the engine, since we shall then determine the total unbalanced moments acting on the engine about this plane, and the answer will therefore be in a form in which we can immediately apply it to determine the effect of the moment. If we are dealing with abstract engines in which the centre of gravity is unknown we can reasonably choose as a reference plane one near the middle of the crankshaft; the plane of symmetry if one exists.

$$\left.\begin{array}{ll}\text{total primary force} & = \omega^2 \sum mr \cos \theta \\[4pt] \text{total secondary force} & = \dfrac{1}{q} \cdot \omega^2 \sum mr \cos 2\theta \\[4pt] \text{total primary moment} & = \omega^2 \sum mrd \cos \theta \\[4pt] \text{total secondary moment} & = \dfrac{1}{q} \cdot \omega^2 \sum mrd \cos 2\theta \end{array}\right\} \begin{array}{l}\text{at this instant} \\ \text{in the direc-} \\ \text{tion of the} \\ \text{line of stroke}\end{array}$$

From the point of view of balancing this engine, we are uninterested in ω and q, which will have finite values: we have to ensure that the summed terms are zero not only at this instant but at *any* such instant. Let us consider typically the first, $\sum mr \cos \theta$. We know that each term in this summation is the projection along the

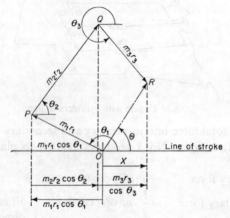

Fig. 3.23 Addition of harmonic terms by vector diagram

line of stroke of a vector of length mr drawn in the direction of the crank. The easiest way to carry out such a summation is to add the vectors* and project the resultant, as indicated in fig. 3.23, where $N (= 3)$ vectors OP, PQ, . . . are added to give the resultant OR, whose projection on the line of stroke is X. So if OR happened at this instant to be perpendicular to the line of stroke X would be for the moment zero; but this would be of no importance. As the crankshaft rotates the whole of the diagram rotates, so the resultant force X can *remain* zero only if the vector diagram is a closed polygon. A similar argument holds for each summation.

We therefore arrive at a very simple conclusion. In order that

* Cf. § 0.14.

such an engine should be balanced, these four vector polygons must close. We can express this *either* for analytical purposes thus:

$$\left.\begin{array}{ll}
\sum mr \ \cos \theta \ = 0 & \sum mr \ \sin \theta \ = 0 \\
\sum mr \ \cos 2\theta = 0 & \sum mr \ \sin 2\theta = 0 \\
\sum mrd \cos \theta \ = 0 & \sum mrd \sin \theta \ = 0 \\
\sum mrd \cos 2\theta = 0 & \sum mrd \sin 2\theta = 0
\end{array}\right\} \tag{3.5}$$

where each term such as $mr \cos \theta$ is a scalar quantity,
or for graphical purposes thus:

$$\left.\begin{array}{l}
\sum_\theta \ mr \ = 0 \\
\sum_{2\theta} mr \ = 0 \\
\sum_\theta \ mrd = 0 \\
\sum_{2\theta} mrd = 0
\end{array}\right\} \tag{3.6}$$

where each term such as mr is a vector, and the symbol $\sum_\theta$, for instance, of course implies that each vector has to be drawn in the direction of the corresponding crank, and so on.

From this point we can pursue our consideration of the problem in several ways. In general terms we can argue that we have to satisfy the eight equations (3.5), and that associated with each crank we have three variables (mass-radius mr, distance from reference plane d, and crank angle θ) at our disposal: a total of $3N$. But we are concerned with relative magnitudes and positions, so the variables associated with one cylinder are not at our disposal: we have only $3(N - 1)$ to play with. On this basis, to be able to balance the engine we must have $3(N - 1) \not< 8$ or $N \not< 4$. This sort of argument can be helpful in moderation, but it leaves out of account the rather important consideration that we are probably hoping to build the engine to do a job of work and not simply to end up as a perfectly balanced bit of machinery: we are fairly certainly not going to have all this freedom of manoeuvre. For instance, we probably want to have the cylinder centre-lines as closely pitched as their dimensions will allow, and to have uniform intervals of time between the instants at which each cylinder fires. Again, the possibility of serious torsional vibration of the crankshaft may rule out of consideration some arrangements which could be attractive from the point of view of balancing alone. However, with the number $N = 4$ in mind let us consider briefly a few scraps from the history of engine-building, restricting ourselves for the time being to engines with cylinders in line.

The single-cylinder engine we have already dealt with.

A double-acting steam engine with two cylinders and 'cranks' at right angles is extremely simple, is self-starting, and has excellent torque characteristics (cf. § 1.10). It is not surprising that it was chosen for the early locomotives and persisted so long. As a vehicle for the practitioner of balancing it is a write-off. The rotating masses can, of course, be balanced very easily by adding masses in the road wheels. The reciprocating masses cannot be balanced at all. The unbalanced horizontal primary forces, which to some degree will reach the train coupling, can to any desired extent be transformed into vertical forces on the track by increasing the magnitude of the rotating 'balancing' masses, and the 'yawing' primary moments transformed correspondingly into 'rolling' moments, and this procedure was included under the generic title 'balancing', but the word is a misnomer. When the minimal two cylinders were increased to three or four the primary objective was an increase in power. No doubt consideration was given to obtaining some degree of self-balance, but this must have been a minor consideration. In fact, rotating masses were almost invariably added to the road wheels greatly in excess of those required to balance the rotating masses, but whether this was done to decrease the yawing moment or merely to annoy the Chief Civil Engineer by damaging his track must remain uncertain.

In marine steam reciprocating machinery space was less restricted and economy more important, so development followed different lines. Simple 'single-expansion' engines soon gave way to double-, triple-, and even occasionally quadruple-expansion engines, and triple-expansion engines might well use two low-pressure cylinders; so three- and four-cylinder engines were common, and in such engines the mass necessarily associated with each cylinder would be far from identical. This provided the engine-builder at least with an opportunity to exercise his ingenuity, and one of the more interesting examples of a design to minimise out-of-balance forces, the 'Schlick symmetrical engine', will be discussed in some slight detail in § 3.3.6. The steam reciprocator may be dead or dying, but this as an exercise in balancing is well worth brief study.

The internal-combustion engine provides an entirely different picture. Whereas a double-acting single-cylinder steam engine gives two working strokes per revolution, the most common type of internal-combustion-engine cylinder gives one working stroke in two revolutions, so that for a similarly uniform torque characteristic four times as many cylinders are required. Moreover, the characteristics of combustion in a 'spark-ignition' type of engine severely restrict the absolute size of a cylinder. For these and many other reasons development has been in the direction of multi-cylinder fast-running engines with symmetrically orientated cranks. In § 3.3.7 we shall investigate the state of balance of a few typical internal-combustion engines: the methods are perfectly straightforward and can easily be applied to other designs.

In later paragraphs we shall deal with one or two examples of engines whose cylinders are not in one bank.

3.3.6 Schlick 'symmetrical' engine

As mentioned in the previous paragraph, the Schlick 'symmetrical' engine is selected as an interesting example of the methods which can be adopted to obtain a reasonable degree of balance in an engine (or other machine) with a limited number of reciprocating parts of varying masses.

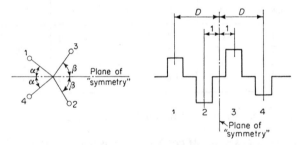

Fig. 3.24 Crankshaft of Schlick 'symmetrical' engine

The chosen layout of the engine is indicated by the end and side diagrammatic views of the crankshaft in fig. 3.24. The slightly unusual sense in which the word 'symmetrical' is used will be clear from this diagram, but in addition it should be stated that all the crank radii are made identical (and r_1 may therefore be taken as the unit for this dimension) and that the reciprocating masses in cylinders 1 and 4 are made the same (and therefore taken as unity) and in 2 and 3 the same (and therefore may be indicated as $M = m_2/m_1$). The distances from the central plane to the four cylinders may similarly be designated $-D$, -1, $+1$, $+D$ respectively, d_3 being taken as unity.

When all these choices have been made it becomes impracticable to balance the engine completely, and no attempt is made to eliminate the secondary moments, which are probably the least likely to cause distress. Let us then investigate the state of balance in other respects: we can tackle this graphically or analytically, and shall choose the latter alternative. The equations (3.7) we wish to satisfy are:

$$
\begin{array}{llll}
\text{Primary forces:} & \sum mr \cos \theta = 0 & (1) \\
& \sum mr \sin \theta = 0 & (2) \\
\text{Secondary forces:} & \sum mr \cos 2\theta = 0 & (3) \\
& \sum mr \sin 2\theta = 0 & (4) \\
\text{Primary moments:} & \sum mrd \cos \theta = 0 & (5) \\
& \sum mrd \sin \theta = 0 & (6)
\end{array}
\quad (3.7)
$$

As when dealing with rotating balancing, it is very helpful to display our data in tabular form, thus:

Table 3.3

PLANE	mr	d	mrd	θ	2θ
1	1	$-D$	$-D$	$(\pi - \alpha)$	-2α
2	M	-1	$-M$	$-\beta$	-2β
3	M	$+1$	$+M$	$+\beta$	$+2\beta$
4	1	$+D$	$+D$	$(\pi + \alpha)$	$+2\alpha$

Inserting these values in our equations we have

$$
\begin{array}{lll}
(1) & -\cos \alpha + M \cos \beta + M \cos \beta - \cos \alpha = 0 \\
(2) & \sin \alpha - M \sin \beta + M \sin \beta - \sin \alpha = 0 \\
(3) & \cos 2\alpha + M \cos 2\beta + M \cos 2\beta + \cos 2\alpha = 0 \\
(4) & -\sin 2\alpha - M \sin 2\beta + M \sin 2\beta + \sin 2\alpha = 0 \\
(5) & D \cos \alpha - M \cos \beta + M \cos \beta - D \cos \alpha = 0 \\
(6) & -D \sin \alpha + M \sin \beta + M \sin \beta - D \sin \alpha = 0
\end{array}
$$

Thus we see that thanks to the choice already made, three of our equations (2, 4, and 5) are identically satisfied, while the others require that

$$
\begin{array}{ll}
(7) & \cos \alpha = M \cos \beta \\
(8) & \cos 2\alpha + M \cos 2\beta = 0 \\
(9) & D \sin \alpha = M \sin \beta
\end{array}
$$

From equations (8) and (7)

$$2 \cos^2 \alpha - 1 + M(2 \cos^2 \beta - 1) = 0$$
$$\therefore 2M \cos^2 \alpha - M + 2 \cos^2 \alpha - M^2 = 0$$
(10) $\therefore 2 \cos^2 \alpha(M + 1) = M(M + 1)$ or $2 \cos^2 \alpha = M$

So our equations reduce to

(10) $2 \cos^2 \alpha = M$
(7) $\cos \alpha = M \cos \beta$
(9) $D \sin \alpha = M \sin \beta$

and we have only three equations to satisfy with four variables, M, D, α, and β at our command. One more arbitrary choice – probably D – can still be made.

The components of the unbalanced secondary moment vector are found from the appropriate equations of (3.5)

$$\sum mrd \cos 2\theta = -D \cos 2\alpha - M \cos 2\beta + M \cos 2\beta + D \cos 2\alpha = 0$$
$$\sum mrd \sin 2\theta = D \sin 2\alpha + M \sin 2\beta + M \sin 2\beta + D \sin 2\alpha$$
$$= 2(D \sin 2\alpha + M \sin 2\beta)$$

The first component being zero, the second gives, as it happens, the total length of the vector and hence the amplitude of the force, which at a crankshaft speed ω therefore works out to

$$2(D \sin 2\alpha + M \sin 2\beta) \, . \, m_1\omega^2 r_1 d_3/q$$

where m_1, r_1, and d_3 are, of course, the *actual* values of these quantities which for brevity in the analysis have been assumed to be unity.

Note that this unbalanced moment has been measured about the plane of symmetry (i.e. the approximate centre of gravity) of the engine, as suggested in the footnote on p. 211. In this particular engine, however, there are no unbalanced forces so the moment is unaffected by our choice of reference plane; in cases where an unbalanced force is left this is not so. See, for instance, § 3.3.7(a).

3.3.7 Balance of typical 'in-line' internal-combustion engines

We shall select as typical a four-cylinder and a six-cylinder four-stroke petrol engine, and a ten-cylinder marine oil engine. Partly because we tackled the Schlick engine (§ 3.3.6) by the analytical approach, but mainly because of the greater physical insight provided by the graphical solution, we shall in these cases adopt the latter.

In each case we display our data in tabular form and then (if necessary) draw the four vector polygons as described in § 3.3.5.

(a) *Four-cylinder four-stroke petrol engine*

The crankshaft is arranged with the two outer throws in line and the two inner throws in line but at 180° to the outer: in other words, if the direction of the left-hand crank be taken as the reference direction 0°, the others, in order, are at 180°, 180°, and 0°. This, of

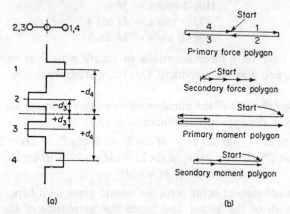

(a)   (b)

Fig. 3.25 Degree of unbalance in conventional four-cylinder 'in-line' engine

course, allows four evenly spaced working strokes in two revolutions of the crankshaft. Symmetry about the centre line is assumed. End and side views of the shaft in this position are not very informative: end and plan views are therefore given in fig. 3.25a.

Table 3.4

PLANE	MASS-RADIUS	DISTANCE FROM REFERENCE PLANE	mrd	$\theta°$	$2\theta°$
1	mr	$-d_4$	$-mrd_4$	0	0
2	mr	$-d_3$	$-mrd_3$	180	0
3	mr	d_3	$+mrd_3$	180	0
4	mr	d_4	$+mrd_4$	0	0

Rough sketches (fig. 3.25b) are, of course, adequate to show that in this engine primary forces and moments, and secondary moments about the centre line, are inherently balanced. The secondary forces, on the other hand, add up to produce a total vector $4mr$ and a corresponding reciprocating force F where

$$F = \pm 4mr \cdot \omega^2/q \qquad (3.8)$$

Note that this is of the same order of *magnitude* as the primary force in a single-cylinder engine, but since it is applied at twice the frequency to a much larger mass (the four-cylinder engine), its effects will be quite different (see chapter 5).

(b) *Six-cylinder four-stroke engine*

The crankshaft is arranged with throws at $0°$, $120°$, $240°$, $240°$, $120°$, $0°$. Symmetry is again assumed as in fig. 3.26a.

Table 3.5

PLANE	MASS-RADIUS	d	$mr\ d$	θ	2θ
1	mr	$-d_6$	$-mrd_6$	0	0
2	mr	$-d_5$	$-mrd_5$	120	240
3	mr	$-d_4$	$-mrd_4$	240	120
4	mr	$+d_4$	$+mrd_4$	240	120
5	mr	$+d_5$	$+mrd_5$	120	240
6	mr	$+d_6$	$+mrd_6$	0	0

Force Polygons

At this stage, or perhaps before if one is particularly bright, it may seem that we have been wasting our time in drawing these force polygons as in fig. 3.26b, since a glance at the symmetrical disposition of the cranks may be enough to show that the primary forces in each half of the engine will be in balance, while doubling the crank angles (cf. the 2θ column) gives an equally symmetrical array. Even so, it is not a bad exercise to draw the polygons until the methods are fully understood.

Moment Polygons

Certainly before bothering to draw these, we may pause and consider the situation from the physical point of view. In an engine such as this (and the four-cylinder one) in which the two halves of the crankshaft are, so to speak, mirror images of each other in the plane of symmetry, the force, whether primary or secondary, required to accelerate a piston on one side of this plane must be exactly equal to the force required to accelerate its opposite number. No *moment* can therefore be produced.

Whether this argument carries conviction or not, the moment polygons should in this instance be drawn as in fig. 3.26c.

So we reach the conclusion that no unbalanced primary or secondary forces remain. Furthermore, we see that each *half* of the engine

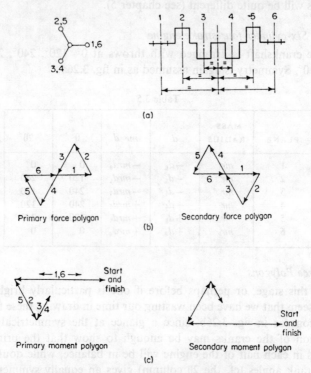

(a)

Primary force polygon

Secondary force polygon

(b)

Primary moment polygon

Secondary moment polygon

(c)

Fig. 3.26 Six-cylinder in-line engine: 'perfect' balance

is independently balanced in relation to both primary and secondary forces, and that the moments arising are relatively small, so no large forces have to be transmitted through the frame. Finally, we might note as a matter of interest that although we have accepted the balance of primary and secondary forces as the limit of our reasonable ambitions, the next item in the infinite series for the acceleration of a piston involves (see § 1.3) not 3θ but 4θ, and it is an easy matter to check that the corresponding forces will similarly balance. The next term involves 6θ, so the forces would add up, but this is of no importance, since the coefficient is so small that unavoidable practical imperfections in manufacture would lead to larger forces. We may therefore reasonably conclude that in so far as balance is concerned, we are unlikely ever to find a more nearly perfect machine than a six-cylinder engine.

Nevertheless, we may require more than six cylinders. Eight cylinders in line is about the limit for the smallish high-speed type of engine: the difficulty of avoiding torsional oscillation of the crankshaft thereafter becomes too great (see chapter 4). Other arrangements we shall be discussing in subsequent paragraphs. Marine oil engines may have more than eight cylinders in line, and as a final example we shall look briefly at a

(c) *Ten-cylinder oil engine, symmetrical about the mid-plane with cranks at 0°, 144°, 288°, 72°, 216°, 216°, 72°, 288°, 144°, 0°*

Having learnt our lesson, we need not consider moments which will automatically balance. It is therefore hardly worth while making out a table: indeed, it is again easy to deduce from the symmetry of

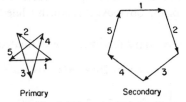

Primary   Secondary

Fig. 3.27 Five- (or ten-) cylinder in-line engine: balance of forces

the angular arrangement of the cranks that all the forces will be in balance. However, the force polygons will be as sketched in fig. 3.27 for one half of the engine, to confirm that the engine will be 'completely balanced'.

3.3.8 'V', 'fan', and radial engines

(a) *V-engines*

We may consider first a V-twin engine as sketched in fig. 3.28, with the plane of symmetry horizontal. The connecting-rods operate on a common crank OC: the angle between the cylinders is α as shown. The crank angle θ is measured from OX: as measured from T.D.C. of piston A it will therefore be $(\theta - \alpha/2)$ and of piston B $(\theta + \alpha/2)$.

The primary force on the frame due to piston A along OA

$$= m\omega^2 r \cos(\theta - \alpha/2)$$

The primary force due to piston B along OB

$$= m\omega^2 r \cos(\theta + \alpha/2)$$

Therefore total primary force along XX

$$= m\omega^2 r [\cos(\theta - \alpha/2) \cos \alpha/2 + \cos(\theta + \alpha/2) \cos(-\alpha/2)]$$
$$= m\omega^2 r \cos \alpha/2 \ (2 \cos \theta \cos \alpha/2)$$
$$= m\omega^2 r \cos \theta \cdot 2 \cos^2 \alpha/2 \tag{3.9a}$$

Total primary force along YY

$$= m\omega^2 r [\cos(\theta - \alpha/2) \sin \alpha/2 + \cos(\theta + \alpha/2) \sin(-\alpha/2)]$$
$$= m\omega^2 r \sin \theta \cdot 2 \sin^2 \alpha/2 \tag{3.9b}$$

and we may note that *in the particular case when $\alpha = 90°$* these reduce to

$$\text{P.F. along } XX = m\omega^2 r \cos \theta$$
$$\text{P.F. along } YY = m\omega^2 r \sin \theta$$

So these are the components of a constant force rotating with the crank, and they can be completely balanced by a mass radius mr opposite the crank. On reflection, this could be considered as a blinding glimpse of the obvious: it is worth while to pause until the reason is clear.

Similarly, the secondary forces can be evaluated

$$\text{S.F. along } XX = \frac{1}{q} m\omega^2 r [\cos 2(\theta - \alpha/2) \cos \alpha/2 + \cos 2(\theta + \alpha/2) \cos(-\alpha/2)]$$
$$= \frac{1}{q} m\omega^2 r \cos 2\theta \cdot 2 \cos \alpha \cos \alpha/2 \tag{3.9c}$$

$$\text{S.F. along } YY = \frac{1}{q} m\omega^2 r \sin 2\theta \cdot 2 \sin \alpha \sin \alpha/2 \tag{3.9d}$$

and if $\alpha = 90°$

$$\text{S.F. along } XX = 0$$

$$\text{S.F. along } YY = \frac{1}{q} m\omega^2 r \sin 2\theta \cdot \sqrt{2}$$

the last quantity being, of course, a harmonic force which can be eliminated only by special means, such as the provision of a Lanchester balancer.

'V-8' engine. The state of unbalance of an eight-cylinder V-engine can be quickly and easily ascertained by an intelligent application of the foregoing paragraphs. If, for instance, the crankshaft is similar to that illustrated in fig. 3.25 we can see at once that each bank will be inherently balanced *except* for secondary forces, and that these (since they are simply four times the size of that due to one piston) will add up for the double bank in exactly the same way as those in the V-twin. In the case when $\alpha = 90°$, for instance, the secondary unbalanced force for the engine will be:

$$4\sqrt{2}\frac{1}{q} \cdot m\omega^2 r \sin 2\theta \text{ along } YY \qquad (3.10)$$

The primary forces can, of course, be balanced at the source by adding rotating balance weights opposite each crank.*

In the case of a 90° V-8 engine with cranks set at 0°, 90°, 270°, 180°, the primary forces can be balanced in the same way, and it is very easy to check that the secondary forces and moments are inherently balanced, so the whole engine can be completely balanced.

'*V-12*' *engine*. This gives no anxiety. If the connecting-rods operate on the same cranks and the crankshaft is that illustrated in fig. 3.26 each bank of pistons will be self-balancing, so the whole engine will be balanced.

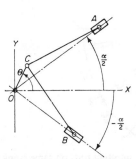

Fig. 3.28 'V-twin' engine

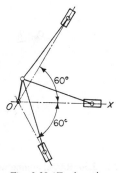

Fig. 3.29 'Fan' engine

(b) '*Fan*' *engines*

This type of engine, in which more than two banks of cylinders operate on a common crankshaft, can be investigated by methods exactly similar to those used for the V-engine. For instance, a three-cylinder engine with centre-lines disposed at 60°, as indicated in fig. 3.29, produces a rotating primary force of magnitude $\frac{3}{2} m\omega^2 r$, which can be eliminated by a balancing mass opposite the crank, but the secondary reciprocating forces of $\frac{1}{2}$ and $\frac{3}{2} \cdot \frac{1}{q} m\omega^2 r$ along XX and YY respectively are not easily balanced.

* See S. H. Grylls, *loc. cit.*

(c) '*Radial*' engine

In a very similar fashion one can extend the analysis to a radial engine with N cylinders, fig. 3.30. If we measure the crank angle θ from some arbitrary datum, such as OX, we can write down:

Primary force due to piston 1 along $01 = m\omega^2 r \cos(\theta + \alpha)$

Primary force due to piston 2 along $02 = m\omega^2 r \cos\left(\theta + \alpha + \dfrac{2\pi}{N}\right)$

and so on.

Hence, resolving along datum OX:

Total P.F. along OX

$$= m\omega^2 r \,|\, \cos(\theta + \alpha)\cos\alpha + \cos\left(\theta + \alpha + \frac{2\pi}{N}\right)\cos\left(\alpha + \frac{2\pi}{N}\right) + \ldots$$

$$\text{to } N \text{ terms } |$$

$$= \tfrac{1}{2}m\omega^2 r \,|\, \{\cos\theta + \cos(\theta + 2\alpha)\} + \left\{\cos\theta + \cos\left(\theta + 2\alpha + \frac{4\pi}{N}\right)\right\}$$

$$+ \ldots \text{to } N \text{ terms}$$

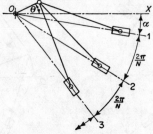

Fig. 3.30 Radial engine with all connecting rods on same crank:
cf. fig. 3.19

Clearly, the first term in each double bracket adds up directly: what about the second? The standard series is summed as follows:

Let $S_N = \cos A + \cos(A + B) + \ldots + \cos\{A + (N - 1)B\}$

then $2S_N \sin B/2 = 2\cos A \sin B/2 + 2\cos(A + B)\sin B/2 + \ldots$

$$= -\sin(A - B/2) + \sin(A + B/2)$$

$$- \sin(A + B/2) + \sin(A + 3B/2)$$

$$- \ldots \text{etc.}$$

$$+ \sin\{A + (2N - 1)B/2\}$$

$$= -\sin(A - B/2) + \sin\{A + (2N - 1)B/2\}$$

$$= 2\cos\{A + (N - 1)B/2\}\sin(NB/2) \qquad (3.11)$$

Now we see that in our particular series sin $NB/2$ has a value $\sin\left(\dfrac{N}{2}\cdot\dfrac{4\pi}{N}\right) = 0$, so unless sin $B/2$ which has a value sin $(2\pi/N)$ is also zero (which can happen only if $N < 3$ and a two-cylinder engine is not usually called a 'radial' engine), the whole series sums up to zero.

We conclude that the total unbalanced primary force along OX is

$$\text{P.F.} = \frac{N}{2}\cdot m\omega^2 r \cos\theta \qquad (3.12)$$

But note that this is along an *arbitrary* axis with which the crank makes an angle θ at the instant considered. In other words, the force is *rotating* with the crank and can be balanced by a rotating mass opposite the crank.

It is very tempting to extend the analysis to secondary forces, which, with the same assumptions, disappear altogether provided $N \not< 4$. Unfortunately such calculations, however gratifying, are utterly unrealistic, since it has not yet been found possible to allow all the connecting-rods to operate on a common crank. A master rod and articulated rods (see fig. 3.19) give a very different value of the secondary force (see § 3.3.3c): moreover, the master rod is usually much heavier than the articulated rods, so even the primary force may not be perfectly balanced.

3.3.9 Other types of engine

Clearly the field, if not inexhaustible, is far from exhausted by our discussions so far. Other types which the student may wish to consider for himself include the 'opposed twin', with cranks at 180°,

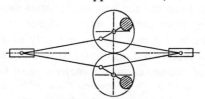

Fig. 3.31 'Lanchester' balanced engine

which would give complete balance if the pistons could be placed opposite each other; the 'flat four'; many different types of opposed-piston engine, the 'swash-plate' engine, and so on. It is perhaps a pity for historical as well as educational reasons to omit altogether an early engine designed by F. W. Lanchester,* which is sketched in

* *Proc. I.A.E.*, Vol. 8, p. 222.

fig. 3.31. It employs two crankshafts geared together and double connecting-rods: if torque is taken equally from the two shafts the primary and secondary forces *and* the torque reaction in the frame are all zero: there is even no side-thrust on the cylinder walls!

§ 3.4 Other Machinery

It would be quite wrong to leave the impression that the science or art of balancing is or should be applied only to prime movers. All that we have learnt in this chapter can be applied directly in this field, and most of the discussion of reciprocating balance has been heavily biased in this direction. The reason is primarily that engines offer a wide choice of simple well-known mechanisms to which the principles can be applied, whereas to discuss textile machinery or printing machinery or shoe-making machinery would be to limit severely the number of those able immediately to appreciate the examples chosen; and secondarily, that these other mechanisms may in themselves be rather too complicated to be so immediately appreciated.

It may be worth while in these circumstances to describe briefly and without detail particular examples of two very different machines in which balancing plays a useful part, to show how such problems can be tackled. The first machine is basically very simple. A heavy mass M in the form of a cutting head is required to reciprocate as sketched in fig. 3.32. The path of the cutting edge need not be exactly straight, but must be repeatable with extreme accuracy, and no lubricant is permissible in its vicinity: the simplest way of achieving this result is to mount the head on the arms shown. The ground vibration arising from running such machines at low speeds was barely tolerable; and it was desired not only to put some on an upper floor of a factory but also to speed them up by a factor of 3, which would, of course, cause the magnitude of the forces to increase by a factor of 9. Clearly some form of balancing *had* to be adopted: what was the best policy?

From the discussion in § 3.3 it will be appreciated that many courses are possible. Two different partial solutions were in fact adopted by competing manufacturers. The first added a mass M_1 as sketched in fig. 3.32, driven by additional cranks set at 180° to those driving M. This provided quite a good solution in that (as should be checked) the primary and secondary forces were completely balanced. Less obviously the 'arcuate' (as opposed to straight-line) motion of both masses requires horizontal forces: it is a good exercise to work out their type and magnitude, noting that at the ends of the stroke the mass requires to be given a component horizontal acceleration, while at the middle it requires to be given a horizontal centripetal acceleration. From such simple considerations it can easily be shown that the force is 'secondary' in frequency, and that its amplitude is $M\omega^2 r^2/R$, i.e. r/R times the primary force required to accelerate the mass M, where r is the crank radius and R the length of the arm supporting M.

The second manufacturer, whose course was affected by the

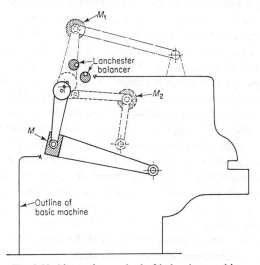

Fig. 3.32 Alternative method of balancing machine

patent situation, added instead a mass M_2 as sketched, driven from the original connecting-rods by articulated rods, and rotating balance masses on the crankshaft. This (cf. § 3.3.8(*a*)) gave primary balance, and the obvious secondary forces (including the effect of the articulated rods) were eliminated by a small Lanchester balancer as sketched (again cf. §§ 3.3.4 and 3.3.8(*a*)). Again the secondary forces due to the arcuate motions were left unbalanced. In fact, this force arising from M_2 was larger than the other similar forces arising from M and M_1, but as it passed closer than the latter to the C.G. of the machine, it was not in effect more troublesome.

Neither of these machines was then perfectly balanced, and it would have been unduly expensive to eliminate the small remaining secondary forces. The remainder of the practical solution consisted of mounting the machines on carefully designed flexible supports – a topic which will be discussed at some length in § 5.5 – when the force transmitted to the floor from the high-speed machines was immensely smaller and less disturbing than that from the original low-speed machines; in fact, it could be accurately described as inappreciable.

The second example is undeniably complicated, but quite typical of the sort of machine with which the engineer may in practice be confronted. It is in fact a textile knitting machine, of the 'flat' type, in which the knitting elements – many thousands of needles, guiding members for the yarn and the completed cloth, and the like – are carried in close array on 'rigid' bars perhaps five or more metres in length, and required to move in precisely controlled paths in order to effect the knitting action. Nothing of this concerns us at the moment, except in so far as it presents a background to our problem of how to achieve in a machine of this type a state of balance which will be sufficiently good to allow the machine to run at the highest speed of which it is capable, without causing annoyance to the operator or others because of any vibration it produces.

The diagrammatic sketches given in fig. 3.33 are in one sense grossly over-simplified, but they show the parts of the machine which affect us. In (*a*) we have an end view or cross-section of the four bars *A*, *B*, *C*, *D*, which carry the knitting elements, simplified to rectangular cross-sections, and a set of driving mechanisms which is repeated at intervals along the length of the bars. Inevitably in this view these mechanisms obscure each other, so in (*b*) a pictorial view of one set is given which may help to clarify the situation. Normally in such a machine the motions of the elements are derived from cams which can, of course, be designed to give, at least at a low machine speed, exactly what the knitting experts want. This particular machine was, however, being designed *de novo* with a high speed in mind. When attention was particularly directed to this situation it was found that an adequate approximation to the required motion of each of these four bars independently could be obtained from a compound harmonic motion involving only the first and second harmonic terms: it was therefore decided to design the machine with two crankshafts *E* and *F*, one running at 'machine speed', the other at twice this speed, and to combine the harmonic

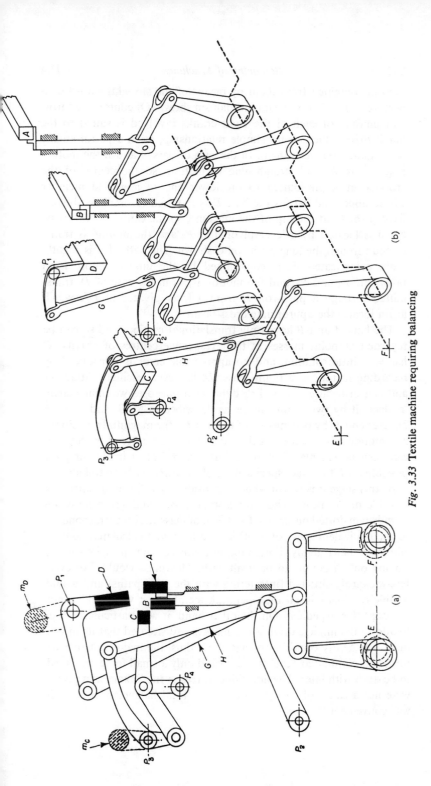

Fig. 3.33 Textile machine requiring balancing

motions obtained from them by means of the straightforward connecting-rod and cross-beam mechanism shown. Of course, the throw
and phasing of each of the eight cranks involved is suited to the
bar it drives: there is no simple relationship, as in a petrol engine
for instance, between them. In the case of bars A and B the motion
required is vertical straight-line oscillation, so the centre of the
cross-beam is pin-jointed to the lower end of a guided member
whose upper end carries the bar. D is required to oscillate about a
fixed pivot P_1 at the top of the machine, so the centre of the cross-
beam is located by a lever pivoted at P_2 and the motion is transmitted through the long push-rod G shown. The path of C is slightly
more complicated: it is constrained by the linkage whose fixed
pivots are P_3 and P_4, and the motion is, as in the case of D, transmitted through a push-rod H from a lever pivoted at P_2', which
again locates the appropriate cross-beam.

The bars A and B have only translational motion, and so can be
regarded as point masses at their respective centres of gravity. C
has very little rotation, and no appreciable error is introduced by
regarding it, too, as a point mass. D, however, oscillates about P_1
and is a distributed mass. For dynamical purposes we immediately
'replace' it by two point masses as described in § 1.9, and exercise
our free choice by placing one of them at P_1: the magnitude, position,
and motion of the second at the corresponding 'centre of percussion'
can then be calculated, and we have immediately in the simplest
possible form the exact specification of the force we have to balance.

At this stage it is worth while to consider what 'local' balancing is
possible or desirable. Thus, for instance, we could add a mass m_D
as shown in dotted outline on fig. 3.33a, of mass-radius corresponding
to the equivalent mass of D which we have just calculated, and so
eliminate the (roughly) horizontal unbalanced force near its source,
leaving only a couple to be dealt with. This mass would, however,
be awkwardly situated (interfering with the yarn paths) and would
double the loads in all the operating members: it was decided not
to adopt this expedient. On the other hand, it was particularly easy
to add to the linkage driving C a mass m_C so situated that its movement was nearly opposite to that of C: thus, the major unbalanced
force was eliminated near its source and only a small force remained
to be dealt with later. The additional forces in the linkage and pivots
were not a cause of embarrassment. No local balancing of A or B
was convenient.

We can now take stock of the situation. Each bar, and indeed each element of the machine, has a motion of the type

$$x = a \cos (\theta + \alpha) + b \cos (2\theta + \beta)$$

where θ is the primary crankshaft angle, measured from some arbitrary datum, and α and β are the phase angles required to give the appropriate movement to the element. For our present purposes we convert these to the form

$$x = c \cos \theta + d \sin \theta + e \cos 2\theta + f \sin 2\theta$$

hence $\qquad \ddot{x} = -\omega^2(c \cos \theta + d \sin \theta + 4e \cos 2\theta + 4f \sin 2\theta)$

The force $m\ddot{x}$ to accelerate the element will in general have horizontal and vertical components (though, of course, in the case of masses A and B the former are zero) and a moment about some arbitrary datum we can choose, preferably near the centre of gravity of the whole machine. There will therefore be altogether twelve terms in the whole expression, but although this may sound rather a lot, it is only the coefficients which concern us, and these may be tabulated under the headings

	HORIZONTAL FORCES				VERTICAL FORCES	MOMENTS
	$\cos \theta$	$\sin \theta$	$\cos 2\theta$	$\sin 2\theta$	$\cos \theta$, etc.	$\cos \theta$, etc.
Component						

The same procedure is followed for each bar, partial balance mass, and element of the driving machinery. Each of these twelve components of the unbalanced force can then be added up by simple arithmetic to give the resultant for the whole machine.

So much for the science; the rest belongs at least in part to the realms of art. To balance completely all these twelve components would require the provision of more independently driven masses than the results could be expected to justify. The compromise adopted was to eliminate the relatively small horizontal forces by 'conceptual' rotating masses on each shaft (we shall return to this again), which, of course, introduced corresponding vertical forces (see § 3.3.4); and to add to the machine two vertically moving balance masses driven, one from the primary shaft and one from the secondary shaft, by special cranks set at the appropriate phase angle to give the correct ratio between cosine and sine terms. Thus, all horizontal and vertical forces were completely eliminated. It would, of course, have been possible to combine these motions, but the separate-mass solution was mechanically simpler. It would further have been possible to set the balance masses eccentrically to the C.G. of the machine and so to produce an effective moment, but space was limited and the possible gains were small, since the phasing was, of course, inappropriate. Thus, the moments were in fact not balanced, but since they were small in relation to the moment of inertia of the machine, their effect was regarded as likely to be inappreciable. Finally, the rotational balance of each shaft, including the 'conceptual' masses referred to above, was achieved by the simple methods outlined in § 3.2.5.

No doubt other compromises could have been adopted; the one described was chosen from several possibilities which were considered. This is what is meant by the reference to the 'art' of balancing a machine. Further, it ought to be admitted – or pointed out – that the fact that all the motions were of the two-component harmonic type was an immense simplification; had the motions been less regular, so good a state of balance could not easily have been achieved. Although in fact design modifications in the machine after the crank angles for the balancing masses had been settled made the final compromise somewhat less good than that suggested by this description, the eventual result was a machine on which it was literally possible to balance coins on edge while it was operating at full speed.

Examples

3.1 Three masses, of 5, 7, and 12 kg, are attached to a shaft in the same transverse plane at radii 30, 45 and 25 mm and at angular positions 0, 52°, and 220° respectively. The shaft is supported in a bearing on each side of this plane, distant 200 mm and 225 mm from it. If the shaft rotates at 300 rev/min, find the out-of-balance force on each bearing.

Find the magnitude and angular setting of the single mass at 40 mm radius which will balance this shaft.

3.2 A shaft carries five pulleys each 1 m apart. The pulleys are out of balance to the following extent:

     (1) 5 kg at 10 mm rad. at 0°
     (2) 6   20   45°
     (3) 7   10   90°
     (4) 2   20   120°
     (5) 6   10   240°

Find:

 (*a*) the out-of-balance forces on bearings 0·5 m outside the extreme pulleys when the shaft speed is 250 rev/min;

 (*b*) the masses required to balance the shaft if placed in pulleys (1) and (5);

 (*c*) the masses required to balance the shaft if placed in pulleys (2) and (4).

3.3 Four pulleys are equally spaced along a shaft, and each has an out-of-balance mass at the same radius. The out-of-balance mass in the second pulley is 3 kg, and the third and fourth out-of-balance masses are at 72° and 220° to it.

Determine the masses in the first, third, and fourth pulleys, and also the angle of the first mass relative to the second if complete balance is to be obtained.

3.4 A rotor is to be balanced by the four-vector method. When it is pivoted about one reference plane XX the vibration amplitude observed is 1·55 mm. A mass of 54 g at 120 mm radius is then added in the YY reference plane in three successive positions inclined at 120° to one another, and in each state the amplitude of vibration is measured. The amplitudes are 1·96, 2·30, and 0·755 mm. Find the value of the initial out-of-balance mass radius in the YY plane and its direction relative to the first position in which the additional mass was fixed.

3.5 Particulars of a four-cylinder marine steam engine are tabulated below. Determine the crank angles and fourth reciprocating mass required to balance all primary forces and moments, and find the value of the unbalanced secondary forces and moments at a crankshaft speed of 100 rev/min. The stroke is 1 200 mm and the connecting-rod/crank ratio is 3·9.

CYLINDER	RECIPROCATING MASS (TONNES)	DISTANCE FROM H.P. CYLINDER
L.P. 1	6·6	12 800 mm
L.P. 2	7·0	8 530 mm
I.P.	6·3	4 570 mm
H.P.	?	0

3.6 A Schlick 'symmetrical' marine steam engine is to have a reciprocating mass of 3·4 tonnes in each outer cylinder. The pitch of the inner pair of cylinders is 4·5 m and of the outer pair 8·4 m. Find the crank angles and the reciprocating mass in each inner cylinder in order that the primary and secondary forces and primary moments may be balanced.

3.7 A four-cylinder two-stroke compression-ignition engine has cranks, taken in order, set at 0°, 180°, 90°, and 270°, and runs at 300 rev/min. Each piston has a mass of 145 kg and each connecting-rod, of which two-thirds may be assumed to be balanced as a revolving mass, 120 kg. The stroke of the engine is 380 mm, the connecting-rods are 950 mm long, and the cylinders are spaced 500 mm apart. Find the unbalanced forces and moments on the foundations.

3.8 Compare the balance of an ordinary six-cylinder petrol engine and an eight-cylinder engine in which the cranks are set at 0°, 90°, 180°, 270°, 270°, 180°, 90°, 0°. The cylinders are all equally spaced.

3.9 A five-cylinder oil engine has cranks at 0°, 144°, 288°, 72°, and 216°. The reciprocating parts for each cylinder weigh 3 kgf, the stroke is 120 mm, and the connecting-rods are 260 mm long. The distance between the cylinder centre lines is 125 mm. Find the unbalanced forces and moments in the engine at a crankshaft speed of 2 200 rev/min.

3.10 A V-twin engine with the cylinders in the same plane has an angle between the banks of 90°. The mass of the reciprocating parts is 0·68 kg per cylinder, the crank radius 45 mm, and the connecting-rod 150 mm.

If the engine speed is 2 500 rev/min, find from first principles the unbalanced forces acting on the frame of the engine, show how these forces may be completely balanced and find the value of each balancing mass-radius required.

3.11 An eight-cylinder engine is built up of two banks of four cylinders, which are inclined to one another at an angle of 90°. The crank is 57 mm, connecting rod 228 mm, the reciprocating mass per cylinder 1·6 kg, and the pitch of the cylinders is 120 mm. Find the unbalanced forces and couples at a speed of 2 500 rev/min if the crankshaft throws are arranged (a) at 0°, 180°, 180°, 0°, and (b) at 0°, 90°, 270°, 180°.

3.12 In a three-cylinder radial engine all the connecting-rods operate on the same crank. The mass of the reciprocating parts is m per cylinder, the crankshaft speed is ω rad/s, the stroke is $2r$, and the connecting-rods are qr in length. Opposite the crank is a balance weight equivalent to $1·5m$ at crank radius. Accepting that the primary inertia forces are balanced, and considering the secondary forces only, find the magnitude of the unbalanced force in a plane at an angle α to one of the cylinders.

3.13 A projected design of a two-stroke oil engine envisages opposed pistons in each cylinder driving separate gear-coupled crankshafts which are to rotate at the same speed in opposite directions but (for the sake of the valve port timing) with a phase difference of 20°. Determine, in terms of the reciprocating mass M associated with each piston, the crank throw r, the connecting-rod/crank ratio q, and the crankshaft speed ω, the unbalanced forces in one cylinder

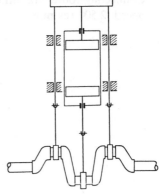

Fig. 3.34 Doxford engine

unit. If three such units are used in the complete engine, the cranks being arranged at 120° to each other, what are the total unbalanced forces and moments?

3.14 A 'Doxford' opposed-piston two-stroke oil engine, as sketched in fig. 3.34, has four cylinders in line. The lower pistons are driven by main cranks, in order along the shaft, at 0°, 90°, 270°, 180°; the upper pistons by cranks at 180° to the corresponding main cranks. The cylinders are at 2 286 mm pitch; the engine speed is 240 rev/min. Other particulars of the engine are given below. Investigate the balance of a single-cylinder unit, and of the complete engine.

	LOWER PISTON	UPPER PISTON
Stroke (mm)	1042	762
Connecting-rod/crank ratio	3·75	7
Reciprocating mass (tonnes)	3·43	4·70

3.15 A three-cylinder engine is arranged with all the pistons operating on a common crank, and the angle between the cylinder centre lines 60°. Show, with proof, how the primary inertia forces due to the reciprocating parts may be balanced, and find the maximum value of the secondary force in the case of the engine for which particulars are given.

Reciprocating parts per cylinder, 3 kg
Stroke, 125 mm
Connecting-rod/crank ratio, 3·5
Speed, 2 500 rev/min

4. Free Vibrations

§ 4.1 Introduction

Although practical difficulties attributable to mechanical vibrations have become of major importance only relatively recently, the mathematical techniques for dealing with vibrations have long been established: the main theoretical foundations of the subject were laid down in a book by Lord Rayleigh entitled *Theory of Sound*, published in 1877. As machine speeds and design stresses increase, so the dangers of serious trouble due to mechanical vibrations are intensified, and successful design becomes even more dependent on the understanding, measurement, and partial (if not complete) elimination of vibrations.

Any system in which masses are coupled to each other by non-rigid members is capable of vibrating. If relative motion between these masses gives rise to restoring forces which are linear functions of the relative displacement, such as occur with elastic coupling members or which can arise as a result of gravitational forces as in a pendulum, then vibration may occur at one or more frequencies. A vibration such as this is described as a *free vibration*, and the frequencies are called the *natural frequencies of vibration* of the system. Forces such as 'viscous' or 'frictional' forces which occur in the system as the result of the motion, and which oppose the motion, are called *damping* or *dissipative* forces.

The vibrations which give rise to trouble in practice can be divided into two groups, which are described as *forced* and *self-excited*. For example, forced vibrations occur when a harmonic disturbance is imposed on a system: if the frequency of this imposed or 'exciting' disturbance approximates to any one of the natural frequencies of the system the magnitude of the resulting vibration increases. If the damping forces are sufficiently small and the exciting frequency coincides with a natural frequency – a critical condition which is termed *resonance* – the vibration will grow to a catastrophically large value. The elimination of all major resonances is one of the main objectives of a designer, since, even if the damping forces are such that failure, either in a short time or more usually by fatigue, does not occur, the resonance will frequently be responsible for the generation of excessive noise or intolerable vibration of the system or surroundings.

In a self-excited vibration the exciting force is not imposed as a harmonic disturbance independent of the motion, but itself arises as a consequence of, and varies with, the motion. It follows fairly obviously that a self-excited vibration occurs at one of the natural frequencies of vibration of the system. As examples of such vibration we may quote 'flutter' in aircraft, where the exciting force is aerodynamic in origin, 'oil-whip' in shafting, where it is hydrodynamic, and 'stick–slip' vibrations, where it is frictional.

In most vibration problems one of the first steps to be taken is therefore to determine, either analytically or experimentally, the natural frequencies.

Usually the stiffnesses in the system being considered give, accurately or to a first approximation, a linear relationship between force and displacement. There are, however, many examples where this is not even approximately true, and in these cases the analysis is more difficult. In such 'non-linear' systems the natural undamped frequency depends on the amplitude, but it is found that the effect of resonance is less serious.

In this chapter we shall consider only the natural undamped

frequencies of various linear systems. The effects of damping on free vibrations, and the vibration of systems subjected to damping and harmonic exciting forces, or to damping forces and harmonic displacement, will be considered in chapter 5. Sources of harmonic excitation and examples of self-excited vibration will also be briefly discussed in that chapter. No attempt will be made to deal with nonlinear systems, which are considered in more advanced books to which references are given at the end of chapter 5.

§ 4.2 Degrees of Freedom

The number of degrees of freedom of an elastic system is the number of parameters required to define its configuration. As outlined in § 1.2, a rigid body in space has six degrees of freedom, three of translation and three of rotation. If such a body is attached to an 'infinite' mass, that is to a body which may be regarded as 'fixed', by a flexible connection which elastically constrains the body in all six degrees, then the body can vibrate in six distinct ways or *modes* or any combination of these modes. For example, fig. 4.1a shows a simple system in which a mass is connected to a rigid foundation by a massless flexible member. The arrangement shown has been selected to provide a degree of symmetry sufficient to avoid 'coupling' between the possible modes of vibration. Thus, the mass may vibrate in the direction of any one of three mutually perpendicular axes (figs. 4.1b, c, d) without having any component of motion in either of the other directions or any rotational motion; or rotationally about any one of these axes (figs. 4.1 e, f, g) without any translational or other rotational motion.

Usually, in order to make a problem more tractable, it is assumed that there is no interaction or coupling between the vibrations occurring along and about the three axes. In some problems, however, this assumption is not even approximately true, and it is then necessary to introduce 'coupling constants' into the equation of motion. An example of this will be considered later.

In a system such as this, or indeed in a much more complicated system, only three types of these motions are usually of much practical significance; the modes of vibration in the z direction, which are known as *axial* or *longitudinal vibrations*, rotational

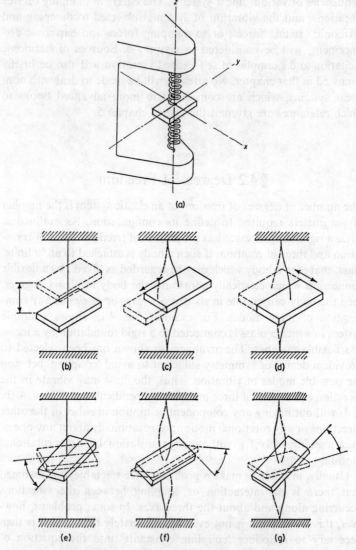

Fig. 4.1 Modes of vibration of a system with six degrees of freedom

modes (e.g. about the axis of a shaft), which are termed *torsional vibrations*, and the translational modes in the x or y direction, which are referred to as *transverse vibrations*. We shall soon appreciate that it is a waste of time to consider axial and torsional vibrations as separate problems, since the equations governing the motion of the masses in either case are similar. Consequently, in the following paragraphs we shall normally consider in detail *either* the axial *or* the torsional vibrations of the system under examination, and write down the corresponding solution for the other case. In general, though there are many exceptions, axial vibrations encountered in engineering occur relative to the earth, which may be regarded as an infinite mass belonging to the system. When such a mass is involved we shall therefore choose the axial case for detailed consideration. Torsional vibrations, however, more frequently occur in systems such as that illustrated in fig. 4.5(a), in which the bearings effectively isolate the system from the earth, so that when no infinite mass is introduced we shall consider preferentially the torsional case. With transverse vibrations of beams the problem is basically different from torsional or axial vibrations, as will be explained in § 4.11.

In a general system with n finite masses elastically coupled in series by $(n - 1)$ stiffnesses, $6n$ degrees of freedom are possible for the masses. Six of these arise from the need to define the position of one of these masses in space, and the others from the need to define the relative position of the other masses. Consequently, only $6(n - 1)$ degrees of freedom involve elastic constraints, so there are only $6(n - 1)$ modes of vibration. These modes are equally divided among the three translational and three rotational motions.

§ 4.3 Vibration of a Two-mass System, One Mass being Infinite

Since one mass is infinite, it may be regarded as a fixed support to which the other is attached by an elastic member. This is a problem dealt with in elementary mechanics, but it is of such fundamental importance that it will be repeated here.

Fig. 4.2a shows a spring-mass system in which a mass, m, is suspended from an infinite mass by a spring in which a force λ causes unit deflection; λ is termed the *stiffness* of the spring. It is assumed that the mass is so guided that only motion in the direction of the axis of the spring is possible. Under equilibrium conditions the gravitational force on the mass is balanced by the force in the spring as shown in fig. 4.2b. If the mass is displaced upwards (which following normal mathematical conventions, see § 0.2, will be chosen as positive) and released, then at some instant it will be in the position shown in fig. 4.2c, and the forces acting on the mass will be a downward gravitational force, mg, and an upward spring force of $mg - \lambda x$; these forces are shown in fig. 4.2d. The net force is $-\lambda x$, the negative sign indicating that the force acts in a downward

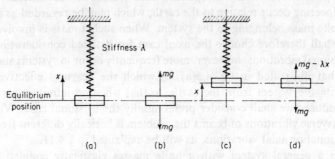

Fig. 4.2 Spring-mass system

direction. Since such a force tends to restore the mass to its equilibrium position, it is usually called a *restoring force*. This restoring force is clearly unaffected by the constant gravitational force. Applying Newton's second law to this system we have

$$m\ddot{x} = -\lambda x$$

or

$$\ddot{x} = -\frac{\lambda}{m}x \tag{4.1}$$

(λ and m being essentially positive quantities) from which we see that the acceleration of the mass is a negative quantity times its displacement from its equilibrium position. This may be recognised immediately as the characteristic of simple harmonic motion (cf. § 0.14), and the 'radiancy' * of the system, or the angular speed of the vector in the auxiliary circle, may be written down as

$$\omega = \sqrt{\frac{\lambda}{m}} \tag{4.2}$$

In dealing with vibrations we shall be involved with so many angular speeds, and this value is of such fundamental importance, that we shall give it the particular symbol ω_n, which stands for the 'natural radiancy of vibration' of a system.

* See § 0.14 for the definition of this 'coined' word.

Alternatively, we may recognise equation (4.1) as a linear differential equation of the second order with constant coefficients and write its general solution either as

$$x = C_1 \sin \omega_n t + C_2 \cos \omega_n t$$

in which ω_n is as before $\sqrt{(\lambda/m)}$ and C_1 and C_2 are arbitrary constants, which depend upon starting conditions, such as displacement and velocity at zero time, or as

$$x = R \cos (\omega_n t + \phi) \tag{4.3}$$

where R and ϕ are now the arbitrary constants. It is easy to see that

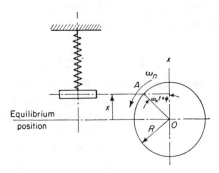

Fig. 4.3 Auxiliary circle

equation (4.3) represents a simple harmonic motion in which the amplitude or maximum displacement is R, and ω_n is the angular velocity of the vector in the auxiliary circle which is shown in fig. 4.3.

From the value of ω_n the frequency f_n and periodic time t_p can, of course, be immediately obtained, since the frequency or *number of cycles per unit time* is the same as the number of revolutions of the vector in the auxiliary circle. The periodic time or period, i.e. the time for one complete vibration cycle, is the reciprocal of the frequency, so

$$f_n = \frac{1}{2\pi} \cdot \omega_n = \frac{1}{2\pi} \sqrt{\frac{\lambda}{m}} \tag{4.4}$$

$$t_p = \frac{2\pi}{\omega_n} = 2\pi \sqrt{\frac{m}{\lambda}}$$

These relationships can be expressed in terms of the static deflection, δ_s, of the spring caused by the gravitational force on the mass, and the gravitational constant g. Such a form of the results is frequently convenient, and it helps to drive home a simple but vitally important physical fact. This deflection is

$$\delta_s = \frac{mg}{\lambda}$$

$$\therefore \ \omega_n = \sqrt{\frac{g}{\delta_s}}; \quad f_n = \frac{1}{2\pi}\sqrt{\frac{g}{\delta_s}}; \quad t_p = 2\pi\sqrt{\frac{\delta_s}{g}}$$

Similar equations for the frequency of other modes of vibration in a suitable system, such as the translational modes in fig. 4.1, can be deduced in an identical manner: in each case the radiancy and frequency are given by equations (4.2) and (4.4), where λ is the total restoring force acting on m for unit displacement. Again the radiancies and frequencies of the rotational modes can be written down by analogy as

$$\omega_n = \sqrt{\frac{k}{I}}; \quad f_n = \frac{1}{2\pi}\sqrt{\frac{k}{I}} \qquad (4.5)$$

where I is the appropriate moment of inertia of the mass and k is the torsional stiffness of the element, that is the restoring torque acting on the mass for an angular displacement of one radian.

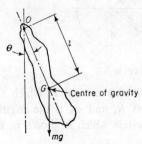

Fig. 4.4 Compound pendulum

We may also briefly consider a case in which the restoring force or couple is the result of a gravitational force, and an elementary example is afforded by the compound pendulum shown in fig. 4.4. In the position shown the gravitational force gives rise to a restoring torque about O of $mgl \sin\theta$, which if θ is small reduces to $mgl\theta$ as $\sin\theta \simeq \theta$. Applying Newton's second law to the rotational motion about O gives

$$I\ddot{\theta} = -mgl\theta$$

where I is the moment of inertia about an axis through O and can be expressed as $mk_G^2 + ml^2$ when k_G is the radius of gyration about the centre of gravity. This is a similar equation to (4.1), and we can therefore write down the following equations for radiancy, frequency, and periodic time:

$$\omega_n = \sqrt{\frac{gl}{k_G^2 + l^2}}, \quad f_n = \frac{1}{2\pi}\sqrt{\frac{gl}{k_G^2 + l^2}}$$

$$T = 2\pi\sqrt{\frac{k_G^2 + l^2}{gl}}$$

The corresponding values for a simple pendulum will be remembered as

$$\omega_n = \sqrt{\frac{g}{l}}, \quad f_n = \frac{1}{2\pi}\sqrt{\frac{g}{l}}, \quad T = 2\pi\sqrt{\frac{l}{g}}$$

From any one of these corresponding pairs it is at once seen that the equivalent simple pendulum, that is a massless rod with a concentrated mass at its end having the same period, will have a length of

$$\frac{k_G^2 + l^2}{l}$$

(cf. § 1.9).

§ 4.4 Vibration of a Two-mass System, Both Masses Finite

Consider the system shown in fig. 4.5a, which consists of two flywheels, I_1 and I_2, coupled by a uniform massless shaft of torsional stiffness k. Such a system might represent, for instance, a marine engine, shafting, and propeller. The flywheel I_1 represents the engine crankshaft and flywheel, plus, perhaps, an arbitrary allowance of part of the moment of inertia of the propeller shaft, and the other flywheel I_2 is the moment of inertia of the propeller plus part of that of the shaft.

Although in practice such a system would normally be rotating, the vibration is unaffected by this rotation, which is superimposed on it. We could, in fact, consider the motion of each element of the system relative to axes rotating with it, but it is more straightforward to consider the vibration of a non-rotating shaft.

Just as we assumed in § 4.3 that the single mass was given an initial displacement and then released, so in this system we may consider that the shaft is initially twisted with both flywheels at rest, and that they are simultaneously released. At this instant the angular momentum of the system is zero, and as no external torque is applied, we know from the principle of conservation of angular momentum that it must remain zero. This implies that the two flywheels must at all times be moving in opposite directions; they will come to rest at the same instant; in other words, the radiancy and frequency of vibration of each flywheel must be the same. We can therefore regard them as those for the system, and use the previous notation ω_n and f_n for these quantities.

If we now measure time and the angular rotations of the fly-wheels θ_1 and θ_2 from the instant at which the shaft is untwisted we can write down the momentum equation

$$I_1\dot{\theta}_1 + I_2\dot{\theta}_2 = 0$$

and integrating this equation with respect to time

$$I_1\theta_1 + I_2\theta_2 = 0$$

the arbitrary constant being zero. Thus, at any time

$$\frac{\theta_1}{\theta_2} = -\frac{I_2}{I_1}$$

that is, the ratio of the angular displacements is constant (the negative sign of course indicating opposite directions of rotation). This

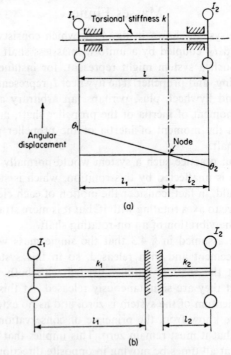

Fig 4.5 Two-mass system

implies that some transverse section of the shaft is stationary: this section is referred to as a *node*. It is therefore permissible to regard this as a rigidly fixed section of the shaft as indicated in fig. 4.5b and to consider each part of the system as in § 4.3. Consequently, we can write down

$$\omega_n = \sqrt{\frac{k_1}{I_1}} = \sqrt{\frac{k_2}{I_2}}; \quad \text{so} \ \frac{k_1}{I_1} = \frac{k_2}{I_2}$$

where k_1 and k_2 are the stiffnesses of the two lengths l_1 and l_2 of the shaft, which are not explicitly known.

In a uniform shaft subjected to a torque the twist is proportional to the length; in other words, the stiffness is inversely proportional to the length. We can therefore write

$$k_1 l_1 = k_2 l_2 = kl$$

where k is the stiffness of the whole shaft.

$$\therefore \ \frac{l_2}{l_1} = \frac{k_1}{k_2} = \frac{I_1}{I_2}; \quad \text{so} \ \frac{l}{l_1} = \frac{l_1 + l_2}{l_1} = \frac{I_1 + I_2}{I_2}$$

$$\therefore \ \omega_n = \sqrt{\frac{k_1}{I_1}} = \sqrt{\frac{kl}{I_1 l_1}} = \sqrt{\left[k \frac{(I_1 + I_2)}{I_1 I_2} \right]} = \sqrt{\left[k \left(\frac{1}{I_1} + \frac{1}{I_2} \right) \right]}$$

and

$$f_n = \frac{1}{2\pi} \sqrt{\left[k \left(\frac{1}{I_1} + \frac{1}{I_2} \right) \right]} \tag{4.6}$$

If the shaft between the two flywheels is made up of several shafts or flexible couplings with individual stiffnesses $k_1, k_2, \ldots k_n$, then these shafts may be replaced by a single shaft of equivalent stiffness k_e. This equivalent shaft must have the same overall stiffness; i.e. the angular deflection caused by a torque T must be the same for both the actual and equivalent shafts; this leads to

$$\frac{T}{k_e} = \frac{T}{k_1} + \frac{T}{k_2} + \cdots \frac{T}{k_n}$$

or

$$\frac{1}{k_e} = \frac{1}{k_1} + \frac{1}{k_2} + \cdots \frac{1}{k_n} = \sum \frac{1}{k_n}$$

The system can then be considered in exactly the same way as the above problem, except that k is replaced by k_e.

Correspondingly, the natural radiancy and frequency of a two-mass axial system can be written down as

$$\omega_n = \sqrt{\left[\lambda \left(\frac{1}{m_1} + \frac{1}{m_2} \right) \right]} \tag{4.7}$$

$$f_n = \frac{1}{2\pi} \sqrt{\left[\lambda \left(\frac{1}{m_1} + \frac{1}{m_2} \right) \right]}$$

§ 4.5 Vibration of Multi-mass Systems, General Considerations

In the case of a two-mass system it was easy to demonstrate mathematically that the masses must be vibrating in opposite phases with the same frequency. Indeed, such a demonstration might well be regarded as superfluous, since it is physically obvious that if any other condition obtained the vibration would progress in such a way that the masses would in time be moving in phase with each other; the vibration would no longer exist, and the conservation of angular momentum would not have been maintained. In the case of a multi-mass system such a simple mathematical demonstration is not possible, since the total motion may well be a combination of several modes. It is nevertheless equally obvious physically that for the case of any persistent single mode of vibration the frequency of each mass must be the same, and that the phase of each is either the same or exactly opposite, a condition which is covered by a change in sign of the amplitude. We shall therefore deal with multi-mass systems by considering a single mode at a time, when we can legitimately make the assumption that the frequency of vibration of each element is that of the system as a whole; but we must remember that the various modes may in practice occur simultaneously, when the individual motions will be superimposed.

§ 4.6 Three-mass System, All Masses Finite

The system is shown diagrammatically in fig. 4.6. We assume that it is vibrating torsionally in a single mode and that the radiancy of the system is ω_n. At some instant let the deflections from the equilibrium position of the three flywheels I_1, I_2, and I_3 be θ_1, θ_2, and θ_3.

The only torque to which the first flywheel is subjected is the torque in the first shaft, of stiffness k_1, but the angle of twist in this shaft is $(\theta_1 - \theta_2)$ and we know that the torque must be a restoring torque: we can therefore apply Newton's second law and write down

$$I_1 \ddot{\theta}_1 = -k_1 (\theta_1 - \theta_2)$$

The second flywheel is subjected to an equal and opposite torque from this shaft, and to a torque from the second shaft of stiffness

k_2 which is twisted through an angle $(\theta_2 - \theta_3)$. We can therefore write

$$I_2\ddot\theta_2 = +k_1(\theta_1 - \theta_2) - k_2(\theta_2 - \theta_3)$$

and, since the third flywheel is subjected only to the torque in the second shaft,

$$I_3\ddot\theta_3 = +k_2(\theta_2 - \theta_3)$$

Now if the radiancy of vibration of the system is ω_n and the amplitudes of vibration of the three flywheels are a_1, a_2, and a_3 respectively,

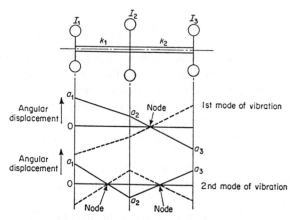

Fig. 4.6 Three-mass system; all masses finite

and if we measure time from the instant at which the shaft is in the undistorted condition we can write

$$\theta_1 = a_1 \sin \omega_n t$$
$$\theta_2 = a_2 \sin \omega_n t$$
$$\theta_3 = a_3 \sin \omega_n t$$

and substituting these values in the three equations of motion and simplifying:

$$-I_1a_1\omega_n{}^2 = -k_1(a_1 - a_2)$$
$$-I_2a_2\omega_n{}^2 = +k_1(a_1 - a_2) - k_2(a_2 - a_3)$$
$$-I_3a_3\omega_n{}^2 = +k_2(a_2 - a_3)$$

These equations apparently contain four unknowns: $\omega_n{}^2$, a_1, a_2, and a_3. However, on further consideration it will be realised that the absolute value of the amplitudes is dependent solely on the starting conditions, and that what we require to determine are the ratios of the amplitudes which define the shape of the dynamic deflection curve. From the first and third of these equations

$$a_1 = \frac{k_1 a_2}{k_1 - I_1 \omega_n^2}$$

$$a_3 = \frac{k_2 a_2}{k_2 - I_3 \omega_n^2}$$

and by substituting these into the second and rearranging the result

$$\omega_n^4 - \omega_n^2 \left\{ k_1 \left(\frac{1}{I_1} + \frac{1}{I_2} \right) + k_2 \left(\frac{1}{I_2} + \frac{1}{I_3} \right) \right\}$$

$$+ k_1 k_2 \left\{ \frac{1}{I_1 I_2} + \frac{1}{I_2 I_3} + \frac{1}{I_3 I_1} \right\} = 0 \quad (4.8)$$

From the form of this equation it is clear that we can obtain two real values of ω_n^2, which agrees with the argument put forward in § 4.2 that this system would have two degrees of torsional freedom and so two modes of vibration. If we now substitute these values of ω_n^2 in the equations above for a_1 and a_3 we shall find two different values of these amplitudes in terms of a_2, and from the sign of these relative amplitudes we shall find that the lower value of ω_n^2 corresponds to a mode of vibration with a single node, and the higher to a vibration with a node in each shaft. These two modes are illustrated in fig. 4.6.

It is perhaps worth repeating that in practice the different modes of vibration may, and often do, occur simultaneously, and if the system were disturbed in a random fashion this would be the case. Only if the starting conditions were carefully arranged to agree with those of either of the modes, for instance by adjusting the relative amplitudes of the flywheels before release, would the system vibrate in a single mode. When we come to consider forced vibrations we shall find that other considerations determine the behaviour of the system.

Again it is wholly unnecessary to go through the similar calculations for a three-mass axial system. The results can be written down by analogy; for instance, to find the natural radiancy of vibration we have to solve the equation

$$\omega_n^4 - \omega_n^2 \left\{ \lambda_1 \left(\frac{1}{m_1} + \frac{1}{m_2} \right) + \lambda_2 \left(\frac{1}{m_2} + \frac{1}{m_3} \right) \right\}$$

$$+ \lambda_1 \lambda_2 \left\{ \frac{1}{m_1 m_2} + \frac{1}{m_2 m_3} + \frac{1}{m_3 m_1} \right\} = 0 \quad (4.9)$$

§ 4.7 Three-mass System, One Mass Infinite

This case can be solved by inspection of equations (4.8) or (4.9). If the centre mass were infinite we should, of course, have two simple isolated systems, which is a trivial case. If one of the other masses,

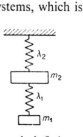

Fig. 4.7 Three-mass system; one mass infinite

say I_3 or m_3, is infinite we obtain a result which is of importance, particularly in the axial case sketched in fig. 4.7.

If, then, in equation (4.9) we make m_3 infinite

$$\omega_n^4 - \omega_n^2 \left\{ \lambda_1 \left(\frac{1}{m_1} + \frac{1}{m_2} \right) + \frac{\lambda_2}{m_2} \right\} + \frac{\lambda_1 \lambda_2}{m_1 m_2} = 0 \qquad (4.10)$$

which again clearly gives two real values for ω_n^2.

§ 4.8 The Vibration of Multi-mass Systems

The method of solution adopted in the previous section can be used for systems with a larger number of masses and stiffnesses, but the solution of the simultaneous equations becomes progressively more tedious unless a computer is available, and the chances of an arithmetical mistake are considerable. The natural frequencies of a system with a large number of masses can be much more readily found by using a numerical method frequently attributed to Holzer (1921), though first suggested by Gümbol (1901).

Fig. 4.8a shows a four-mass system and the associated notation which will be used to demonstrate the essential features of Holzer's method. First, it is necessary either to guess a value for the natural radiancy for whichever of the modes of vibration is to be determined, or, better, to obtain some approximate value by 'simplifying' the system, as in the numerical example which follows this outline of the method. (It is often possible to reduce a system, however crudely, to an equivalent two-mass system, the frequency of which can be easily calculated; this frequency can then be used for the first estimate of the frequency of the first mode of vibration of the original multi-mass system.) It is also necessary to choose some value for the deflection at any particular instant of one mass: for convenience let us say the amplitude of the first mass. But as the frequency of vibration is unaffected by amplitude (as long as the system is *assumed* to remain elastic), the value taken for this amplitude is of no importance and may conveniently be chosen as unity, i.e. one radian.

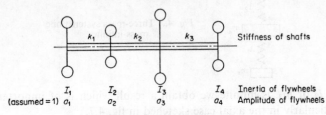

Fig.4.8a Four-mass system

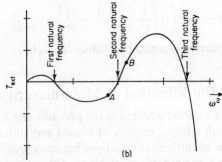

Fig. 4.8b Variation of 'external torque' with assumed value of ω^2

We consider the vibrating system at the instant when the flywheels are in their fully deflected position and assume our first estimated radiancy, say ω_1, to be correct. As in every case of simple harmonic motion, we know that acceleration is $-\omega_1^2$ times the displacement, and therefore that the torque T_1 in the first shaft, which is the only torque acting on I_1, is numerically,

$$T_1 = I_1(\omega_1^2 \times a_1) = I_1\omega_1^2$$

since the amplitude a_1 of I_1 has been assumed to be 1 radian. Consequently, the twist in the first shaft will be T_1/k_1; but we know that the twist in the shaft is $(a_1 - a_2) = (1 - a_2)$, so we can determine a_2:

$$a_2 = 1 - \frac{T_1}{k_1}$$

and knowing a_2, we can now say that the torque acting on I_2 to give it its acceleration is $I_2\omega_1^2 a_2$. But the torque in the second shaft must be capable of accelerating I_2 *and* providing the torque T_1 in the first shaft, so

$$T_2 = T_1 + I_2\omega_1^2 a_2$$

From T_2 the twist in the second shaft can be calculated, and hence the amplitude of vibration of the third flywheel

$$a_3 = a_2 - \frac{T_2}{k_2}$$

This enables us to calculate the torque to accelerate I_3. The torque in the third shaft must be capable of accelerating I_3 and providing the torque T_2 in the second shaft, so

$$T_3 = T_2 + I_3\omega_1^2 a_3$$

and
$$a_4 = a_3 - \frac{T_3}{k_3}$$

It is now possible to calculate the torque required to accelerate I_4. If it can be imagined that there is an external torque T_{ext}, then this must be capable of accelerating I_4 and providing the torque T_3 in the third shaft, so

$$T_{ext} = T_3 + I_4\omega_1^2 a_4$$

If the assumed value of ω_1 coincides with one of the natural frequencies, then T_{ext} will be zero, and the mode of vibration can be readily established by observing the changes in sign of the amplitude of vibration (i.e. a). For instance, if there is one change of sign, then there is only one node, and the mode of vibration is the simplest or fundamental. If T_{ext} is not zero, then ω_1 must be incorrect and another value of radiancy must be assumed and the process repeated. The finding of a sufficiently accurate approximation to ω_n can be speeded up if the variation of T_{ext} with ω^2, which is shown in fig. 4.8b, is realised. From the graph it can be seen that T_{ext} is zero at the origin and at the three natural frequencies of the system. If, for instance, two fairly close values of ω^2 give the points A and B, then a reasonable value of the third guess can be obtained by interpolation, assuming a linear relationship between ω^2 and T_{ext}, and checked if necessary, when an almost perfect final interpolation can be made.

An explanatory description of the Holzer method such as the

foregoing makes it appear somewhat lengthy and complicated. Once
the principles have been understood, however, the calculations can
be presented in the form of a table and carried out rapidly and with
ease. Essentially the method is arithmetical, and it is best illustrated
by a numerical example taken from practice, such as the idealised

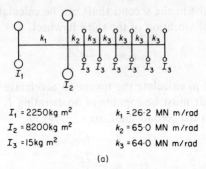

$I_1 = 2250\,\text{kg m}^2$ $k_1 = 26\cdot2\,\text{MN m/rad}$
$I_2 = 8200\,\text{kg m}^2$ $k_2 = 65\cdot0\,\text{MN m/rad}$
$I_3 = 15\,\text{kg m}^2$ $k_3 = 64\cdot0\,\text{MN m/rad}$

(a)

Fig. 4.9a Idealized engine load system

system shown in fig. 4.9a, which represents a six-cylinder engine and
flywheel coupled to a load with a moment of inertia I_1. By inspection
it can be seen that the six small 'flywheels' which represent the crank
throws and associated masses are coupled by relatively stiff shafts

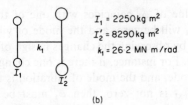

$I_1 = 2250\,\text{kg m}^2$
$I_2' = 8290\,\text{kg m}^2$
$k_1 = 26\cdot2\,\text{MN m/rad}$

(b)

Fig. 4.9b Equivalent two-mass system

and that the shaft connecting them to I_2 is also relatively stiff. It is
therefore not unreasonable, as a first approximation, to reduce the
system to a two-flywheel system, as shown in fig. 4.9b, in which I_1
and k_1 are unchanged and I_2' has a moment of inertia equal to
$(I_2 + 6I_3)$. $\omega_n{}^2$ for this simplified system has been considered in
§ 4.4 and is given, from equation (4.6), by

$$\omega_n{}^2 = k_1 \left(\frac{1}{I_1} + \frac{1}{I_2'} \right)$$

and substituting in this equation, we obtain

$$\omega_n^2 = 26 \cdot 2 \times 10^6 \left\{ \frac{1}{2\,250} + \frac{1}{8\,290} \right\}$$

$$= 14\,805 \ (\text{rad/s})^2$$

This value of ω_n^2 will serve as an approximate value ω_1^2 for the first mode of vibration of the system shown in fig. 4.9a. Table 4.1 sets out the computation for this assumed value of ω_1^2, and no difficulty should be experienced in following through the scheme, step by step. For convenience columns 1, 2, and 6, which consist essentially of the display of data, can be filled in straight away. We then have to tackle the computation row by row. The first entry in

Table 4.1

1 Inertia Number n	2 $I_n \omega_1^2$ Nm	3 a_n rad	4 $I_n \omega_1^2 a_n$ Nm	5 $T = \Sigma I_n \omega_1^2 a_n$ Nm	6 $n k_{n+1}$ Nm/rad	7 Twist $(a_n - a_{n+1})$ $= T/k$ rad
1	$33 \cdot 3 \times 10^6$	1	$33 \cdot 3 \times 10^6$	$33 \cdot 3 \times 10^6$	$26 \cdot 2 \times 10^6$	$1 \cdot 271$
2	$121 \cdot 4 \times 10^6$	$-0 \cdot 271$	$-32 \cdot 90$	$0 \cdot 4 \times 10^6$	$65 \cdot 0 \times 10^6$	$\cdot 006$
3	$0 \cdot 22 \times 10^6$	$-0 \cdot 277$	$-0 \cdot 06$	$0 \cdot 34 \times 10^6$	$64 \cdot 0 \times 10^6$	$\cdot 005$
4	$0 \cdot 22 \times 10^6$	$-0 \cdot 282$	$-0 \cdot 06$	$0 \cdot 28 \times 10^6$	$64 \cdot 0 \times 10^6$	$\cdot 004$
5	$0 \cdot 22 \times 10^6$	$-0 \cdot 286$	$-0 \cdot 06$	$0 \cdot 22 \times 10^6$	$64 \cdot 0 \times 10^6$	$\cdot 003$
6	$0 \cdot 22 \times 10^6$	$-0 \cdot 289$	$-0 \cdot 06$	$0 \cdot 16 \times 10^6$	$64 \cdot 0 \times 10^6$	$\cdot 003$
7	$0 \cdot 22 \times 10^6$	$-0 \cdot 292$	$-0 \cdot 06$	$0 \cdot 10 \times 10^6$	$64 \cdot 0 \times 10^6$	$\cdot 002$
8	$0 \cdot 22 \times 10^6$	$-0 \cdot 294$	$-0 \cdot 06$	T_{ext} $= +0 \cdot 04 \times 10^6$		

column 3 is our assumed amplitude of I_1, in column 4 the torque required to accelerate I_1, in column 5 the torque which must exist in the shaft 'behind it' or 'to its right', in column 6 the stiffness of this shaft, and in column 7 the twist which will occur in this shaft. If now we subtract this from a_1 we find a_2 (negative already, because we have passed through a node, as we might have expected), and we can now start on the second row. The torque in the second shaft is equal to the sum of that in the first shaft and the torque to accelerate I_2, that is the sum of $I\omega^2 a$ up to and including the second flywheel. Knowledge of this torque enables the twist in the second shaft to be computed and so on.

If the correct value of ω^2 had been assumed, T_{ext}, the torque after the last flywheel would be zero. It can be seen that T_{ext} is very small compared with most of the other values in column 5, and it would require an extremely small increase in the assumed value of ω^2 to reduce the external torque to zero. The amplitudes of vibration are given in column 3, and, as might be expected, there is only one change of sign which indicates that there is only one node, i.e. that the mode of vibration is the first.

In this particular example the reduction of the problem to a two-mass system to determine an approximate value of ω_n gives a value very close indeed to the correct one, but this is, of course, exceptional, and frequently it is not possible to make such a good simplification. For example, there is no obvious assumption which can be made to determine a reasonable value of ω^2 for the second mode of vibration, and this value could only be obtained by guesswork and then trial and error.

In addition to serving as a means of determining the natural frequencies, the Holzer table is extremely useful in the interpretation of results obtained from measurements made of vibrations in practice. For instance, in the example which has just been considered the amplitude of vibration at the lowest natural frequency at some point in the system may have been measured under running conditions; if so, it would be possible from column 3 to deduce by simple proportion the amplitude at any other point in the system, and from columns 3 and 5 to determine the torque at any section of the shaft due to vibration. Alternatively, if the vibrational strain is measured at any section, then the vibrational torque can be deduced, and from columns 5 and 3 the torques and displacements throughout the system can be found.

In the example shown in fig. 4.9a the masses at each end are finite, but in some systems, more usually associated with axial vibrations, as stated at the end of § 4.2, one or both of the ends of the system may be regarded as 'fixed'. The conditions to be satisfied are then (in the latter case, for instance) that the amplitude of vibration at each end shall be zero. As an example of this, consider the problem shown in fig. 4.10 for which the Holzer table 4.2 has been worked out for a value of ω^2 found to this degree of approximation by previous trial and error. It should be noted that the data given in the example have been converted to SI units. To start the table for this system the deflection of the first mass is assumed to be $+1$ m, which implies that the force in the first spring is $-\lambda_1$ (i.e. $-21 \cdot 6 \times 10^6$ N) or alternatively the deflection in the first spring is -1 m.

The end condition to be satisfied in this problem is that the amplitude at the fixed end of the last spring shall be zero; for the particular value of ω^2 assumed, however, the amplitude is $+0.004^m$, which is very small. From the table it will be seen that there is no change of sign in column 3, so there are no nodes except those constrained

Table 4.2

	$\omega^2 = 517\ 500$ (rad/sec)^2					
1	2	3	4	5	6	7
Mass No. n	$m_n\omega^2$ kg/s^2	a_n m	$m_n\omega^2 a_n$ N	$\Sigma m_n\omega^2 a_n$ = F, N	$n\lambda_{n+1}$ N/m	F/λ m
	Fixed end	0		$-21{\cdot}6\ \times\ 10^6$	$21{\cdot}6 \times 10^6$	-1
1	20·70	$+1$	$20{\cdot}70 \times 10^6$	$-0{\cdot}9\ \times\ 10^6$	$33{\cdot}3 \times 10^6$	$-0{\cdot}027$
2	11·64	$+1{\cdot}027$	$11{\cdot}95 \times 10^6$	$+11{\cdot}05 \times 10^6$	$10{\cdot}8 \times 10^6$	$+1{\cdot}023$
	Fixed end	$+0{\cdot}004$				

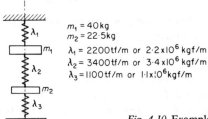

$m_1 = 40$ kg
$m_2 = 22{\cdot}5$ kg
$\lambda_1 = 2200$ tf/m or $2{\cdot}2 \times 10^6$ kgf/m
$\lambda_2 = 3400$ tf/m or $3{\cdot}4 \times 10^6$ kgf/m
$\lambda_3 = 1100$ tf/m or $1{\cdot}1 \times 10^6$ kgf/m

Fig. 4.10 Example on axial vibration

constrained at each end, so that this solution is for the first or simplest mode of vibration, and its frequency will be

$$\frac{1}{2\pi} \sqrt{517\ 500} = 114{\cdot}5 \text{ Hz*} = 6\ 868 \text{ c/min}$$

§ 4.9 Vibration of Systems involving Gearing

Many cases of severe torsional vibration arise in transmission systems, and not infrequently these systems incorporate a step-up or reduction gear. In the following analysis it is assumed that the gear-teeth are infinitely stiff in bending, and that either the mean torque is greater than the amplitude of the vibration torque or there is no backlash, so that no non-linearity is caused by 'lost motion'. Usually, in practice, there is backlash in a geared system, and it is clearly desirable that the amplitude of the vibration torque should not

*Note that in the SI system the unit of frequency, c/s, has been named after Hertz and given the abbreviation Hz.

exceed the mean torque, so that the gear is not subjected to torque reversal, which although it would effectively raise the damping in the system, would also lead to impact loads on, and serious wear of, the gear-teeth, as well as a considerable amount of noise.

In fig. 4.11a the gear ratio is $-nx/x = -n$, where nx and x are the number of teeth on gears I_2 and I_3 respectively, the negative sign indicating the change in the sense of rotation. It must be realised that although this ratio affects the torque and angular motion transmitted, the frequency is constant throughout the system. The equa-

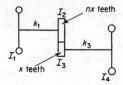

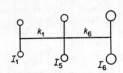

Fig. 4.11a Geared system   *Fig. 4.11b* Equivalent simple system

tion of motion of the first flywheel can be written down as for the first flywheel in the system considered in § 4.6:

$$I_1\ddot{\theta}_1 = -k(\theta_1 - \theta_2)$$

The gears I_2 and I_3 vibrate together with no backlash, and the torques to accelerate them are $I_2\ddot{\theta}_2$ and $I_3\ddot{\theta}_3$ about their respective axes of rotation. In addition, I_2 is acted on by a disturbing torque in the first shaft of $k_1(\theta_1 - \theta_2)$ and I_3 is acted on by a torque of $-k_3(\theta_3 - \theta_4)$. To simplify the equation of motion for I_2 and I_3 it is best to convert all the torques to one or other of the axes of rotation. For instance, let us consider all the torques as acting about the axis of the first shaft, then the torque about this axis to accelerate I_2 and I_3 will be

$$I_2\ddot{\theta}_2 + (-n)(I_3\ddot{\theta}_3)$$

Similarly, the torques acting on these flywheels will become

$$k_1(\theta_1 - \theta_2) + (-n)\{-k_3(\theta_3 - \theta_4)\}$$

or equating

$$I_2\ddot{\theta}_2 - nI_3\ddot{\theta}_3 = k_1(\theta_1 - \theta_2) + nk_3(\theta_3 - \theta_4)$$

We can, however, note that

$$\frac{\theta_2}{\theta_3} = -\frac{1}{n}$$

so $$\theta_3 = -n\theta_2 \quad \text{and} \quad \ddot{\theta}_3 = -n\ddot{\theta}_2$$

Consequently, the above equation of motion reduces to

$$I_2\ddot\theta_2 + n^2I_3\ddot\theta_2 = k_1(\theta_1 - \theta_2) + nk_3(-n\theta_2 - \theta_4)$$

where θ_4 is the actual displacement of I_4,

or $$(I_2 + n^2I_3)\ddot\theta_2 = k_1(\theta_1 - \theta_2) - n^2k_3(\theta_2 - \theta_6)$$

where we define θ_6 $(=-\theta_4/n)$ as the equivalent angular displacement of I_4 referred to the first shaft.

The equation of motion for the last flywheel, which in its simplest form can be written

$$I_4\ddot\theta_4 = +k_3(\theta_3 - \theta_4)$$

can be expressed as

$$-nI_4\ddot\theta_6 = -nk_3(\theta_2 - \theta_6)$$

or by multiplying by n as

$$n^2I_4\ddot\theta_6 = n^2k_3(\theta_2 - \theta_6)$$

In this form it can be seen that all the torques are the effective torques acting about the axis of the first shaft.

These three equations of motion would be identical to those deduced in § 4.6 for a simple flywheel system as shown in fig. 4.11b if

$$\left.\begin{aligned} I_5 &= I_2 + I_3n^2 \\ I_6 &= I_4n^2 \\ k_6 &= k_3n^2 \end{aligned}\right\}   \text{(4.11)}$$

A system with a larger number of masses can be similarly reduced to an equivalent simple system with no gear, or alternatively a Holzer table can be used without simplifying the system. As an example of the latter method, consider the system shown in fig. 4.12

Table 4.3

$f = 333$ c/min $= 5\cdot55$ Hz $\omega^2 = 1\ 215$ (rad/s)^2						
1 Inertia number n	2 $I_n\omega^2$ Nm	3 a_n rad	4 $I_n\omega^2 a_n$ Nm	5 $\Sigma I_n\omega^2 a_n$ $= T$, Nm	6 nk_{n+1} Nm/rad	7 T/k rad
1	25 520	1	25 520	25 520	$7\cdot845 \times 10^5$	0·0325
2	25 520	0·9675	24 690	50 210	$7\cdot845 \times 10^5$	0·0640
3	25 520	0·9035	23 060	73 270	$7\cdot845 \times 10^5$	0·0934
4	25 520	0·8101	20 670	93 940	$1\cdot961 \times 10^5$	0·4790
5	10 940	0·3311	3 620	97 560		
		$\div -2\cdot5$		$\times -2\cdot5$		
				$= -243\ 900$		
6	58 320	$-0\cdot1324$	$-7\ 720$	$-251\ 620$	$6\cdot178 \times 10^5$	$-0\cdot4073$
7	917 300	$+0\cdot2749$	$+252\ 170$	$T_{\text{ext}} = +550$		

and the final Holzer tabulation for the first mode of vibration, which is given in table 4.3. Again the data given in the example have in the table been converted into SI units. The procedure is the same as that previously described up to finding the amplitude of vibration of the gear pinion I_5. The amplitude of I_5 being known, the torque to accelerate I_5 can be computed, and the torque transmitted to the gear-wheel I_6 will then be equal to the sum of this torque and the torque in the fourth shaft. This torque is then referred to the propeller shaft by multiplying by the gear ratio, $-2\frac{1}{2}$, the minus sign

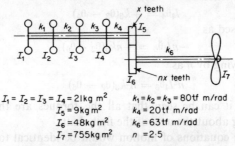

$$I_1 = I_2 = I_3 = I_4 = 21\,\text{kg m}^2 \qquad k_1 = k_2 = k_3 = 80\,\text{tf m/rad}$$
$$I_5 = 9\,\text{kg m}^2 \qquad k_4 = 20\,\text{tf m/rad}$$
$$I_6 = 48\,\text{kg m}^2 \qquad k_6 = 63\,\text{tf m/rad}$$
$$I_7 = 755\,\text{kg m}^2 \qquad n = 2.5$$

Fig. 4.12 Practical example of geared systems

indicating the change in sense of the torque. The amplitude of the gear-wheel I_6 is also found by dividing the amplitude of vibration of the gear pinion I_5 by $-2\frac{1}{2}$, and the torque to accelerate I_6 can then be readily calculated. The torque in the shaft of stiffness k_6 is the sum of the torque transmitted through the gear and the torque to accelerate I_6, and the rest of the table is then as before. Neglecting the change in sign of a_n due to the gear, there is only one other change of sign, so the mode of vibration is the first.

The vibration torque being transmitted through the gear can be deduced if the vibration amplitude or stress at any point in the system is known. For instance, if the vibration amplitude of I_1 were found to be 0·05 rad, then the amplitude of the vibration torque between the gears, referred to the first shaft, would be

$$0.05 \times 97\,560 = 4\,878 \text{ Nm}$$

§ 4.10 Vibration of Systems with a Distributed Mass and Stiffness

So far the natural frequency of vibration of systems consisting of discrete masses separated by stiffnesses has been considered. A system in which the mass is distributed can be regarded as one in which the number of masses is infinite, and as might be expected, it will be found to have an infinite number of modes of vibration. As an example, the *axial* vibration of a uniform shaft fixed at one end

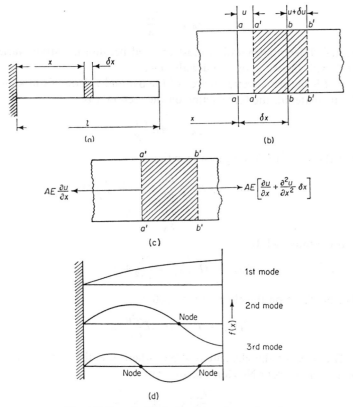

Fig. 4.13 Longitudinal vibrations of a uniform bar

and free at the other will be considered; this system is shown in fig. 4.13a. In the solution of this problem the lateral contraction and expansion of the bar are neglected, which is justifiable only if the cross-sectional dimensions of the bar are small compared with the wavelength of the vibration or the distance between nodes.

Consider an element of the bar δx long and at a distance x from one end of the bar; this element is shown in detail in fig. 4.13b. At rest, let the element be in the position denoted by *aa* and *bb*. When the bar is vibrating axially the section *aa* will at some instant be displaced a distance u from its position of rest to $a'a'$, and the section bb will be displaced $u + \delta u$ to $b'b'$. The increment δu is then the change in length of the element and the mean strain in the element $\delta u/\delta x$. In the limit the strain at section $a'a'$ will be

$$\varepsilon_{a'a'} = \frac{\partial u}{\partial x}$$

Partial differential notation must be used because the strain varies both with the value of x and with time.

If A is the cross-sectional area of the bar and E Young's modulus for its material, the force acting on the section x is

$$F_{a'a'} = AE\frac{\partial u}{\partial x}$$

For the section $b'b'$ we can now write

$$\text{Strain} = \varepsilon_{b'b'} = \frac{\partial u}{\partial x} + \frac{\partial}{\partial x}\left(\frac{\partial u}{\partial x}\right)\delta x$$

$$= \frac{\partial u}{\partial x} + \frac{\partial^2 u}{\partial x^2}\delta x$$

and the force will be

$$F_{b'b'} = AE\left\{\frac{\partial u}{\partial x} + \frac{\partial^2 u}{\partial x^2}\delta x\right\}$$

The net force on the element is therefore

$$F_{b'b'} - F_{a'a'} = AE\frac{\partial^2 u}{\partial x^2}\delta x$$

The mass of the element is $A\delta x\rho$ (where ρ is the density of the material), so from Newton's second law

$$AE\frac{\partial^2 u}{\partial x^2}\delta x = A\delta x\rho\frac{\partial^2 u}{\partial t^2}$$

or
$$\frac{E}{\rho}\frac{\partial^2 u}{\partial x^2} = \frac{\partial^2 u}{\partial t^2} \qquad\qquad (4.12)$$

where $\dfrac{E}{\rho}$ is essentially a positive quantity.

We can solve this equation by the application of common sense. It is reasonable to assume that any single mode of vibration will be simple harmonic, and that the amplitude of vibration will vary with x, and be some function of x, say $f(x)$. We can therefore write

$$u = f(x)C_1\sin(\omega_n t + \phi)$$

and if we substitute this value of u in equation (4.12) we obtain

$$\frac{E}{\rho}\frac{d^2f(x)}{dx^2}C_1 \sin(\omega_n t + \phi) = -f(x)\omega_n^2 C_1 \sin(\omega_n t + \phi)$$

or

$$\frac{d^2}{dx^2}f(x) = -\omega_n^2 \frac{\rho}{E}f(x)$$

and by analogy with the equations of simple harmonic motion (cf. § 0.14) or by inspection we see that the solution of this equation is

$$f(x) = C_2 \sin\left(\omega_n \sqrt{\frac{\rho}{E}} . x + \beta\right)$$

so

$$u = R \sin\left(\omega_n \sqrt{\frac{\rho}{E}} . x + \beta\right) \sin(\omega_n t + \phi) \qquad (4.13)$$

where $R \,(= C_1 C_2)$, ϕ and β are arbitrary constants.

Now at the fixed end of the bar vibrational motion must be zero; i.e. when $x = 0$, $u = 0$. At the free end of the bar there can be no force and therefore no strain; i.e. when $x = l$, $\frac{\partial u}{\partial x} = 0$. These statements are true for all values of t. Substituting the first of these values in equation (4.13) we have

$$0 = R \sin\beta \sin(\omega_n t + \phi)$$

for all values of t, and since R cannot be zero (or the bar would not be vibrating),

$$\sin\beta = 0 \text{ or } \beta = 0$$

Differentiating (4.13) and then substituting the second condition,

$$\frac{\partial u}{\partial x} = R\,\omega_n \sqrt{\frac{\rho}{E}} \cos\left(\omega_n \sqrt{\frac{\rho}{E}} . x\right) \sin(\omega_n t + \phi)$$

$$\therefore 0 = R \cos\left(\omega_n \sqrt{\frac{\rho}{E}} . l\right) \sin(\omega_n t + \phi) \text{ for all values of } t$$

$$\therefore \cos\left(\omega_n \sqrt{\frac{\rho}{E}} . l\right) = 0$$

$$\therefore \omega_n \sqrt{\frac{\rho}{E}} . l = \frac{2n - 1}{2}\pi \text{ where } n \text{ is any integer}$$

or

$$\omega_n = \frac{2n - 1}{2}\frac{\pi}{l}\sqrt{\frac{E}{\rho}}$$

or

$$f_n = \frac{\omega_n}{2\pi} = \frac{2n - 1}{4l}\sqrt{\frac{E}{\rho}}$$

Thus we have obtained, as we expected, an infinite number of possible frequencies of vibration, and a glance at equation (4.13) shows that each is associated with a different mode. Taking as examples the values $n = 1$, 2, and 3, we have

$$u_1 = R_1 \sin \frac{\pi}{2l} x \sin \left(\frac{\pi}{2l} \sqrt{\frac{E}{\rho}} \cdot t + \phi_1 \right) \quad f_{n_1} = \frac{1}{4l} \sqrt{\frac{E}{\rho}}$$

$$u_2 = R_2 \sin \frac{3\pi}{2l} x \sin \left(\frac{3\pi}{2l} \sqrt{\frac{E}{\rho}} \cdot t + \phi_2 \right) \quad f_{n_2} = \frac{3}{4l} \sqrt{\frac{E}{\rho}}$$

$$u_3 = R_3 \sin \frac{5\pi}{2l} x \sin \left(\frac{5\pi}{2l} \sqrt{\frac{E}{\rho}} \cdot t + \phi_3 \right) \quad f_{n_3} = \frac{5}{4l} \sqrt{\frac{E}{\rho}}$$

and fig. 4.13d illustrates these three modes. If the free end of the bar were subjected to a small axial force at any one of these frequencies the corresponding vibration would be excited: if this free end were subjected to an axial blow all the modes of vibration might be excited simultaneously, each with, in general, its own amplitude R.

To determine the natural frequencies of longitudinal vibration of a bar with other end conditions involves only substituting the appropriate end conditions into equation (4.13). Similar equations can be deduced for the torsional vibrations of a uniformly distributed moment of inertia and stiffness.

When Rayleigh's approximate method of solving such problems is considered in a later section the case of a uniform bar such as that shown in fig. 4.13, but with a concentrated mass M at its free end, will be considered. It is consequently of some interest to obtain the frequency of such a system in order to compare it with the value deduced by the approximate method. The only effect that the addition of mass M will have on the solution of equation (4.13), which applies to the simple uniform shaft, will be that instead of satisfying zero force condition at $x = l$, this force will now have to be such as to accelerate the mass M. The force must be a restoring force, so the end condition at $x = l$ will be

$$-AE \left(\frac{\partial u}{\partial x} \right)_{x=l} = M \left(\frac{\partial^2 u}{\partial t^2} \right)_{x=l}$$

As before, the condition that $u = 0$ at $x = 0$ gives $\beta = 0$. Differentiating (4.13) and then substituting the condition at $x = l$ gives

$$-AER \, \omega_n \sqrt{\frac{\rho}{E}} \cos \left(\omega_n \sqrt{\frac{\rho}{E}} \cdot l \right) \sin \left(\omega_n t + \phi \right) =$$

$$-MR \, \omega_n{}^2 \sin \left(\omega_n \sqrt{\frac{\rho}{E}} \cdot l \right) \sin(\omega_n t + \phi)$$

or
$$\frac{AE}{M}\sqrt{\frac{\rho}{E}} = \omega_n \tan \omega_n \sqrt{\frac{\rho}{E}} \cdot l$$

If we multiply through by $l\rho$ and substitute for $Al\rho$ the total mass of the shaft, say m, then we can rearrange the above equation to give

$$\frac{m}{M} = \omega_n \sqrt{\frac{\rho}{E}} \cdot l \tan \omega_n \sqrt{\frac{\rho}{E}} \cdot l \qquad (4.14)$$

For any value of m/M the values of $\omega_n \sqrt{\frac{\rho}{E}} \cdot l$ which satisfy this equation can be found graphically or by trial and error. Considering only the frequency of the first mode of vibration, the solutions for several values of m/M are given in table 4.4.

Table 4.4

m/M	0·1	1	10	100	∞ (i.e. $M = 0$)
$\omega_n \sqrt{\frac{\rho}{E}} \cdot l$	0·311	0·860	1·429	1·555	$\pi/2 = 1\cdot571$

Considering, for example, the solution for $m/M = 0\cdot1$, then

$$\omega_n \sqrt{\frac{\rho}{E}} \cdot l = 0\cdot311$$

or
$$\omega_n = \frac{0\cdot311}{l} \sqrt{\frac{E}{\rho}}$$

which, alternatively, can be expressed as

$$\omega_n = 0\cdot311 \sqrt{\frac{EA}{l^2 A\rho}} = 0\cdot311 \sqrt{\frac{EA}{lm}} = 0\cdot311 \sqrt{\frac{\lambda}{m}}$$

where $\frac{EA}{l} = \lambda$, the stiffness of the bar in the axial direction. In general, $\omega_n = A' \sqrt{\frac{\lambda}{m}}$, where A' is the value of $\omega_n \sqrt{\frac{\rho}{E}} \cdot l$ appropriate to the value m/M.

§ 4.11 Transverse Vibration of Beams

4.11.1 Note on sign conventions in beam theory

It is much to be regretted that no satisfactory sign conventions have been universally adopted in the subject we have now to discuss, namely simple beam theory. Much of the confusion arises from the carelessness mentioned in § 0.2, and some is undoubtedly due to the preference of engineers for dealing with positive quantities. Thus, a beam loaded with weights tends usually to deflect downwards, so the downward y direction is often accepted as 'positive for deflection', and the x-axis along the beam taken to the left or right as occasion arises. It is not surprising that the average student is utterly confused.

Let us run through the elements of a simple example, without

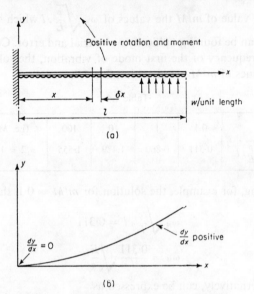

Positive rotation and moment

w/unit length

(a)

$\frac{dy}{dx} = 0$

$\frac{dy}{dx}$ positive

(b)

Fig. 4.14 Sign convention in beam theory

proofs, merely to establish the signs. For this example consider a cantilever which is subjected to a positive loading of w/unit length. If we take our standard directions for the axes with the origin at the fixed end (since the origin must remain fixed), then the situation is as illustrated in fig. 4.14a. The shear force F on the beam at section x (i.e. the force exerted by the material beyond x) is in the direction of the y-axis and therefore positive.

$$\therefore F = w(l - x)$$

and at $x + \delta x$ it will be

$$F + \delta F = w \{l - (x + \delta x)\}$$

$$\therefore \delta F = -w\delta x, \quad \text{or} \quad w = -\frac{dF}{dx}$$

The bending moment at section x (i.e. the moment exerted by the material beyond x) is

$$M = \tfrac{1}{2}w(l - x)^2$$

and at $x + \delta x$

$$M + \delta M = \tfrac{1}{2}w\,\{l - (x + \delta x)\}^2$$
$$= \tfrac{1}{2}w\,\{(l - x) - \delta x\}^2$$
$$\therefore \ \delta M = -w(l - x)\delta x = -F\delta x$$

or

$$F = -\frac{dM}{dx}$$

Both the shearing force and the bending moment have a positive value at any section in this beam. The former produces a positive shearing deflection which is usually neglected in comparison with the bending deflection illustrated in fig. 4.14b. Here we see that the slope of the beam, dy/dx is zero at the origin and positive elsewhere, growing in magnitude as x increases, so that d^2y/dx^2 is positive throughout. We have no anxiety, therefore, in writing down

$$M = +EI\frac{d^2y}{dx^2}$$

We can now write down our important relationships

$$\text{Deflection} \quad = y$$

$$\text{Slope} \quad = \frac{dy}{dx}$$

$$\text{Bending moment} = EI\frac{d^2y}{dx^2}$$

$$\text{Shear force} \quad = -\frac{dM}{dx} = -EI\frac{d^3y}{dx^3}$$

$$\text{Rate of loading} \quad = -\frac{dF}{dx} = +EI\frac{d^4y}{dx^4}$$

Note that we can rotate the figures in (4.14) through any angle, say 90° or 180°, and still have a self-consistent system of signs, with anti-clockwise rotation and bending moment positive.

4.11.2 *Preliminary remarks on the general problem*

In the previous paragraphs free torsional and axial vibrations (of single mass, multi-mass, and distributed mass systems) have been considered, and the same analytical techniques are applicable in each case. Why, then, are these methods not applicable to problems of transverse vibrations other than the single-mass case? Let us compare, for instance, the torsional and lateral vibrations of a uniform shaft of known dimensions and properties carrying a number of discrete masses. In the torsional case if at any instant we know the angular displacements of the masses we can immediately write down the 'rate of twist', $d\theta/dx$, in each part of the shaft $\Big($ if the mass of the shaft is negligible, then $\dfrac{d\theta}{dx} = \dfrac{\theta_{n+1} - \theta_n}{l_n}\Big)$, and from this deduce the torque therein, and hence the torque applied to each mass and its angular acceleration. In other words, we can write down the equation of motion for each mass in terms of the first derivative of the angular deflection with respect to the length of the

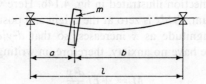

Fig. 4.15 Transverse vibrations of a single mass

shaft, and of course an analogous statement could be made in relation to axial vibrations. In the case of lateral vibrations no correspondingly simple situation exists. The restoring force on each deflected mass arises directly from the shear force in the shaft treated as a beam, and the shear force can be written down, not in terms of the first derivative of the deflection, dy/dx (which is the slope of the beam), but in terms of d^3y/dx^3.

To simplify the analysis of transverse vibrations we shall neglect the shear deflections in the beam, which are very small compared with the bending deflections as long as the cross-sectional dimensions are small compared with the distance between nodes. It will also be assumed that the moment of inertia of each mass about an axis perpendicular to the plane of vibration is negligible. The reason for this is apparent from fig. 4.15, in which it can be seen that when the beam is deflected the masses have rotational as well as transla-tional motion, so that when the beam is vibrating not only lateral restoring forces but also restoring moments are necessary. A more complete analysis of the problem allowing for rotation is given by Lord Rayleigh and Timoshenko.

4.11.3 Beam with single concentrated mass

In a system such as that illustrated in fig. 4.15, where a single mass is mounted on a light flexible beam on supports, no difficulty of analysis arises. If the mass is deflected transversely from its equilibrium position the beam will exert a restoring force which might equally well be applied by any other form of spring, and we can without further thought write (cf. § 4.3):

$$\omega_n = \sqrt{\frac{\lambda}{m}} = \sqrt{\frac{g}{\delta_s}}$$

where λ is the stiffness of the beam, or δ_s the static deflection of the beam under the weight of m. (From elementary beam theory we know that, with the usual notation $\delta_s = \dfrac{Wa^2b^2}{3EIl}$ or $\lambda = \dfrac{W}{\delta_s} = \dfrac{3EIl}{a^2b^2}$ and $W = mg$.)

4.11.4 Beam with two concentrated masses: exact solution

Let us now consider the case (fig. 4.16a) in which a massless beam supports two concentrated masses, m_1 and m_2. At some instant when

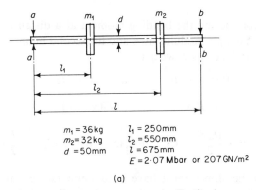

$m_1 = 36\,\text{kg}$	$l_1 = 250\,\text{mm}$
$m_2 = 32\,\text{kg}$	$l_2 = 550\,\text{mm}$
$d = 50\,\text{mm}$	$l = 675\,\text{mm}$
	$E = 2 \cdot 07\,\text{Mbar}$ or $207\,\text{GN/m}^2$

(a)

Fig. 4.16a Example on transverse vibration

the beam is vibrating let the deflections under these masses be y_1 and y_2, and let the forces acting on the beam at these points and causing these deflections be F_1 and F_2 as in fig. 4.16b. The reactions on the masses will be $-F_1$ and $-F_2$, and we can as usual assume that the motion is simple harmonic with a radiancy ω_n, so we can write down these (restoring) forces as

$$-F_1 = -m_1\omega_n^2 y_1 \qquad -F_2 = -m_2\omega_n^2 y_2$$

Clearly we can determine y_1 and y_2 in terms of F_1 and F_2 by applying the methods normally used in 'strength of materials',* and so obtain equations which, in conjunction with those for F_1 and F_2, will enable us to determine the quantity ω_n.

This is probably best illustrated by a worked example, and we shall take the data given in fig. 4.16a.

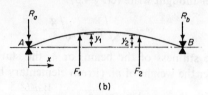

Fig. 4.16b Equivalent static system

Working first in general terms and taking moments about B, we determine the reaction at A as

$$R_a = \frac{F_1(l - l_1) + F_2(l - l_2)}{l}$$

Now considering the bending moment at a distance x from A and taking Young's modulus as E and the second moment of area about the neutral axis as I, we get

$$
\begin{array}{cccc}
 & 0 < x < l_1 & l_1 < x < l_2 & l_2 < x < l \\
EI\dfrac{d^2y}{dx^2} = & -R_a x & +F_1(x - l_1) & +F_2(x - l_2) \\
EI\dfrac{dy}{dx} = -\dfrac{R_a x^2}{2} + C_1 & +\dfrac{F_1}{2}(x - l_1)^2 + C_2 & +\dfrac{F_2}{2}(x - l_2)^2 + C_3
\end{array}
$$

but as the slopes must have a common tangent at the points of application of the loads, the constants C_2 and C_3 must be zero; hence

$$EI\frac{dy}{dx} = -\frac{R_a x^2}{2} + C_1 \quad + \frac{F_1}{2}(x - l_1)^2 \quad + \frac{F_2}{2}(x - l_2)^2$$

$$EIy = -\frac{R_a x^3}{6} + C_1 x + C_4 + \frac{F_1}{6}(x - l_1)^3 + 0 \quad + \frac{F_2}{6}(x - l_2)^3 + 0$$

---

* See A. Morley, *Strength of Materials*, 11th edn. Longmans, 1954; or J. Case and A. H. Chilver, *Strength of Materials*, Arnold.

At $x = 0$ the deflection $y = 0$, therefore $C_4 = 0$. Also $y = 0$ at $x = l$, so

$$0 = -\frac{R_a l^3}{6} + C_1 l + \frac{F_1}{6}(l - l_1)^3 + \frac{F_2}{6}(l - l_2)^3$$

or

$$C_1 = +\frac{R_a l^2}{6} - \frac{F_1}{6l}(l - l_1)^3 - \frac{F_2}{6l}(l - l_2)^3$$

We can now insert numerical values.

Thus

$$R_a = \frac{0 \cdot 425 F_1 + 0 \cdot 125 F_2}{0 \cdot 675}$$

and hence

$$C_1 = 0 \cdot 02885 F_1 + 0 \cdot 01356 F_2$$

and we can now substitute for C_1 and for R_a in terms of F_1 and F_2. The arithmetic is somewhat laborious, but perfectly straightforward, and well worth doing. We obtain

$$EIy_1 = 0 \cdot 005575 F_1 + 0 \cdot 002913 F_2 \text{ Nm}^3$$
$$EIy_2 = 0 \cdot 002913 F_1 + 0 \cdot 002334 F_2 \text{ Nm}^3$$

Now $F_1 = 36\omega_n^2 y_1$ and $F_2 = 32\omega_n^2 y_2$.

Inserting these values we get

$$EIy_1 = 0 \cdot 2007\omega_n^2 y_1 + 0 \cdot 09322\omega_n^2 y_2 \text{ Nm}^3$$
$$EIy_2 = 0 \cdot 1049\omega_n^2 y_1 + 0 \cdot 07469\omega_n^2 y_2 \text{ Nm}^3$$

As the absolute magnitudes of y_1 and y_2 do not affect the natural frequency, we are left with two unknowns y_1/y_2 and ω_n^2, and two equations. Eliminating the amplitude ratio from these two equations gives the frequency equation

$$0 \cdot 005212\omega_n^4 - 0 \cdot 27539\omega_n^2 EI + E^2 I^2 = 0$$

and

$$EI = 207 \times 10^9 \times \frac{\pi}{4} \times 0 \cdot 025^4 = 63\,506 \text{ Nm}^2$$

Solving the frequency equation gives the two values

$$\omega_n = 499 \cdot 1 \text{ rad/s and } 1\,759 \cdot 2 \text{ rad/s}$$
$$f_n = 794 \cdot 5 \text{ Hz and } 2800 \text{ Hz}$$
$$\text{or } 4\,767 \text{ c/min and } 16\,799 \text{ c/min}$$

The corresponding amplitude ratios can be found from either of the simultaneous equations above to be

$$y_1/y_2 = +1.7196 \text{ and } -0.5174$$

The minus sign for the amplitude ratio given by the higher of the two natural frequencies indicates that there is a node between the two masses, in addition to the two constrained at the ends, whereas the lower frequency mode has no node except those constrained at the ends.

4.11.5 Beam with uniformly distributed mass: exact solution

The analysis employed to solve the problem of fig. 4.16a can be extended to systems with a larger number of masses and supports, but the arithmetic becomes progressively more laborious, and usually Rayleigh's method, which is discussed in the next section, would be used. A system with a uniformly distributed mass is, however, more readily amenable to solution. Consider, for instance, a uniform shaft of mass m and length l, between supports whose nature we shall discuss later. When this shaft is vibrating transversely let the displacement from the equilibrium position at a distance x from one support be at some particular instant y. If an element of shaft δx in length, and therefore of mass $\frac{m}{l}\delta x$, is considered, then the force to accelerate this element must be

$$\text{mass} \times \text{acceleration} = \left(\frac{m}{l}\delta x\right)\frac{\partial^2 y}{\partial t^2}$$

and this force must be provided by the shaft. The reaction to this force can be considered as a distributed load on the beam of magnitude

$$-\frac{m}{l}\frac{\partial^2 y}{\partial t^2} \text{ per unit length}$$

From the consideration of the theory of bending of loaded beams (as in § 4.11.1) we can therefore write down

$$EI\frac{\partial^4 y}{\partial x^4} = -\frac{m}{l}\frac{\partial^2 y}{\partial t^2}$$

or

$$\frac{\partial^4 y}{\partial x^4} = -\frac{m}{lEI}\frac{\partial^2 y}{\partial t^2} \tag{4.15}$$

As outlined in the solution of the axial vibration of a uniform bar in § 4.10, we can make an intelligent guess that the solution of this fourth-order partial differential equation will be of the form

$$y = f(x)\, C \sin(\omega_n t + \phi)$$

Substituting this solution into (4.15) and simplifying gives

$$\frac{d^4}{dx^4} f(x) = \frac{m\omega_n{}^2}{lEI} f(x) \text{ where } \frac{m\omega_n{}^2}{lEI} \text{ is essentially positive} \quad (4.16)$$

and it is convenient to rewrite this as

$$\frac{d^4}{dx^4} f(x) = \alpha^4 f(x) \text{ where } \alpha^4 = \frac{m\omega_n{}^2}{lEI} \quad (4.17)$$

and so the solution is to this extent satisfactory, but we still have to solve equation (4.16) before we can determine ω_n.

There is no difficulty in obtaining solutions of (4.16): we require only to find functions which when differentiated four times return to a positive constant times the original form, and those available include $e^{\alpha x}$, $e^{-\alpha x}$, $e^{j\alpha x}$, $e^{-j\alpha x}$, $\sin \alpha x$, $\cos \alpha x$, $\sinh \alpha x$, $\cosh \alpha x$. To obtain a complete solution we require four arbitrary constants; thus we can choose the four as below

$$f(x) = C_1 \sin \alpha x + C_2 \cos \alpha x + C_3 \sinh \alpha x + C_4 \cosh \alpha x$$

(though if the last two functions are unfamiliar $e^{\alpha x}$ and $e^{-\alpha x}$ can equally well be used), and the adequacy of this solution can be deduced by substitution in equation (4.16).

The arbitrary constants C_1, C_2, C_3, and C_4 depend, of course, on the end conditions, that is, the nature of the supports, of which there are four well-defined cases, two of which we now consider in detail.

Case 1. Beam freely supported at each end: If the beam is freely supported at each end, then at all values of time the deflection $(= y)$ at each support and the bending moment $\left(= EI \dfrac{d^2 y}{dx^2} \right)$ at each support must be zero, i.e. $f(x) = 0$ at $x = 0$ and $x = l$, and $\dfrac{d^2 f(x)}{dx^2} = 0$ at $x = 0$ and $x = l$.

For $x = 0$

$$f(x) = 0 = C_2 + C_4$$

and

$$\frac{d^2 f(x)}{dx^2} = 0 = -C_2 + C_4$$

from which we deduce that $C_2 = C_4 = 0$.

For $x = l$

$$f(x) = C_1 \sin \alpha l + C_3 \sinh \alpha l = 0$$

and

$$\frac{d^2 f(x)}{dx^2} = -C_1 \sin \alpha l + C_3 \sinh \alpha l = 0$$

which give $C_1 \sin \alpha l = C_3 \sinh \alpha l = 0$. As $\sinh \alpha l \left(= \frac{e^{\alpha l} - e^{-\alpha l}}{2} \right)$ cannot be zero for a finite value of l the constant C_3 must be zero. If C_1 were also zero we would have $f(x) = 0$; i.e. no vibration; so this is not a solution of interest, and we are forced to conclude that C_1 is not zero, but that $\sin \alpha l = 0$. This gives

$$\alpha l = \pi, 2\pi, 3\pi, \ldots n\pi$$

where n is any integer, and substituting this result in equation (4.17)

$$\omega_n = n^2 \pi^2 \sqrt{\frac{EI}{ml^3}} \text{ and } f_n = \frac{n^2 \pi}{2} \sqrt{\frac{EI}{ml^3}}$$

As in the case of axial vibration (cf. § 4.10), each of this infinite number of frequencies corresponds to a different mode, as is clear from substitution of the various values of α in the equation for $f(x)$, which gives the deflected form

$$f(x) = C_1 \sin \frac{n\pi x}{l}$$

Fig. 4.17 indicates the form of the first three modes.

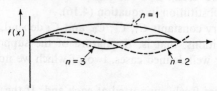

Fig. 4.17 Natural modes of vibration of a simply supported beam

Case 2. Beam directionally constrained at each end: If the beam is built in at each end, then at all values of time the deflection and slope $(= dy/dx)$ will be zero at each end, i.e.

$$f(x) = 0 \text{ at } x = 0 \text{ or } x = l \text{ and } \frac{d}{dx} f(x) = 0 \text{ at } x = 0 \text{ or } x = l$$

For $x = 0$

$$f(x) = 0 = C_2 + C_4$$

$$\frac{d}{dx} f(x) = 0 = C_1 + C_3$$

which give $C_4 = -C_2$ and $C_3 = -C_1$.

For $x = l$

$$f(x) = C_1 \sin \alpha l + C_2 \cos \alpha l + C_3 \sinh \alpha l + C_4 \cosh \alpha l = 0$$

$$\frac{d}{dx} f(x) = C_1 \alpha \cos \alpha l - C_2 \alpha \sin \alpha l + C_3 \alpha \cosh \alpha l + C_4 \alpha \sinh \alpha l = 0$$

Substituting for C_4 and C_3 in the last two equations and simplifying we get

$$\cos \alpha l \cosh \alpha l = 1$$

This equation can be solved by plotting $\cosh \alpha l \left(= \dfrac{e^{\alpha l} + e^{-\alpha l}}{2} \right)$ against $\sec \alpha l$. The first two consecutive roots are

$$\alpha l = 4 \cdot 73 \text{ and } 7 \cdot 853$$

and substituting these in equation (4.17) we get

$$\omega_n = 4 \cdot 73^2 \sqrt{\frac{EI}{ml^3}} \text{ or } 7 \cdot 853^2 \sqrt{\frac{EI}{ml^3}}$$

and

$$f_n = 3 \cdot 56 \sqrt{\frac{EI}{ml^3}} \text{ or } 9 \cdot 82 \sqrt{\frac{EI}{ml^3}}$$

Case 3. If the beam is freely supported at one end and directionally constrained at the other the solution obtained is $\cot \alpha l = \coth \alpha l$, and the first value of αl which satisfies this is $3 \cdot 92$, which gives

$$\omega_n = 3 \cdot 92^2 \sqrt{\frac{EI}{ml^3}}$$

Case 4. If the beam is supported as a cantilever, then the solution obtained is $\cos \alpha l \cosh \alpha l = -1$ and the first solution is $\alpha l = 1 \cdot 875$, which gives

$$\omega_n = 1 \cdot 875^2 \sqrt{\frac{EI}{ml^3}}$$

§ 4.12 Rayleigh's Method

4.12.1 General approach: 'potential' and 'strain' energy

In the previous section the transverse vibration of a beam with two concentrated masses was considered, and it can be appreciated that the solution of more complicated systems would be exceedingly laborious. Rayleigh suggested for computing the frequency of complex systems an approximate method which has been extensively applied to problems of transverse vibrations and many other systems.

Rayleigh showed that if a *reasonable* dynamic deflection curve were *assumed* for the first mode of vibration the frequency which could then be easily computed would not be seriously in error. If an inaccurate curve is assumed the beam must be subjected to constraints in order that it may vibrate in this unreal form, and this implies that the *calculated frequency of vibration will always be too high*, but a considerable error in the assumed form leads usually to a relatively small error in the result. In any case it is possible to use the first computations to improve the assumed deflection curve, so that a more accurate answer can be established. Although Rayleigh's method is more frequently applied to the determination of the first mode, it can be applied to higher-order modes.

Normally the method of computing the natural frequency after guessing a reasonable deflection curve is to equate the change in 'potential' energy, when the system is fully deflected from its equilibrium position, to the maximum kinetic energy of the system when vibrating with simple harmonic motion. The use of the adjective 'potential' in this situation is in itself unfortunate, since some authorities regard potential energy as including the energy stored, for instance, in the distortion of a spring, while others restrict the term to gravitational (and atomic and nuclear) potential energy. In fact, as we shall see, we are not really concerned with the gravitational potential energy at all, which is eliminated from our equations, but with the 'distortional' or 'strain' energy in the deflected elastic members. The confusion which troubles most beginners is made worse by the fact that we frequently find it convenient to assume as a dynamic deflection curve one which is identical to the static deflection curve, only because the strain energy associated with this assumed deflection curve can then be readily expressed in terms of the change of gravitational energy due to the displacement of the beam as it is loaded statically.

4.12.2 *Vibration of simple spring-mass system: energy method*

In an attempt to clarify the matter we shall first again consider the simple spring-mass system dealt with in § 4.3. Although it was there shown that gravity, which exerts a constant force on the mass, had no influence on the vibration, it is now worth while to repeat this demonstration, tackling the problem from energy considerations. In fig. 4.18a the mass is shown vibrating horizontally in frictionless guides, and in fig. 4.18b it is shown vibrating vertically in a gravitational field.

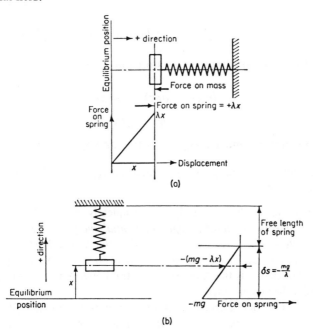

Fig. 4.18

(1) *Horizontal motion*

Energy stored in spring in equilibrium position $= 0$
„   „   „   „   „   deflected position
$\qquad = $ Area under force extension graph
$\qquad = \frac{1}{2} . x . \lambda x$
$\qquad = \frac{1}{2}\lambda x^2$

∴ Increase in 'potential' energy (wholly in the form of strain energy stored in system) $= \frac{1}{2}\lambda x^2$.

(2) *Vertical motion*

In equilibrium position force on mass due to gravity $= -mg$
„   „   „   „   „   „   „   „   spring $= +mg$
$\qquad$ and the force on the spring $= -mg$
In deflected position force on spring $= -(mg - \lambda x)$.

Energy stored in spring in equilibrium position $= \frac{1}{2} \frac{(-mg)^2}{\lambda}$

$$= \frac{1}{2} \frac{(mg)^2}{\lambda}$$

„ „ „ „ „ deflected position $= \frac{1}{2} \frac{(mg - \lambda x)^2}{\lambda}$

Reduction of strain energy stored in deflected spring

$$= \frac{1}{2\lambda} | (mg)^2 - (mg)^2 + 2mg\lambda x - \lambda^2 x^2 |$$

$$= mgx - \tfrac{1}{2}\lambda x^2$$

Gain of potential energy (gravitational) by $m = mgx$.

$\therefore$ Net gain change of potential energy in deflected system $= \frac{1}{2}\lambda\, x^2$ as before.

Again we see that we can ignore completely the gravitational field and its effects: *the total change in potential energy in the system due to a deflection x of the spring is equal to the change in strain energy stored in an unloaded spring due to deflection x*. It is very important to realise that it is this 'potential' or strain energy which is required in using Rayleigh's method.

When a particle is executing simple harmonic motion (c.f. §0.14) its velocity is given by $-\omega r \sin(\omega t + \phi)$, and therefore its maximum velocity as it passes through the mid-position by ωr, and its maximum kinetic energy by $\frac{1}{2}m(\omega r)^2$. In this case the maximum KE is $\frac{1}{2}m(\omega_n x)^2$, and by equating this to the reduction in 'potential' energy in the mid-position we obtain

$$\omega^2_n = \frac{\frac{1}{2}\lambda x^2}{\frac{1}{2}mx^2} = \frac{\lambda}{m}; \ \text{or} \ \omega_n = \sqrt{\frac{\lambda}{m}}$$

4.12.3 Spring-mass system in which the mass of the spring matters

A very simple example of the application of Rayleigh's method is afforded by the case of mass M on the end of a spring whose mass m is not negligible. Since we know that the presence of a gravitational field does not affect the motion, we shall ignore it. We shall first make the simplest possible assumption about the shape of the displacement curve of the spring, namely that the amplitude of motion of an element δl distant l from the fixed end of the spring as shown in fig. 4.19 is

$$\frac{l}{L} X$$

where L is the whole length of the spring and X the amplitude of M. This is clearly not correct, because it implies that the force required

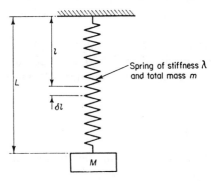

Fig. 4.19

to accelerate the material of the spring does not distort it, but it makes the calculation of the energies very easy.

The strain energy stored in the spring in the deflected position is

$$\tfrac{1}{2}\lambda X^2$$

where λ is the stiffness of the spring.

The kinetic energy of the mass M at the mid-position is as before

$$\tfrac{1}{2}M(\omega_n X)^2$$

The mass of the element δl is $m\dfrac{\delta l}{L}$, so its kinetic energy at the mid position is $\tfrac{1}{2}\left(m\dfrac{\delta l}{L}\right)\left(\omega_n \dfrac{l}{L} X\right)^2$, and of the whole spring is therefore

$$\tfrac{1}{2}m\omega_n^2 X^2 \frac{1}{L^3}\int_0^L l^2 dl$$

$$= \tfrac{1}{2}\frac{m}{3}\,\omega_n^2 X^2$$

So the total kinetic energy in the system is

$$\tfrac{1}{2}\left(M + \frac{m}{3}\right)\omega_n^2 X^2$$

and equating this to the strain energy we get

$$\omega_n = \sqrt{\frac{\lambda}{M + \frac{1}{3}m}} \qquad (4.18)$$

and

$$f_n = \frac{1}{2\pi} \sqrt{\frac{\lambda}{M + \frac{1}{3}m}}$$

From equation (4.2) it can be seen that the radiancy of the system is the same as that of a mass $M + \frac{1}{3}m$ supported by a massless spring of stiffness λ.

It is of interest to compare this solution with the exact solution which has been considered in § 4.10. In order to carry out this comparison it is necessary to put equation (4.18) in the form

$$\omega_n = A'' \sqrt{\frac{\lambda}{m}}$$

where $A'' = 1/\sqrt{(M/m + \frac{1}{3})}$.

This approximate coefficient A'' can then be compared directly with the similar 'correct' coefficient $A' = \omega_n \sqrt{\dfrac{\rho}{E}} \cdot l$ given at the end of § 4.10. The values are given for various values of m/M in table 4.5.

Table 4.5

m/M	0·1	1	10	100	∞
A''	0·311	0·866	1·519	1·707	1·8157
A'	0·311	0·860	1·429	1·555	1·5708
% error	Negligible	+0·7	+6·3	+9·8	+15·6

It will be seen that unless the spring mass is greater than the attached mass, the errors involved in assuming a linear deflection curve are less than 1 per cent. This table shows that we can make a large error in our assumptions as to the shape of the deflection curve without necessarily getting a large error in the frequency, but that in the extreme case of an unloaded spring the initial assumption that the mass of the spring had no effect on the dynamic deflection curve was too far from the truth.

4.12.4 Vibration of an 'unloaded' spring

A *less* erroneous (though clearly far from accurate) assumption in
this case would be *that the deflection was the same as that which*

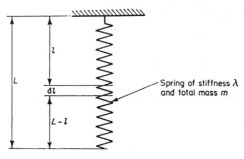

Spring of stiffness λ
and total mass m

Fig. 4.20

would occur in a gravitational field. The force acting on the element is
now that due to the unsupported part of the spring as shown in fig.
4.20, namely

$$\left(\frac{L-l}{L}\right) mg$$

The stiffness of this element $= \frac{L}{\delta l}\lambda$

So the deflection in it, (force/stiffness) $= \frac{(L-l)}{L^2\lambda} mg\delta l$

and the deflection from the fixed end to this point

$$= \frac{mg}{\lambda L^2}\int_0^l (L-l)\delta l$$

$$= \frac{mg}{\lambda L^2}(Ll - \tfrac{1}{2}l^2)$$

If we assume that the dynamic deflection curve is identical to this
static deflection curve, then the kinetic energy of the element in
mid-position is

$$\tfrac{1}{2}\left(\frac{\delta l}{L}m\right) \omega_n^2 \left\{\frac{mg}{\lambda L^2}(Ll - \tfrac{1}{2}l^2)\right\}^2$$

and of the whole spring

$$\frac{m^3\omega_n^2 g^2}{2\lambda^2 L^5}\int_0^L (L^2l^2 - Ll^3 + \tfrac{1}{4}l^4)dl$$

$$= \frac{1}{15}\frac{m^3\omega_n^2 g^2}{\lambda^2}$$

The work done by gravity in deflecting the element is

$\frac{1}{2}$ maximum force on element $\times$ deflection

$$= \frac{1}{2}\left(\frac{\delta l}{L}mg\right)\frac{mg}{\lambda L^2}(Ll - \frac{1}{2}l^2)$$

and for the whole spring

$$\frac{m^2g^2}{2\lambda L^3}\int_0^L (Ll - \frac{1}{2}l^2)dl = \frac{1}{6}\frac{m^2g^2}{\lambda}$$

and this work must be stored as strain energy in the spring.

Equating energies

$$\omega_n{}^2 = \frac{15\lambda^2}{m^3g^2}\frac{m^2g^2}{6\lambda}$$

$$= 2\cdot 5\frac{\lambda}{m}$$

or $\omega_n = 1\cdot5811\ \sqrt{(\lambda/m)}$, which has to be compared with the accurate answer from § 4.10 of $\omega_n = \frac{\pi}{2}\sqrt{\frac{\lambda}{m}} = 1\cdot5708\ \sqrt{\frac{\lambda}{m}}$. The error is well under 1 per cent.

4.12.5 Transverse vibration of a beam: general case

In the case of a beam we need not repeat the previous argument, in that it is equally obvious that the total strain stored in the deflected beam is a function of its deflected shape, and is totally unaffected by the presence or absence of a gravitational field. If we assume any such shape we can calculate this energy from the elementary principles of strength of materials. Thus, if an element, length δx, of an initially straight beam is subjected to a gradually applied bending moment M, and the corresponding change of slope in δx is $\delta\phi$ the strain energy stored in the element is

$$\frac{1}{2}M\delta\phi$$

and $\delta\phi = \delta x/R$, where R is the radius of curvature of this element.
The equation relating M and R is well known

$$\frac{M}{I} = \frac{E}{R}$$

so the strain energy stored in the element $= \frac{1}{2}M\frac{M}{EI}\delta x$

and the strain energy stored in the whole beam

$$= \tfrac{1}{2} \int_0^l \frac{M^2}{EI} \, dx \qquad (4.19)$$

Suppose we consider first a simply supported uniform beam of total mass m. We have said that we can assume any *reasonable* deflection curve. This, however, is not as simple as it might appear. Suppose, for instance, that we were ingenuous enough to say 'a parabola looks a reasonable curve' and to choose, therefore, the expression

$$y = 4y_0 \frac{x}{l}\left(1 - \frac{x}{l}\right)$$

(which is the standard parabola $y = 4ax^2$ transferred to give $y = 0$ at $x = 0$ and $x = l$, and $y = y_0$ at $x = l/2$. From symmetry this gives $dy/dx = 0$ at $x = l/2$).

We could proceed as follows

$$\frac{d^2y}{dx^2} = -\frac{8y_0}{l^2}$$

$$\therefore \quad M = EI\frac{d^2y}{dx^2} = -8EI\frac{y_0}{l^2}$$

and the change in potential energy is

$$\frac{EI}{2}\int_0^l \frac{64y_0^2 dx}{l^4} = \frac{32EI\,y_0^2}{l^3}$$

The kinetic energy of the system in the equilibrium position will be

$$\frac{m\omega_n^2}{2l}\int_0^l \left\{4y_0\frac{x}{l}\left(1 - \frac{x}{l}\right)\right\}^2 dx = \frac{4}{15}\,m\omega_n^2 y_0^2$$

Equating the change of potential energy and the kinetic energy gives

$$\omega_n = 10\cdot 95 \sqrt{\frac{EI}{ml^3}}$$

which is $10\cdot9$ per cent higher than the exact solution given in § 4.11. The reason for this rather large discrepancy is that we have not satisfied the condition that the bending moment is zero at $x = 0$ and $x = l$, which would have implied that d^2y/dx^2 should have been zero, whereas for the curve assumed it has a constant value of $-8y_0/l^2$. In other words, the assumed curve cannot be correct, as it does not give simply supported end conditions. If we had appreciated this to begin with we might have realised why the 'static deflection curve' is so good a choice, since it can easily be made to suit *any* end conditions which are applied to the vibrating beam.

Let us now consider the same problem, but assuming a dynamic deflection curve identical to the static deflection curve. To avoid confusion both in relation to 'potential' energy and in relation to sign conventions, it is probably easiest and best to think of the beam in the vertical position as shown in fig. 4.21a. It is, of course, perfectly permissible to deal with it in the horizontal position, with axes as in either fig. 4.21b or 4.21c, as suggested at the end of

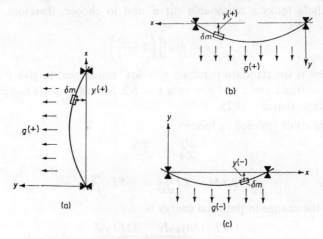

Fig. 4.21

§4.11.1, since these axes follow the convention recommended in §0.2, that anti-clockwise rotation takes us from x to y axis, though the negative gravitational field and deflections in (c) do not help to clarify the argument. In this case we shall choose the easy way shown in fig. 4.21a: in subsequent cases where we are merely applying the general result there is no need for uniformity.

To determine the strain energy associated with the 'static deflection curve' we can assume that a 'gravitational field' acts in the positive y direction. Let this field grow *gradually* from zero until it reaches an intensity g. Any element of the beam of mass δm will experience a gradually increasing force, reaching a maximum value δmg, while suffering a gradually increasing deflection which reaches a maximum value, say y. The work done on the element by the gravitational field will be $\frac{1}{2}(\delta mg)y$, and on the whole beam

$$\int_0^l \frac{1}{2} gy\,dm$$

and this work must be stored in the beam in the form of strain energy. If the gravitational field is now suddenly removed the beam will vibrate about the x axis, the amplitude of δm being y, the kinetic energy of the element in the mean position $\frac{1}{2}\delta m(\omega_n y)^2$, and of the whole beam

$$\int_0 \tfrac{1}{2}\,\omega_n^2 y^2 dm$$

Equating these energies we have

$$\omega_n{}^2 = g\frac{\displaystyle\int_0^l y\,dm}{\displaystyle\int_0^l y^2\,dm} \tag{4.20}$$

Even if the beam is loaded separately or additionally with discrete masses, the argument is completely unaffected, but we should normally write the summations in the form

$$\omega_n{}^2 = g\frac{\sum y\,\delta m}{\sum y^2 \delta m} \tag{4.21}$$

Before we leave the general case it is worth while to note that a simply supported beam with an 'overhung' mass as in fig. 4.22 would

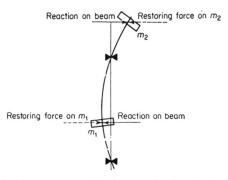

Reaction on beam   Restoring force on m_2

m_2

Restoring force on m_1 Reaction on beam

m_1

Fig. 4.22 Beam with overhung mass vibrating in first mode

fairly obviously have a first mode of vibration in which the masses were moving in opposite phases, so the strain energy we have to calculate is that for a beam deflected by a 'gravitational field' acting, say, to the left on the central mass and to the right on the overhung mass. This is perhaps a shade less easy to understand if the beam is thought of in the horizontal position, though of course no less true.

We can now consider a few specific problems to show how the
Rayleigh method and equation (4.20) or (4.21) are applied. We
deliberately choose cases we have already dealt with by more rigorous
methods so that we can check the degree of accuracy of the result.

4.12.6 Frequency of transverse vibration of a uniform beam of total mass m, assuming that static and dynamic curves are identical

Let us this time consider a simply-supported beam with axes as in
fig. 4.23. In a gravitational field g the rate of loading on the beam is

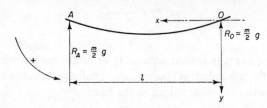

Fig. 4.23

mg/l, so we can immediately write down the bending moment at x
as

$$BM = -\frac{mg}{2}(l - x) + \frac{mg}{l}\frac{(l - x)^2}{2} = \frac{mg}{2}\left(\frac{x^2}{l} - x\right)$$

$$\therefore EI\frac{d^2y}{dx^2} = \frac{mg}{2}\left(\frac{x^2}{l} - x\right)$$

$$\therefore EIy = \frac{mg}{2}\left(\frac{x^4}{12l} - \frac{x^3}{6}\right) + C_1 x + C_2$$

When $x = 0$ $y = 0$ $\therefore C_2 = 0$

When $x = l$ $y = 0$ $\therefore C_1 = \frac{1}{24}mgl^2$

$$\therefore y = \frac{mg}{24EIl}(x^4 - 2lx^3 + l^3 x)$$

hence $$\int_0^l y\,dx = \frac{mg}{24EIl}\cdot\frac{l^5}{5}$$

and $$\int_0^l y^2\,dx = \frac{m^2g^2}{576E^2I^2l^2}\cdot\frac{31l^9}{630}$$

Substituting these values into equation (4.20) leads to

$$\omega_n = 9{\cdot}8768\sqrt{\frac{EI}{ml^3}}$$

as compared with the exact solution for the first mode (i.e. $n = 1$) given in case 1 under § 4.11.5

$$\omega_n = 9\cdot8696 \sqrt{\frac{EI}{ml^3}}$$

showing that the error in the approximate method is less than $0\cdot1$ per cent.

It is equally easy to deal with the other end conditions considered in § 4.11.5.

4.12.7 Transverse vibration of a beam with concentrated masses

In general terms, there is nothing to add to § 4.12.5. We shall therefore take as a specific, or worked, example the massless shaft carrying two concentrated masses which was illustrated in fig. 4.16a, for which the exact solution was calculated in § 4.11.4.

It is convenient to calculate the static deflections by first working out in general terms the deflection at *any* point in a shaft due to

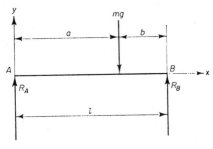

Fig. 4.24

a *single* concentrated mass as sketched in fig. 4.24. There is no difficulty about this. If we take our axes as indicated we can write down the numerical value of the bending moment at section x as

$$x < a; \text{ adding when } x > a$$
$$M = R_a \cdot x \qquad\qquad -mg(x - a)$$

and since the bending moment in the beam at $x = a$ is $+R_b \cdot b$ our signs are clearly correct, and will lead to a negative value of y in the negative gravitational field. Since we are interested only in the product of $g \cdot y$ or of y^2, we can possibly avoid some awkwardness by deliberately altering this sign, for instance by assuming the gravitational field to act upwards, so that the beam will be regarded as deflecting as was in fact sketched in fig. 4.16b. We then have

$$x < a; \text{ adding when } x > a$$

$$EI \frac{d^2y}{dx^2} = -mg \frac{b}{l} \cdot x \qquad\qquad + mg(x - a)$$

$$EIy = -\frac{mgbx^3}{6l} + Ax + B \qquad + \frac{1}{6}mg(x - a)^3$$

whence, for $x < a,\; y = \dfrac{mg}{6EI} \cdot \dfrac{bx}{l}(l^2 - b^2 - x^2)$

We can now calculate the total deflection y_1 under m_1 in fig. 4.15 by adding the deflections due to m_1 and to m_2. Inserting numerical values

$$y_1 = 0 \cdot 8779 \times 10^{-7} \times (36 \times 9 \cdot 81) + 0 \cdot 4587 \times 10^{-7} \times (32 \times 9 \cdot 81)$$
$$= 31 \cdot 003 \times 10^{-6} + 14 \cdot 400 \times 10^{-6} = 45 \cdot 403 \times 10^{-6} \,\text{m}$$

Similarly, the deflection y_2 under m_2 will be

$$y_2 = 0 \cdot 3676 \times 10^{-7} \times (32 \times 9 \cdot 81) + 0 \cdot 4587 \times 10^{-7} \times (36 \times 9 \cdot 81)$$
$$= 11 \cdot 539 \times 10^{-6} + 16 \cdot 200 \times 10^{-6} = 27 \cdot 739 \times 10^{-6} \,\text{m}$$

We have now only to insert these values in equation (4.21) to find the approximate value of ω_n: since, however, we are going on to find a better approximation let us call this one ω_1. Hence

$$\omega_1{}^2 = g \cdot \frac{\sum y \delta m}{\sum y^2 \delta m}$$

$$= \frac{9 \cdot 81 \,\{36 \times 45 \cdot 403 \times 10^{-6} + 32 \times 27 \cdot 739 \times 10^{-6}\}}{36 \times 45 \cdot 403^2 \times 10^{-12} + 32 \times 27 \cdot 739^2 \times 10^{-12}}$$

$$= 250\,340$$

$$\therefore\; \omega_1 = 500 \cdot 3 \text{ rad/s and } f_1 = 796 \cdot 3 \text{ Hz} = 4\,778 \text{ c/min}$$

4.12.8 *Improving by successive approximation the first frequency computed using Rayleigh's method*

The value of ω_1 we have calculated for the natural radiancy of the beam discussed in the preceding paragraph will, for the reasons given in § 4.12.1, be too high, and had we not previously determined the correct value, we might have had some anxiety about the magnitude of the error. It is possible to improve the answer and dispel this anxiety by a successive approximation method.

It may be helpful in promoting understanding if we repeat the reason for the existence of an error when we assume a dynamic deflection identical to the static deflection. In the case of a single mass we equate energies and so obtain

$$\omega_n{}^2 = \frac{gm\delta_s}{m(\delta_s)^2} = \frac{g}{\delta_s}$$

and no error is introduced because the mass can certainly vibrate with an amplitude δ_s. In the multi-mass case we equate energies and obtain

$$\omega_1{}^2 = g\,\frac{\sum m\delta_s}{\sum m(\delta_s)^2}$$

but ω_1 will be exactly equal to ω_n only if the system can vibrate in a stable fashion with each individual amplitude equal to the static deflection. This is not in general the case: if such a deflection curve were initially imposed some or all of the other modes of vibration would be excited. It is, of course, not the absolute amplitude of any one mass which is incompatible, but the relative amplitudes, i.e. in the two-mass system we are discussing, the ratio of the amplitudes. It is this *ratio* which we shall now attempt to improve, but it is probably easier to follow the argument, the units and the arithmetic if we discuss the problem in terms of amplitudes similar to the static deflection. The more sophisticated can delete $\omega_1{}^2$ from the expressions we obtain or substitute $\omega_1 = 1$.

The method of finding the natural frequency of vibration, starting again from the beginning, can be summarised as follows:

(a) We assume gravity to act on the masses, i.e. that the deflecting forces are m_1g and m_2g.

(b) We calculate the deflection under *each* mass due to *both* forces and obtain

$$y_1 (= \delta_{s1}) \text{ and } y_2 (= \delta_{s2})$$

(c) We calculate the kinetic and strain energies and equate to find the value ω_1 as in § 4.12.7.

(d) We now calculate the forces which the shaft must apply to each mass to give it the necessary acceleration in the fully deflected state, namely

$$-m_1\omega_1{}^2 y_1 \text{ and } -m_2\omega_1{}^2 y_2$$

These individual forces are not the ones first assumed, but much more importantly they will not be in the same ratio as m_1g/m_2g, which implies that the assumed shape of the deflection curve is wrong.

(*e*) (Repeating (*b*)) we calculate the deflection under each mass due to both the forces $m_1\omega_1^2 y_1$ and $m_2\omega_1^2 y_2$, the forces which are acting on the beam, and obtain

$$y_1' \text{ and } y_2'$$

(*f*) (Repeating (*d*)) we calculate the forces which the shaft must apply . . . and so on.

With each repetition we progress towards the situation in which the two forces applied by the deflected shaft and similarly the amplitudes reach a constant ratio. In other words, when this ratio becomes constant, perfect agreement between the shape of the assumed and actual curves will have been achieved, and we can calculate the accurate value of ω_n.

It will probably help to clarify this argument if we go through an outline of the arithmetic for the beam we have been discussing:

(*a*) The assumed forces were

$$m_1 g = 36 \times 9.81, \quad m_2 g = 32 \times 9.81$$
$$= 353.2 \text{ N} \qquad = 313.9 \text{ N}$$

(*b*) The calculated deflection under each mass was

$$y_1 = 45.403 \times 10^{-6}, \quad y_2 = 27.739 \times 10^{-6} \text{ m}$$

and we note that $y_1/y_2 = 1.637$

(*c*) We calculate and equate the energies and find

$$\omega_1^2 = 250\ 340 \ (\text{rad/s})^2$$

(*d*) Assuming this value we calculate

$$m_1 \omega_1^2 y_1 = 36 \times 250\ 340 \times 45.403 \times 10^{-6} \text{ kg m/s}^2$$
$$= 409.2 \text{ not } 353.2 \text{ N}$$

and similarly $m_2 \omega_1^2 y_2 = 222.2$ not 313.9 N

We should note that we could have equally well assumed $\omega_1 = 1$, but this would have made the forces appear unrealistically small.

(*e*) The deflections due to these loads would be

$$y_1' = 0.8779 \times 10^{-7} \times 409.2 + 0.4587 \times 10^{-7} \times 222.2$$
$$= 46.115 \times 10^{-6} \text{ m}$$
$$y_2' = 26.938 \times 10^{-6} \text{ m}$$

and we note that $\dfrac{y_1'}{y_2'} = 1.712$

We can now go on to the third and if necessary fourth approximation. Table 4.6 gives the result which shows the rapid convergence towards the correct value of y_1/y_2 of 1·7196 as given in § 4.11.4.

Table 4.6

APPROXIMATION	1	2	3	4
F_1 N	353·2	409·2	415·6	418·0
F_2 N	313·9	222·2	215·8	215·8
$y_1 \times 10^6$ m	45·40	46·12	46·38	46·60
$y_2 \times 10^6$ m	27·74	26·94	27·00	27·11
y_1/y_2	1·637	1·712	1·718	1·719

At any stage we can calculate the corresponding approximate value of ω_n^2, which converges even more rapidly towards its correct value. If we wait until the fourth approximation has been obtained, by which time we see that the ratio y_1/y_z is within the limits of accuracy of our calculation constant, we can proceed:
Strain energy in deflected beam is

$$\tfrac{1}{2}\{418\cdot0 \times 46\cdot60 \times 10^{-6} + 215\cdot8 \times 27\cdot11 \times 10^{-6}\}\ \text{Nm}$$
$$= 12\cdot66 \times 10^{-3}\ \text{Nm}$$

Kinetic energy in mid position is

$$\frac{\omega_n^2}{2}\{36 \times 46\cdot60^2 \times 10^{-12} + 32 \times 27\cdot11^2 \times 10^{-12}\}\ \text{Nm}$$
$$= 50\cdot84 \times 10^{-9}\ \omega_n^2\ \text{Nm}$$

Equating energies, $\omega_n^2 = 249\ 020$
$$\omega_n = 499\ \text{rad/s}$$

which agrees with the 'exact' answer calculated in § 4.11.4 to an order of accuracy far greater than can really be justified by the calculations. If we had assumed $\omega_1 = 1$ and accepted the inconvenience of F_1 and F_2 becoming much smaller at each successive approximation, we would have obtained an identical answer, and the student is advised to go through the arithmetic and convince himself that this is so.

It might appear from this example that Rayleigh's method is just as complicated as the analytical approach of § 4.9, but the successive approximation method outlined in this paragraph is seldom required. Moreover, for a system with many more masses the analytical method becomes excessively laborious, whereas Rayleigh's method is still relatively easy to apply. Most important of all, there are many cases, for instance when the mass of a loaded shaft cannot be neglected or when its section is not uniform, in which an analytical solution cannot be obtained. It is still possible to obtain easily and quickly a good approximation to the natural frequency by using the energy method.

§ 4.13 Coupled Vibrations

In most vibration problems the effects of coupling between the six principal motions are neglected; for instance, it is frequently assumed that torsional vibration of a shaft causes no axial vibration, and

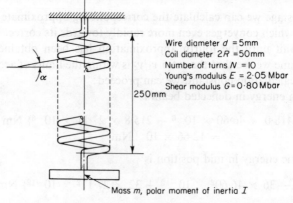

Wire diameter d = 5mm
Coil diameter $2R$ = 50mm
Number of turns N = 10
Young's modulus E = 2·05 Mbar
Shear modulus G = 0·80 Mbar

250mm

Mass m, polar moment of inertia I

Fig. 4.25

vice versa. Nevertheless, there are some problems where coupling cannot be neglected. A simple example of a case in which it may not be safe to do so is afforded by a helical spring which is fixed at one end and supports a mass in the form of a flywheel at the free end. When an axial force is applied to this spring there is a small torsional displacement in addition to the axial deflection, and similarly when a torque is applied about the axis of the spring there is a small axial deflection in addition to the torsional displacement. As a consequence, axial vibration of the mass is accompanied by torsional vibration, which may or may not be trivial in amount, and vice versa.

In the particular case being considered the axial deflection due to a force P is P/λ, and this is accompanied by a torsional displacement which we may designate as $P/C_{\theta x}$, where $C_{\theta x}$ is the 'coupling constant'. Similarly, a torque T causes a torsional displacement of T/k, which is accompanied by an axial deflection of $T/C_{x\theta}$. From Maxwell's reciprocal theorem* in strength of materials, it is deduced that $C_{\theta x} = C_{x\theta} = C$, so that only three stiffness parameters are involved in the problem.

Let the system be disturbed; and then at some particular instant let the axial and torsional displacement be x and θ. In terms of the restoring force and torque in the spring the displacements are

$$x = \frac{P}{\lambda} + \frac{T}{C}$$

and

$$\theta = \frac{T}{k} + \frac{P}{C}$$

These two equations can be rearranged to give the restoring force and torque in terms of x and θ, and this leads to

$$P = \frac{\lambda C}{(C^2 - \lambda k)}(Cx - k\theta)$$

and

$$T = \frac{kC}{C^2 - \lambda k}(C\theta - \lambda x)$$

Considering the equations of motion for rotation about and displacement along the axis of the spring we get

$$m\ddot{x} = -\frac{\lambda C}{(C^2 - \lambda k)}(Cx - k\theta)$$

and

$$I\ddot{\theta} = -\frac{kC}{(C^2 - \lambda k)}(C\theta - \lambda x)$$

Assuming $\theta = A \cos \omega_n t$ and $x = B \cos \omega_n t$, then substituting in these two equations of motion gives

$$-m\omega_n^2 B = -\frac{\lambda C}{(C^2 - \lambda k)}(CB - kA) \qquad (4.22)$$

and

$$-I\omega_n^2 A = -\frac{kC}{(C^2 - \lambda k)}(CA - \lambda B) \qquad (4.23)$$

These two equations contain two unknowns, ω_n^2 and the amplitude ratio A/B, and consequently they can be solved for these two unknowns. As an example, consider the spring shown in fig. 4.25.

* See, for example, Sir R. Southwell, *Introduction to the Theory of Elasticity*, 2nd edn., Oxford University Press, 1941.

From the theory of open coiled springs* the stiffnesses we are interested in are

$$\lambda = 4 \cdot 964 \text{ kN/m}$$
$$k = 3 \cdot 926 \text{ Nm/rad}$$
$$C = 3 \cdot 627 \text{ kN/rad}$$

Substituting these values in equations (4.22 and 4.23), we can obtain the equation

$$B = \frac{5 \cdot 381 A}{4\,971 - m\omega_n{}^2} \tag{4.24}$$

Combining this with equation (4.23) leads to the frequency equation

$$Im\omega_n{}^4 - (4\,971I + 3 \cdot 932m)\omega_n{}^2 + 19\,516 = 0 \tag{4.25}$$

and this can be solved for any particular value of m and I. For instance, if the mass m is in the form of a disk 300 mm diameter and 25 mm thick made from steel with a density of $7 \cdot 85 \times 10^3$ kg/m³, then

$$m = \pi r^2 t \rho = \pi \times \frac{0 \cdot 09}{4} \times 0 \cdot 025 \times 7 \cdot 85 \times 10^3 = 13 \cdot 87 \text{ kg}$$

$$I = m\frac{r^2}{2} = \frac{13 \cdot 87}{2} \times \frac{0 \cdot 09}{4} = 0 \cdot 156 \text{ kg m}^2$$

and solving equation (4.25) gives

$$\omega_n{}^2 = 25 \cdot 0 \text{ or } 358 \cdot 7$$

and

$$f_n = 0 \cdot 7958 \text{ Hz or } 3 \cdot 014 \text{ Hz}$$

From these values we can calculate the ratio of axial to torsional vibration amplitude for both modes of vibration. Improved accuracy is achieved if equation (4.24) is used for the low-frequency

---

* It can be shown (cf. S. Timoshenko, *Strength of Materials*, vol. 2, Van Nostrand, 1948) that

$$\lambda = \frac{\cos \alpha}{2\pi N R^3 \left\{ \dfrac{\cos^2 \alpha}{GI_p} + \dfrac{\sin^2 \alpha}{EI_d} \right\}}$$

$$k = \frac{\cos \alpha}{2\pi N R \left\{ \dfrac{\sin^2 \alpha}{GI_p} + \dfrac{\cos^2 \alpha}{EI_d} \right\}}$$

$$C = \frac{\cos \alpha}{2\pi N R^2 \sin \alpha \cos \alpha \left\{ \dfrac{1}{GI_p} - \dfrac{1}{EI_d} \right\}}$$

where α is the helix angle

N the number of coils

R the radius of the coils

I_p the polar second moment of area of the wire cross-section

I_d the second moment of area about a diameter of the wire cross-section.

mode and the similar equation deduced from (4.23) is used for the high-frequency mode, as this avoids finding the small difference between two large numbers. The ratios found are

$$\frac{B}{A} = +0\cdot0012 \text{ and } -9\cdot667 \text{ m/rad}$$

It will be seen that in this particular case the coupling has a negligible effect, and indeed the two natural frequencies deduced ignoring the coupling constant would be 0·7984 Hz. and 3·011 Hz. On the other hand, an interesting special case of coupled vibrations is afforded by 'Wilberforce's spring', in which a mass suspended from a coiled spring has two protruding screws, as shown in fig. 4.26, with masses which can be adjusted radially so that the polar moment of inertia I can be altered without altering the total mass m. In a suitable design, it is possible to adjust I so that when the mass is displaced vertically and then released an axial vibration occurs initially without a torsional component, but gradually the motion changes to a torsional vibration

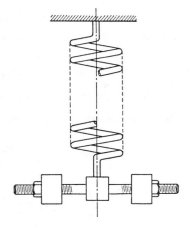

Fig. 4.26 Wilberforce's spring

with no axial component, and so on. This phenomenon can be explained by the existence of two modes of vibration having amplitude ratios of

$$\frac{B}{A} = +x \text{ and } -x$$

At the instant of release both these modes are excited in equal measure in such a way that the axial displacements add up and the torsional displacements cancel one another. But as the frequencies of the modes are different, the two motions do not stay in phase, and ultimately the torsional displacements are superimposed on one another while the axial displacements cancel one another out, so there is no net axial displacement. The frequency of this 'hunting' motion or 'beat' is equal to the difference in the two natural frequencies of the system. As an example let us consider the spring discussed earlier in this paragraph supporting a mass m with moment of inertia I. Substituting for λ, k, and C in equations (4.22) and (4.23) we obtain

$$-m\omega_n^2 B = -4\,971B + 5\cdot381A \text{ N}$$
$$-I\omega_n^2 A = -3\cdot932A + 5\cdot381B \text{ Nm}$$

Substituting $B = -Ax_1$ for one mode of vibration

$$m\omega_n^2 Ax_1 = 4\,971Ax_1 + 5\cdot381A$$
$$I\omega_n^2 A = 3\cdot932A + 5\cdot381Ax_1$$

or

$$\frac{I}{m} = x_1 \frac{(3\cdot932 + 5\cdot381x_1)}{(4\,971x_1 + 5\cdot381)}$$

Substituting $B = +Ax_2$ for the other mode of vibration

$$\frac{I}{m} = x_2 \frac{(3\cdot932 - 5\cdot381x_2)}{(4\,971x_2 - 5\cdot381)}$$

To achieve the condition we want we have to equate x_1 and x_2. Hence, eliminating $\dfrac{I}{m}$ and simplifying, we have

$$x^2 = \frac{3\cdot932 \times 5\cdot381}{5\cdot381 \times 4\,971} = 0\cdot000791$$

or

$$x = 0\cdot02813$$

$$\therefore \quad \frac{I}{m} = 0\cdot02813 \frac{(3\cdot932 + 0\cdot02813 \times 5\cdot381)}{(4\,971 \times 0\cdot02813 + 5\cdot381)}$$

$$= 0\cdot000791m^2 = K^2$$

where K is the radius of gyration to give the required result.

Substituting for $I = 0\cdot000791m$ in equation (4.25) for ω_n^2 gives

$$\omega_n = \frac{69\cdot2}{\sqrt{m}} \text{ or } \frac{71\cdot9}{\sqrt{m}} \text{ rad/s}$$

and

$$f_n = \frac{11\cdot01}{\sqrt{m}} \text{ or } \frac{11\cdot44}{\sqrt{m}} \text{ Hz}$$

Consequently, the frequency of the beat will be

$$\frac{11\cdot44}{\sqrt{m}} - \frac{11\cdot01}{\sqrt{m}} = \frac{0\cdot43}{\sqrt{m}} \text{ Hz}$$

§ 4.14 Free Vibration of a Gyroscope

Gyroscopes are frequently employed in guidance and stabilising systems and their dynamic response, and hence natural frequency, are consequently of considerable interest. With a freely mounted gyroscope the restoring couples are a consequence of what might be termed a 'dynamic coupling' between the precessional motions about two mutually perpendicular axes at right angles to the axis of spin. For the purpose of our analysis let us consider the gyroscope shown in fig. 4.27, which is freely mounted in a gimbal frame to allow the axis of spin, ox, to precess about oy and oz. In the initial position the axis of spin is horizontal and the polar moment of inertia of the flywheel is I_x and its velocity of spin ω_x. The moment of inertia of

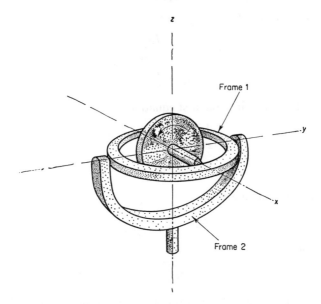

Fig. 4.27 Gimbal-mounted gyroscope

the frame 1 and flywheel about the y axis is I_y, and of the frames 1 and 2 plus the flywheel about the z axis is I_z.

Let us now consider the gyroscope when it is slightly disturbed from its original equilibrium position so that at some instant there are precessional velocities of ω_y and ω_z about the y and z axes respectively. Considering the motion about the y axis and assuming that the angular displacements are small, we can see that the rate of change of angular momentum due to positive rotation about the z axis acts about the y axis, and as no external torques are applied it causes an equal and opposite torque reaction about the y axis which is in a clockwise or negative sense. The equation of motion will be

$$I_y \dot{\omega}_y = -I_x \omega_x \omega_z$$

Similarly if we consider rotation about the z axis we can put down

$$I_z \dot{\omega}_z = +I_x \omega_x \omega_y$$

Rewriting the equations we get

$$I_y \dot{\omega}_y + I_x \omega_x \omega_z = 0$$

and

$$I_z \dot{\omega}_z - I_x \omega_x \omega_y = 0$$

From the second of these equations

$$\omega_y = \frac{I_z \dot{\omega}_z}{I_x \omega_x}$$

or

$$\dot{\omega}_y = \frac{I_z \ddot{\omega}_z}{I_x \omega_x}$$

Substituting these into the first equation

$$\frac{I_y I_z \ddot{\omega}_z}{I_x \omega_x} + I_x \omega_x \omega_z = 0$$

or

$$\ddot{\omega}_z + \frac{I_x^2 \omega_x^2 \omega_z}{I_y I_z} = 0$$

Solving this equation gives

$$\omega_z = C_2 \cos\left(\frac{I_x \omega_x}{\sqrt{(I_y I_z)}} t - \phi_2\right)$$

and similarly

$$\omega_y = C_1 \cos\left[\frac{I_x \omega_x}{\sqrt{(I_y I_z)}} t - \phi_1\right]$$

where C_1, C_2, ϕ_1, and ϕ_2 are arbitrary constants depending on initial conditions. The radiancy of the vibration will be

$$\omega_n = \frac{I_x \omega_x}{\sqrt{(I_y I_z)}}$$

or

$$f_n = \frac{1}{2\pi} \frac{I_x \omega_x}{\sqrt{(I_y I_z)}}$$

When disturbed the gyroscope executes a coupled angular vibration about the oy and oz axes, so the end of the axis of spin will describe an ellipse in a plane parallel to the zy plane.

§ 4.15 Experimental Techniques for Finding Natural Frequencies

Despite the analytical techniques which have been developed for computing natural frequencies, there are many practical problems which are still not readily amenable to analysis, and experimental methods have been established to investigate these more complicated systems. Perhaps one of the best examples is the practical determination of the natural frequencies of a turbine blade which may be of aerofoil cross-section and perhaps twisted from root to tip. This is a problem of very great importance in the development of steam and gas turbines.

The classical, and still frequently employed, method of exciting natural frequencies of fairly small components is to stroke them with a resined violin bow. For these circumstances the coefficient of friction is less under conditions of relative motion than under static conditions. Consequently, when the bow is passed across the member 'stick–slip' excitation of one or more of the natural frequencies tends to occur (see § 5.12). Alternatively, a lump of solid carbon dioxide can be held against a component. If the latter is a good thermal conductor contact causes a sudden generation of gas, which forces the component and the lump of carbon dioxide apart; gas escapes and contact is remade and again vibration is probable. Free vibrations can be similarly excited by holding a compressed-air jet close to the component. All these methods are, however, rather hit and miss, as the mode of vibration excited cannot easily be controlled, except to some minor extent by adjusting the position of application of the exciting force.

Various methods of electrical excitation are now extensively used. One of the methods most frequently employed is to connect a light coil (suspended in a powerful magnetic field as in a conventional loudspeaker) mechanically to the component under examination, and to feed it with an alternating current from a variable-frequency oscillator. This ensures that only a single mode of vibration is excited at any one time, and the frequency at which each resonance occurs can be read directly from the oscillator if this is suitably calibrated.

Even if the frequencies at which resonances occur are determined, it is difficult to know precisely the mode that is being excited at any one frequency. This can be very important, especially if it is desired to correlate failures in service with some particular mode. Visualisation of the mode of vibration can often be readily achieved by sprinkling fine dry sand on the cleaned surface of the component if it is reasonably level. When resonance occurs the sand moves up and down with the surface, and if the acceleration exceeds that due to gravity the sand will 'dance' on the surface and ultimately fall off or accumulate along the line of the nodes where the surface acceleration is zero. Consequently, a 'sand pattern' is produced which shows the position of all the nodal lines. Examples of such patterns are shown in fig. 4.28.

Alternatively, the position of the nodal lines can be found by using, for instance, a proximity meter, an instrument which is sensitive to the distance between a fine probe and the surface of the component. Yet another method involving less-expensive equipment is to place a pencil point lightly against the surface. This is sufficient to damp considerably the resonant vibration, except when the pencil point is at a node. Consequently, the position of the nodal lines can be traced out.

References

See end of chapter 5.

Examples

4.1 A massless rod AB of length 800 mm is pivoted at A and supports a mass of 4 kg concentrated at B. A vertical spring is attached to point C at a distance 500 mm from the pivot, and the spring tension is adjusted so that under equilibrium conditions the rod is horizontal. The stiffness of the spring is 1·5 kN/m.

If the rod is displaced a small distance from its equilibrium position and then released, find the natural frequency.

4.2 A disk of 500 mm diameter, thickness 40 mm, and density 7 850 kg/m³ oscillates torsionally at a frequency of 300 c/min on one end of a shaft which is built in at the other end. A second disk is then bolted to the first and the frequency is reduced to 200 c/min.

Find the moment of inertia of the second disk.

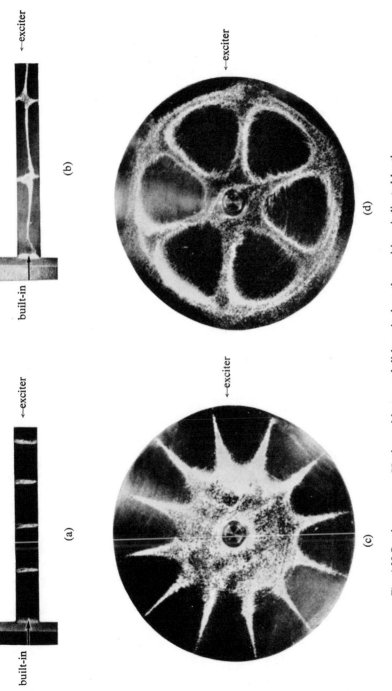

Fig. 4.28 Sand patterns. Horizontal beam and disk excited, at the positions indicated by the arrows, by vertical harmonic forces of different frequencies

4.3 A connecting-rod is 250 mm in length between the centres of the small and big ends, and it has a big-end bearing of 80 mm internal diameter and a small end bearing of 20 mm diameter. When the rod is swung about a knife edge through the small-end bearing the frequency of small oscillations is 1·02 Hz, and when swung about a knife edge through the big end 1·12 Hz.

Find the position of the centre of gravity and the radius of gyration about the centre of gravity.

4.4 Two flywheels with moments of inertia of 5 kg m² and 8 kg m² are separated by a uniform shaft 600 mm long. The stiffness of the shaft is 300 tf m/rad.

Find the position of the node and the natural frequency of torsional vibration.

4.5 Three identical masses are equidistantly mounted on a uniform shaft. Show from first principles that the two natural frequencies of torsional oscillation of the shaft are in the ratio $1 : \sqrt{3}$.

4.6 A Ward–Leonard set for operating a D.C. motor consists of an A.C. motor direct coupled to two D.C. generators in tandem. The three armatures (in order along the shaft) have moments of inertia 0·4, 0·35 and 0·2 kg m², and the stiffness of the shaft between each unit is 5 000 Nm/rad. Find the lower natural frequency of torsional oscillation of the system.

4.7 Fig. 4.29 shows an equivalent torsional system. Find the frequency equation for the system. In the particular case where

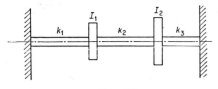

Fig. 4.29

$I_2 = 2I_1$ and $k_1 = k_2 = k_3$ find the ratio of the amplitudes of I_1 and I_2 for the two modes of vibration.

4.8 Write down the equation of motion of each of the masses shown in fig. 4.30, and hence determine the natural frequencies of the system.

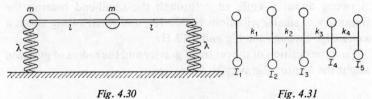

Fig. 4.30   Fig. 4.31

4.9 Fig. 4.31 shows the equivalent torsional system for an engine coupled to a dynamometer on a test bed. Show that the frequencies of the first and second modes of vibration are about 2 120 c/min and 4 465 c/min.

If when the engine is under test the amplitude of vibration of I_4 is measured and found to be 0·75°, when the system is vibrating in the second mode, determine the amplitude of the vibration torque in the shaft between I_4 and I_5.

$$I_1 = 2\text{·}70 \text{ kg m}^2 \qquad k_1 = 1100 \text{ tf m/rad}$$
$$I_2 = 3\text{·}10 \text{ kg m}^2 \qquad k_2 = 60 \text{ tf m/rad}$$
$$I_3 = 5\text{·}60 \text{ kg m}^2 \qquad k_3 = 30 \text{ tf m/rad}$$
$$I_4 = 1\text{·}30 \text{ kg m}^2 \qquad k_4 = 35 \text{ tf m/rad}$$
$$I_5 = 3\text{·}70 \text{ kg m}^2$$

4.10 Fig. 4.32 shows an equivalent torsional system, both ends of which may be considered as built-in. The frequency of the first mode of vibration lies between 3 800 c/min and 4 200 c/min.

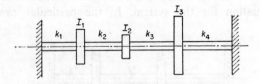

Fig. 4.32

Determine the natural frequency to closer limits.

$$I_1 = 0\text{·}42 \text{ kg m}^2 \qquad k_1 = 95 \text{ kN m/rad}$$
$$I_2 = 0\text{·}30 \text{ kg m}^2 \qquad k_2 = 150 \text{ kN m/rad}$$
$$I_3 = 0\text{·}90 \text{ kg m}^2 \qquad k_3 = 53 \text{ kN m/rad}$$
$$k_4 = 271 \text{ kN m/rad}$$

4.11 Fig. 4.33 shows the equivalent dynamic system of a resonance type fatigue machine. In this machine the mass m_5 is constrained to vibrate with a fixed amplitude at a frequency equal to the natural frequency of the second mode.

Find the value of m_5 so that the natural frequency may be 3 000 c/min. If the amplitude of vibration of m_5 is 1 cm, determine the dynamic force applied to the specimen, and the force applied to the foundations.

$$m_4 = m_3 = 3 \cdot 4 \text{ kg} \qquad \lambda_5 = \lambda_4 = \lambda_3 = \quad 485 \text{ tf/m}$$
$$m_2 = 26 \cdot 5 \text{ kg} \qquad \qquad \lambda_2 = 5350 \text{ tf/m}$$
$$m_1 = 500 \text{ kg} \qquad \qquad \lambda_1 = \quad 224 \text{ tf/m}$$

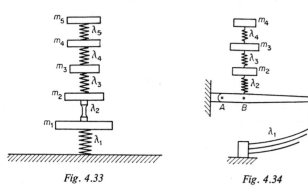

Fig. 4.33   Fig. 4.34

4.12 In order to test a relatively flabby cantilever spring in a resonance-type fatigue machine a simple lever system is used as shown diagrammatically in fig. 4.34. Resonance is excited by applying a harmonic force to mass m_4. If the frequency of this force is 3 000 c/min, find the value of the mass m_4. It may be assumed that the arm ABC and the connecting link CD are rigid.

$$m_2 = m_3 = 3 \cdot 4 \text{ kg}$$
$$\lambda_1 = 3 \cdot 65 \text{ MN/m}$$
$$\lambda_2 = \lambda_3 = \lambda_4 = 4 \cdot 62 \text{ MN/m}$$
Moment of inertia of arm ABC about $A = 0 \cdot 85 \text{ kg m}^2$
$$AB = 100 \text{ mm}$$
$$BC = 500 \text{ mm}$$

4.13 Fig. 4.35 gives information about a geared engine system. Show that the frequency of the first mode is about 3 147 c/min.

If, while the engine is subjected to torsional vibration at the frequency of the first mode, the maximum acceleration measured at I_5 is 6 000 rad/s² find the vibrational torque transmitted to I_4 by I_3.

$$I_1 = 6\cdot32 \text{ kg m}^2 \qquad k_1 = \ \ 270 \text{ kN m/rad}$$
$$I_2 = 1\cdot26 \text{ kg m}^2 \qquad k_2 = \ \ 340 \text{ kN m/rad}$$
$$I_3 = 0\cdot17 \text{ kg m}^2 \qquad k_4 = 1100 \text{ kN m/rad}$$
$$I_4 = 0\cdot02 \text{ kg m}^2 \qquad n \ = 2$$
$$I_5 = 0\cdot30 \text{ kg m}^2$$

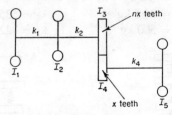

Fig. 4.35

4.14 A uniform steel bar has a length of 1·25 m. Find the frequency of the first mode of longitudinal vibration if

(*a*) both ends are clamped
(*b*) both ends are free

$E = 2\cdot07$ Mbar and $\rho = 7\ 850$ kg/m³.

4.15 A uniform steel tube 600 mm long is mounted so that it is torsionally freely supported. Find the lowest frequency of torsional vibration assuming that the shear modulus is 0·8 Mbar and that the density is 7 850 kg/m³.

4.16 When a nickel bar is subjected to an axial magnetic field its length changes, and this effect can be used to produce axial vibrations. A nickel tube, 12 mm bore, 15 mm outside diameter, and 300 mm long, carries a mass of 0·1 kg at each end. What is the lowest frequency at which resonance will occur?
For nickel, $\rho = 8\ 870$ kg/m³; $E = 2\cdot07$ Mbar.

4.17 A spring has n coils with a mean diameter D and a wire diameter d. The stiffness of a coil spring is given by

$$\frac{Gd^4}{8nD^3}$$

where G is the shear modulus of the material. If this spring is clamped in a compressed state between two rigid platens, find the lowest frequency of longitudinal vibration.

$$n = 20$$
$$D = 35 \text{ mm}$$
$$d = 3 \text{ mm}$$
$$G = 0.8 \text{ Mbar}$$
$$\text{Density of material} = 7\,850 \text{ kg/m}^3$$

4.18 Find the natural frequencies of transverse vibration of a cantilever of length $2l$ carrying two equal masses, m, one at the free end and the other at the centre of the beam. Neglect the mass of the beam.

4.19 One method of measuring the frequency of a vibrating surface is to use a vibration tachometer (see fig. 5.20), which contains a variable length reed which is effectively clamped at one end, the length of the reed being calibrated against frequency. If the reed has a width of 3 mm and a thickness of 0.25 mm, calculate the length of reed required to measure frequencies down to 200 c/min.
$E = 2.07$ Mbar and density 7 850 kg/m^3.

4.20 Find the frequency of free longitudinal vibrations of a vertical spring carrying a mass, taking into consideration the mass of the spring. Calculate the frequency when a mass of 7 kg is suspended from a helical spring with coils formed from 3 m of steel wire 6 mm in diameter, the spring being extended 1 cm under a load of 1.6 kgf. Density of steel is 7 850 kg/m^3.

4.21 A uniform simply supported beam of total mass $M/2$, flexural rigidity EI, and length l carries a concentrated mass M at its centre. Find, using Rayleigh's method, the undamped natural frequency of the first mode of transverse vibration, assuming the dynamic deflection curve to be

    (*a*) that due to a concentrated load at the centre of the beam
    (*b*) that due to a uniformly distributed load
    (*c*) half a sine wave   [over

The deflection of a simply supported beam due to a central load, W, is given by

$$y = \frac{W}{48EI}(3l^2x - 4x^3) \text{ for } 0 < x < \frac{l}{2}$$

The deflection of a simply supported beam due to a uniformly distributed load w/unit length is

$$y = \frac{w}{24EI}(x^4 + l^3x - 2lx^3)$$

4.22 Fig. 4.36 gives details of a steel disk on a tapered steel shaft. Using Rayleigh's method and neglecting the mass of the shaft calculate the frequency of transverse vibration. For steel it may be assumed that $E = 2·07$ Mbar and its density 7 850 kg/m³.

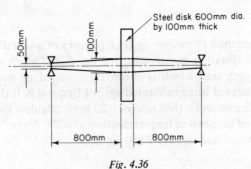

Fig. 4.36

5. Damped and Forced Vibrations

§ 5.1 Introduction

In the cases discussed in the previous chapter it is assumed that a system is disturbed once, and that thereafter no further external force is applied to it, so that it continues to vibrate with constant amplitude. Such a vibration is described as 'free', and the system as a 'conservative' one, since energy is neither gained nor lost by it. In practice, no such systems exist, since forces which oppose the motion, usually called 'damping' or 'dissipative' forces, invariably act; additionally, forces which tend to cause or to increase the vibration, usually called 'exciting' or 'disturbing' forces, may act as well.

In this chapter we discuss damped vibrations and the response of damped systems to exciting forces, in particular to forces which are suddenly applied and to forces which vary harmonically. Consideration will also be given to associated problems such as vibration absorbers, the vibration isolation of machinery, vibration measuring instruments, and the 'whirling' of shafts.

§ 5.2 Damping Forces

Damping forces in mechanical systems can arise from numerous sources, of which the most common are hysteresis in solid materials, resistance due to the motion of fluids, which we may call fluid damping, and friction between 'dry' materials, which we may call 'Coulomb' damping.

(a) *Hysteresis damping.* No material is perfectly elastic even at small stresses, and when a material is taken through a complete stress cycle a hysteresis loop is formed in the stress/strain diagram. This loop represents energy absorbed by the material, which is dissipated as heat to the surroundings. Usually the hysteresis loop is so small as to preclude the possibility of measuring it with normal extensometers used in strength of materials testing, and other methods of measurement have been developed. Hysteresis losses vary considerably from material to material, and also with the level of stress and temperature, and can seldom be estimated with any great degree of accuracy.

(b) *Fluid damping.* In many vibratory systems there is relative motion across oil films, which ideally gives rise to 'viscous' damping forces, that is, forces which are proportional to the shearing velocity. Such forces may arise naturally, for instance in film-lubricated journal bearings or may be deliberately introduced into a system by adding a 'dashpot', as described in § 5.3.

(c) *Coulomb (or solid friction) damping.* If relative vibratory motion occurs between two dry surfaces through which loads are being transmitted an approximately constant force will resist the relative motion. Such forces can, for instance, arise in riveted structures in which small relative movements may occur across the joints.

In (elementary) vibration theory it is customary to assume that all damping forces are 'viscous', because this gives rise to a linear differential equation which is readily amenable to analysis. The energy dissipated per cycle with viscous damping is proportional to the frequency, whereas with hysteresis or Coulomb damping the energy dissipated per cycle at a given amplitude is independent of frequency. At first sight these are serious differences. However, even though in our treatment the physical form of the damping may have been grossly misrepresented, nevertheless provided that the correct value of the energy dissipation is used, the response of a system to a harmonic exciting force, both in relation to amplitude and phase, can usually be found with sufficient accuracy. In this chapter we shall consider viscous damping forces only.

§ 5.3 Damped Vibrations

Consider the vibration of a simple spring-mass system such as we discussed in § 4.3, to which has been added viscous damping. This damping is conventionally represented as shown in fig. 5.1 by a 'dashpot', or loosely fitting piston in an oil-filled cylinder, though

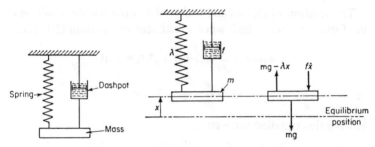

Fig. 5.1 Representation of viscous damping   *Fig. 5.2* Spring-mass system with damping

such a device does not usually give truly viscous damping: equally conventionally we may assume the mass to be guided so that only vertical motion is permitted. We shall take the positive direction of the x-axis upwards.

Let the mass be m, the stiffness of the spring λ, and the *damping constant* of the dashpot (i.e. the force it exerts for unit velocity of the piston in the cylinder) f. Let the mass be displaced from its equilibrium position and at some subsequent time let the displacement and velocity of the mass be $+x$ and $+\dot{x}$ respectively, i.e. both upwards. At this instant, then, the forces acting on m are as shown in fig. 5.2, i.e.

the gravitational force $-mg$ (i.e. downwards)
the total force due to the spring $+(mg - \lambda x)$ (i.e. upwards)
the force due to the dashpot $-f\dot{x}$ (i.e. downwards)

The net force on m is

$$-f\dot{x} - \lambda x$$

and from Newton's second law of motion we can write

$$m\ddot{x} = -f\dot{x} - \lambda x$$

or

$$\ddot{x} + \frac{f}{m}\dot{x} + \frac{\lambda}{m}x = 0$$

From § 4.3 we know that $\frac{\lambda}{m} = \omega_n{}^2$, where ω_n is the natural radiancy of the *undamped* system so we can rewrite this equation as

$$\ddot{x} + \frac{f}{m}\dot{x} + \omega_n{}^2 x = 0 \qquad (5.1)$$

The solution of this second order differential equation will be of the form $x = Ae^{\alpha t}$, which when substituted in equation (5.1) gives

$$A\alpha^2 e^{\alpha t} + \frac{f}{m}A\alpha e^{\alpha t} + \omega_n{}^2 Ae^{\alpha t} = 0$$

or

$$\alpha^2 + \frac{f}{m}\alpha + \omega_n{}^2 = 0$$

Solving this equation for α gives

$$\alpha = -\frac{f}{2m} \pm \sqrt{\left[\left(\frac{f}{2m}\right)^2 - \omega_n{}^2\right]} \qquad (5.2)$$

This solution for α gives rise to three particular cases of some importance, depending on the value of the quantity inside the square-root sign. If this quantity is zero, then the system is said to be 'critically' damped, if positive the system is 'overdamped', and if negative, so that the root is unreal or imaginary, the system is 'underdamped': the meaning of these terms will become apparent. The damping constant f_c for the critically damped system is of considerable importance, as the damping f in a general system is frequently and conveniently expressed non-dimensionally by the *damping ratio* $\frac{f}{f_c} = c$, of which we shall make much use. From equation (5.2) we can at once see that

$$\frac{f_c}{2m} = \omega_n$$

or

$$f_c = 2m\omega_n = 2m\sqrt{\frac{\lambda}{m}} = 2\sqrt{\lambda m} \qquad (5.3)$$

and

$$\frac{f}{m} = \frac{f}{f_c} \cdot \frac{f_c}{m} = 2c\omega_n \qquad (5.4)$$

So equation (5.2) can be rewritten as

$$\alpha = -c\omega_n \pm \omega_n \sqrt{(c^2 - 1)} \qquad (5.5)$$

We can now consider the three particular cases arising from this equation:

Case 1. $c > 1$.   This leads to two real solutions for α:

$$\alpha_1 = -c\omega_n + \omega_n \sqrt{(c^2 - 1)}$$
$$\alpha_2 = -c\omega_n - \omega_n \sqrt{(c^2 - 1)}$$

Consequently the most general solution for the displacement x will be

$$x = A_1 e^{\alpha_1 t} + A_2 e^{\alpha_2 t} \qquad (5.6)$$

where the arbitrary constants A_1 and A_2 depend on the initial conditions. These conditions might, for instance, be the displacement and velocity at zero time.

As an example consider a spring-mass system which has $\omega_n = 16$ rad/s and $c = 2$. Let us start the vibration by deflecting the mass a positive distance X from its equilibrium position and releasing it. We then have at zero time $x = X$, $\dot{x} = 0$. From equation (5.6) we obtain by differentiation

$$\dot{x} = A_1 \alpha_1 e^{\alpha_1 t} + A_2 \alpha_2 e^{\alpha_2 t} \qquad (5.7)$$

Substituting the end conditions into equations (5.6) and (5.7) gives

$$(x)_{t=0} = X = A_1 + A_2$$
$$(\dot{x})_{t=0} = 0 = A_1 \alpha_1 + A_2 \alpha_2$$

where (from 5.5) α_1 and α_2 are $-4 \cdot 29$ and $-59 \cdot 7$ respectively. Solving these two equations gives

$$A_1 = +1 \cdot 0773 X \text{ and } A_2 = -0 \cdot 0773 X$$

So the complete solution for the displacement will be

$$x = +1 \cdot 0773 X e^{-4 \cdot 29t} - 0 \cdot 0773 X e^{-59 \cdot 7t}$$

which is shown plotted in fig. 5.3.

Whatever the starting conditions, the displacement will die away with time as both α_1 and α_2 are negative.

Case 2. $c < 1$.   The solutions for α are now complex

$$\alpha_1 = -c\omega_n + j\omega_n \sqrt{(1 - c^2)}$$
$$\alpha_2 = -c\omega_n - j\omega_n \sqrt{(1 - c^2)}$$

where $j = \sqrt{-1}$, so the most general solution for x will be

$$x = e^{-c\omega_n t} \{A_1 e^{+j\omega_n \sqrt{1-c^2} \cdot t} + A_2 e^{-j\omega_n \sqrt{1-c^2} \cdot t}\}$$

in which A_1 and A_2 are arbitrary constants. Alternatively, this equation can be expressed as

$$x = e^{-c\omega_n t}\{(A_1 + A_2) \cos \omega_n \sqrt{1 - c^2}\, t$$
$$+ j(A_1 - A_2) \sin \omega_n \sqrt{1 - c^2}\, t\}$$

or $x = e^{-c\omega_n t}\{A_3 \cos \omega_n \sqrt{1 - c^2}\, t + A_4 \sin \omega_n \sqrt{1 - c^2}\, t\}$

or $x = A_5 e^{-c\omega_n t} \cos \{\omega_n \sqrt{1 - c^2}\, t + \phi\}$ (5.8)

where A_5 and ϕ (or A_3 and A_4) are arbitrary constants depending on the starting conditions.

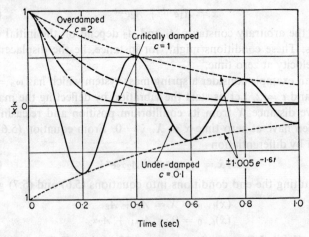

Fig. 5.3 Response of damped spring-mass system released from a displacement at zero time

As an example let us consider the same spring-mass system as in case 1, but now with much less damping so that $c = 0.1$. Substituting the same end conditions as before into equation (5.8) and its derivative gives

$$(x)_{t=0} = X = A_5 \cos \phi$$
$$(\dot{x})_{t=0} = 0 = -c\omega_n A_5 \cos \phi - A_5 \omega_n \sqrt{1 - c^2} \sin \phi$$

so $\tan \phi = -\dfrac{c\omega_n}{\omega_n \sqrt{(1 - c^2)}} = -\dfrac{c}{\sqrt{(1 - c^2)}}$

$$= -\frac{0.1}{\sqrt{0.99}} \simeq -0.1005$$

and $A_5 = \dfrac{X}{\cos \phi} = \dfrac{X}{\sqrt{(1 - c^2)}} \simeq 1.005X$

The complete solution for the displacement is

$$x = 1 \cdot 005 X e^{-1 \cdot 6t} \cos \{15 \cdot 92t - 0 \cdot 1005\}$$

which is shown plotted in fig. 5.3.

From equation (5.8) it will be seen that in general the motion is harmonic with an exponentially decaying amplitude. The radiancy of these oscillations is $\omega_n \sqrt{(1 - c^2)}$, which, for the magnitude of the damping met with in many engineering problems, is very close indeed to the undamped natural radiancy.

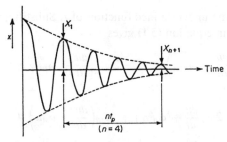

Fig. 5.4 Determination of damping from a die-away vibration record

If these exponentially decaying oscillations are recorded, then it is easy to determine the magnitude of the damping involved. In fig. 5.4 the successive peaks have values of X_1, X_2, . . . X_{n+1}, and from equation (5.8) it can be shown that these maxima occur when $\cos \{\omega_n \sqrt{1 - c^2} \, t + \phi\}$ is unity, and that successive peaks are separated by a time equal to the periodic time

$$t_p = \frac{2\pi}{\omega_n \sqrt{(1 - c^2)}}$$

From equation (5.8) the ratio of X_1 to X_{n+1} will be

$$\frac{X_1}{X_{n+1}} = \frac{A_5 e^{-c\omega_n t}}{A_5 e^{-c\omega_n (t + nt_p)}} = e^{c\omega_n nt_p}$$

or

$$c = \frac{1}{\omega_n nt_p} \log_e \frac{X_1}{X_{n+1}}$$

The ratio of successive peaks is constant and equal to $e^{-c\omega_n t_p}$. The quantity $c\omega_n t_p$ is called the logarithmic decrement.

Case 3. $c = 1$. From cases 1 and 2 it will be seen that critical damping marks the transition from a non-oscillatory die-away motion to a die-away oscillatory type of motion.

For this case equation (5.5) gives only one solution for α

$$\alpha_1 = -\omega_n$$

The corresponding expression for x would be

$$x = A_1 e^{-\omega_n t}$$

This solution, however, contains only one arbitrary constant A_1 instead of two, so it cannot be the most general solution. To obtain the general solution let us tentatively assume that

$$x = u e^{-\omega_n t}$$

where u is some undetermined function of t. Substituting for x and $f/m = 2c\omega_n$ in equation (5.1) gives

$$\left(\frac{d^2u}{dt^2} - 2\omega_n \frac{du}{dt} + \omega_n^2 u \right) e^{-\omega_n t} + 2\omega_n \left\{ \frac{du}{dt} - \omega_n u \right\} e^{-\omega_n t}$$
$$+ \omega_n^2 u\, e^{-\omega_n t} = 0$$

or $\quad \left(\dfrac{d^2u}{dt^2} - 2\omega_n \dfrac{du}{dt} + \omega_n^2 u \right) + 2\omega_n \left(\dfrac{du}{dt} - \omega_n u \right) + \omega_n^2 u = 0$

which reduces to $\dfrac{d^2u}{dt^2} = 0$ or $u = A_1 + A_2 t$, where A_1 and A_2 are arbitrary constants, and the equation for x is

$$x = (A_1 + A_2 t) e^{-\omega_n t}$$

As an example we will consider the same spring-mass system as in cases 1 and 2, but now with $c = 1$

$$(x)_{t=0} = X = A_1$$
$$(\dot{x})_{t=0} = A_2 - \omega_n A_1 = 0$$
so $\qquad\qquad A_2 = \omega_n A_1 = 16X$

The displacement is given by

$$x = X(1 + 16t) e^{-16t}$$

which is shown plotted in fig. 5.3.

§ 5.4 Forced Vibrations

Let us now consider the behaviour of the simple damped spring-mass system which we have been considering in § 5.3, when it is subjected to a disturbing force. This force may be simple or complex in nature, and it is impossible to deal with all the variants. However, we shall select a few typical cases of particular significance which will serve as a guide to the methods used, and which demonstrate some of the important features of the response of such systems to external forces. These typical cases include (*a*) a suddenly applied force of constant magnitude, (*b*) a very rapidly applied force, (*c*) a steadily increasing force, (*d*) a simple harmonic force of constant amplitude, (*e*) a simple harmonic force with an amplitude proportional to the square of its frequency, and (*f*) a complex harmonic force.

Since the vibrating system we are discussing is that shown in fig. 5.2, the forces acting on the mass at the instant when its displacement and velocity are x and $\dot{x}$ will be exactly the same as those considered in § 5.3, except that in addition there is an applied force which will be some function of time, say $f(t)$. We can therefore write down the equation of motion

$$m\ddot{x} = -\lambda x - f\dot{x} + f(t)$$

or rewriting and substituting ω_n^2 for λ/m and $2c\omega_n$ for f/m

$$\ddot{x} + 2c\omega_n\dot{x} + \omega_n^2 x = \frac{1}{m}f(t) \qquad (5.9)$$

This differential equation will have a solution consisting of two parts, the 'complementary function' or 'transient solution', which is the solution of the left-hand side of the equation equated to zero, and the 'particular integral' or 'steady state solution', which is the part which has to be added to the complementary function to allow the total value of the left-hand side to equal the right-hand side. In the previous paragraph we have already solved the equation

$$\ddot{x} + 2c\omega_n\dot{x} + \omega_n^2 x = 0$$

and found that three possible forms of the solution exist depending on the value of c, all of them giving values of x which diminish with time. Thus, the complementary function is of importance only during the early stages of the vibration, and as a consequence it is usually referred to as a 'transient'. In practice, the damping is nearly always sub-critical and the transient vibration will be of the form given in equation (5.8), namely

$$x = A_5 e^{-c\omega_n t} \cos\{\omega_n \sqrt{1 - c^2}\, t + \phi\} \qquad (5.10)$$

Though this will give the *form* of the first part of the complete solution to equation (5.9) whatever the form of applied force, it does not mean that it is independent of it, as the arbitrary constants A_5 and ϕ depend on the forcing function $f(t)$ and the way it is applied at zero time.

We can now consider the particular forcing functions mentioned above.

5.4.1 Response of damped spring-mass system to a suddenly applied force

Let the force F be suddenly applied at $t = 0$ to the mass when it is at rest in its equilibrium position. When $t > 0$

$$\ddot{x} + 2c\omega_n\dot{x} + \omega_n^2 x = \frac{F}{m} \qquad (5.11)$$

and if the damping is sub-critical the transient solution is given by equation (5.10). There are several ways of finding the particular integral, but perhaps the easiest is to consider what will be the state of affairs when the transient vibration has died away. Fairly obviously, we are left with a static system with a force F applied, so the deflection X_s will be F/λ. Alternatively, we can guess a form of x similar to the right-hand side of the equation, which is a constant; thus we could assume

$$x = A$$

If so

$$\dot{x} = \ddot{x} = 0$$

or, substituting into (5.11),

$$\omega_n^2 A = \frac{F}{m}$$

or

$$A = \frac{F}{m\omega_n^2} = \frac{F}{\lambda} = X_s$$

Thus, the complete solution of (5.11) is

$$x = A_5 e^{-c\omega_n t} \cos\{\omega_n \sqrt{1 - c^2}\, t + \phi\} + X_s \qquad (5.12)$$

As the mass started from rest we know that at $t = 0$, $x = \dot{x} = 0$. Consequently

$$(x)_{t=0} = 0 = A_5 \cos\phi + X_s$$

$$(\dot{x})_{t=0} = 0 = -c\omega_n \cos\phi - \omega_n \sqrt{(1 - c^2)} . \sin\phi$$

or

$$\tan\phi = -\frac{c}{\sqrt{(1 - c^2)}}$$

whence we obtain the expression

$$\frac{x}{X_s} = 1 - \frac{e^{-c\omega_n t}}{\sqrt{(1 - c^2)}} \cos\left\{\omega_n \sqrt{1 - c^2}\, t + \tan^{-1}\left(\frac{-c}{\sqrt{(1 - c^2)}}\right)\right\}$$

Though this expression, as is often the case with vibrating systems, appears to be rather clumsy, it can be easily demonstrated that it fits the known facts. When t is large the expression reduces to

$$x = X_s$$

while if c is very small we see that the early stages of the vibration are

$$x = X_s(1 - \cos \omega_n t)$$

which is no more than we could have written down without the slightest difficulty: a weight suddenly added to a spring causes a simple harmonic vibration, the amplitude being equal to the static

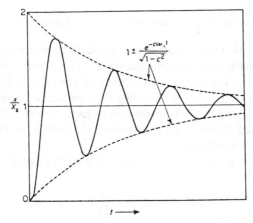

$$1 \pm \frac{e^{-c\omega_n t}}{\sqrt{1 - c^2}}$$

$$\frac{x}{X_s}$$

Fig. 5.5 Response of damped spring-mass system when a force, F, is suddenly applied at zero time. ($c = 0.1$)

displacement. More importantly, however, this analysis shows how the transient solution enables us to satisfy the starting conditions, and why it is impossible in a lightly damped system to avoid 'overshoot' when forces are suddenly applied. For instance, with small values of the damping, the first displacement is nearly twice the final displacement, which we may describe as an overshoot of 100 per cent, while even with damping as great as $\frac{1}{10}$ of the critical value the behaviour illustrated in fig. 5.5 shows that the overshoot is still 72·6 per cent. There is nothing we can do about this except to increase the damping forces, which will make the reaction of the mass more sluggish.

5.4.2 Response of a damped spring-mass system to a rapidly applied force

Although the example in § 5.4.1 is instructive, it is somewhat un-realistic, in that forces are seldom applied with the infinite sudden-ness therein assumed. For instance, even an explosion takes a finite time to occur, and much more time is absorbed during the com-bustion of a propellant in a gun or a charge in an internal-combustion engine. Even more time is required to get steam into a cylinder and so on. We may profitably give brief consideration to a less extreme case than considered in § 5.4.1, where the force function may be taken as

$$0 < t < \pi/\omega \qquad\qquad \pi/\omega < t$$
$$\text{force} = \frac{F}{2}(1 - \cos \omega t) \qquad \text{Force} = F$$

in other words, a force increasing sinusoidally to its maximum value, and then remaining constant. Such a case is of particular interest in instruments for measuring variation of pressure, as for instance engine indicators used for recording pressure against crank angle in steam or oil engines, where the gas pressure exerts a force on a spring-loaded piston, the displacement of which is required to be proportional to pressure.

During the period $0 < t < \pi/\omega$ we can write down the equation of motion as

$$\ddot{x} + 2c\omega_n\dot{x} + \omega_n^2 x = \frac{F}{2m}(1 - \cos \omega t) \qquad (5.13)$$

and the transient solution will again be of the form (5.10). For the steady state solution we may assume a form similar to the right-hand side, for instance,

$$x = A + B \cos (\omega t - \varepsilon) \qquad (5.14)$$

where, of course, A, B, and ε have to be determined. To find these constants we might substitute the assumed value of x in (5.13) and expand the trigonometrical terms, and then collect together the sine, cosine, and constant terms. This would give three simultaneous equations from which A, B, and ε could be found. A much more expeditious and illuminating and equally rigorous solution is found by using a vectorial method, thus:

Given equation (5.14), we can write down each term in the left-hand side of (5.13)

$$\omega_n{}^2x = \omega_n{}^2A + \omega_n{}^2B \cos(\omega t - \varepsilon)$$

$$2c\omega_n\dot{x} = 0 - 2c\omega_n\omega B \sin(\omega t - \varepsilon) = 2c\omega_n\omega B \cos\left(\omega t - \varepsilon + \frac{\pi}{2}\right)$$

$$\ddot{x} = -\omega^2 B \cos(\omega t - \varepsilon) \qquad = +\omega^2 B \cos(\omega t - \varepsilon + \pi)$$

and the sum of these terms must be *identically* equal to the right-hand side of equation (5.13); that is

$$\omega_n{}^2A + B\left\{\omega_n{}^2 \cos(\omega t - \varepsilon) + 2c\omega_n\omega \cos\left(\omega t - \varepsilon + \frac{\pi}{2}\right)\right.$$

$$\left. + \omega^2 \cos(\omega t - \varepsilon + \pi)\right\}$$

$$\equiv \frac{F}{2m} - \frac{F}{2m} \cos \omega t \qquad (5.15)$$

Clearly the terms independent of time must be equal, i.e.

$$\omega_n{}^2 A = \frac{F}{2m} \quad \text{or} \quad A = \frac{F}{2m\omega_n{}^2} = \frac{F}{2\lambda}$$

We have encountered before in § 3.3.5 (and in § 0.14) the conditions set up by the remaining terms, and should be accustomed by now to the vector solution as indicated in fig. 5.6. In this diagram we choose an arbitrary direction to represent the angle $(\omega t - \varepsilon)$ radians and draw a vector length $\omega_n{}^2$ in that direction to represent the first term in the brackets in equation (5.15). The other vectors follow

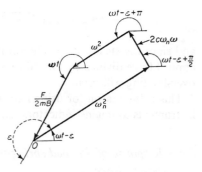

Fig. 5.6 Vector diagram for equation 5.15

obviously enough. In order that the assumed form (5.14) may be correct the closing vector must have a length $F/2mB$ and be in the direction (ωt). Thus we can write, after inspection of fig. 5.6

$$\frac{F}{2mB} = \sqrt{[(\omega_n{}^2 - \omega^2)^2 + (2c\omega_n\omega)^2]}$$

$$\tan \varepsilon = \frac{2c\omega_n\omega}{(\omega_n{}^2 - \omega^2)} = \frac{2c\omega/\omega_n}{(1 - \omega^2/\omega_n{}^2)}$$

and the complete solution to the problem for $0 < t < \pi/\omega$ is

$$x = A_5 e^{-c\omega_n t} \cos(\omega_n \sqrt{1 - c^2}\, t + \phi)$$
$$+ \frac{F}{2\lambda} + \frac{F/2m}{\sqrt{[(\omega_n^2 - \omega^2)^2 + (2c\omega_n\omega)^2]}} \cos(\omega t - \varepsilon)$$

or $\;= A_5 e^{-c\omega_n t} \cos(\omega_n \sqrt{1 - c^2}\, t + \phi)$
$$+ \frac{F}{2\lambda} \left\{ 1 + \frac{\cos(\omega t - \varepsilon)}{\sqrt{\left[\left(1 - \dfrac{\omega^2}{\omega_n^2} \right)^2 + \left(2c\, \dfrac{\omega}{\omega_n} \right)^2 \right]}} \right\}$$

Now let us look at this equation qualitatively. Let us first of all assume that the damping in the system is small, i.e. $c \to 0$ and $\varepsilon \to \pi$. If ω_n and ω are of the same order of magnitude we shall have to go through the arithmetic to find the constants A_5 and ϕ to satisfy the starting conditions (i.e. $x = 0$ and $\dot{x} = 0$ at $t = 0$), and the degree of overshoot, but we need not go to this degree of elaboration to estimate the answer. If the natural radiancy ω_n is low we shall have a situation closely akin to that in fig. 5.5. If, however, ω_n is large in relation to ω we can write

$$x \simeq A_5 e^{-c\omega_n t} \cos(\omega_n \sqrt{1 - c^2}\, t + \phi) + \frac{F}{2\lambda}\{1 - \cos \omega t\}$$

and since the steady state solution is itself at all times equal to the applied force divided by the stiffness, A_5 will be zero and no serious overshooting will occur.

This is the reason for using light masses and stiff springs in such instruments as engine indicators (cf. chapter 1, example 1.3).

5.4.3 *Response of damped spring-mass system to a steadily increasing force*

A steadily increasing force can be represented mathematically by a function, say Kt, where K is some constant. We can therefore write down the equation of motion

$$\ddot{x} + 2c\omega_n\dot{x} + \omega_n^2 x = \frac{K}{m}t \qquad (5.16)$$

As usual the transient solution will be as in equation (5.10). For the steady state solution $\ddot{x}$ will be zero, and consequently the form of x will be

$$x = At + B$$

Substituting this into equation (5.16) we obtain

$$2c\omega_n A + \omega_n^2(At + B) = \frac{K}{m}t$$

from which it follows that

$$A = \frac{K}{m\omega_n^2} = \frac{K}{\lambda}$$

and that

$$2c\omega_n A + \omega_n^2 B = 0$$

so

$$B = -\frac{2c}{\omega_n}\frac{K}{\lambda}$$

The steady state solution will then be

$$x = \frac{K}{\lambda}\left(t - \frac{2c}{\omega_n}\right)$$

and the complete solution will be

$$x = A_5 e^{-c\omega_n t}\cos(\omega_n\sqrt{1 - c^2}\, t + \phi) + \frac{K}{\lambda}\left(t - \frac{2c}{\omega_n}\right)$$

To find the constants A_5 and ϕ we must satisfy the starting conditions. For instance, if this continuously increasing force starts at

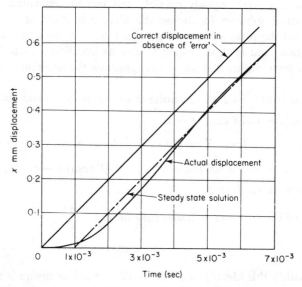

Fig. 5.7 Response of damped spring-mass system when subjected to a steadily increasing force ($c = 0.5$, $\omega_n = 1\,000$ rad/s, $\lambda = 3$ MN/m, $K = 0.3$ MN/s)

$t = 0$ then $x = 0$ and $\dot{x} = 0$ at this instant. A typical example of this kind of response is shown in fig. 5.7. It is worth noting that the displacement is always affected by the damping in the system, and if this is increased in an effort to eliminate the transient vibration quickly, the 'error' in the steady state displacement is increased (see § 6.7).

5.4.4 *Response of a damped spring-mass system to a simple harmonic force of constant magnitude*

The applied force in this case will be represented by a function such as $F \cos \omega t$, where ω is the radiancy of the force and F its amplitude. We can therefore write down the equation of motion

$$\ddot{x} + 2c\omega_n\dot{x} + \omega_n^2 x = \frac{F}{m} \cos \omega t \tag{5.17}$$

Again the transient solution will be as in equation (5.10). Again we choose for the steady state solution a function similar in form to the right-hand side, say

$$x = X \cos (\omega t - \varepsilon)$$

It is, of course, equally possible and indeed convenient for the analytical solution to choose the form $x = A \cos \omega t + B \sin \omega t$, but not the form $x = A \cos \omega t$, which contains only one arbitrary constant, and is clearly going to give us insufficient latitude, since the $\dot{x}$ term implies that $\sin \omega t$ will appear on the left-hand side of the equation.

Just as in § 5.4.2 we substitute this value in (5.17) and obtain

$$\omega_n^2 x = \omega_n^2 X \cos (\omega t - \varepsilon)$$

$$2c\omega\dot{x} = -2c\omega_n\omega X \sin (\omega t - \varepsilon) = 2c\omega_n\omega X \cos \left(\omega t - \varepsilon + \frac{\pi}{2} \right)$$

$$\ddot{x} = -\omega^2 X \cos (\omega t - \varepsilon) \quad = \omega^2 X \cos (\omega t - \varepsilon + \pi)$$

whence we know that

$$X \left[\omega_n^2 \cos (\omega t - \varepsilon) + 2c\omega_n\omega \cos \left(\omega t - \varepsilon + \frac{\pi}{2} \right) \right.$$
$$\left. + \omega^2 \cos (\omega t - \varepsilon + \pi) \right] \equiv \frac{F}{m} \cos \omega t$$

To satisfy this identity at all times all we need to ensure is that the vectors sketched in fig. 5.8a form a closed polygon. Preferably we can divide all the vectors by ω_n^2, which has the big advantage that all the vectors represent non-dimensional quantities as shown in fig. 5.8b, including

$$\frac{F}{Xm\omega_n^2} = \frac{F}{X\lambda} = \frac{F/\lambda}{X} = \frac{X_s}{X}$$

$$= \frac{\text{Static deflection due to a force } F}{\text{Amplitude of dynamic displacement}}$$

From the vector diagram

$$\frac{X_s}{X} = \sqrt{\left[\left(1 - \frac{\omega^2}{\omega_n^2}\right)^2 + \left(2c\frac{\omega}{\omega_n}\right)^2\right]}$$

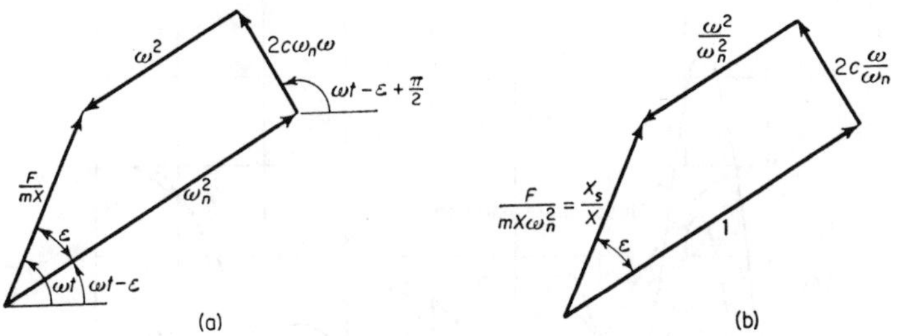

Fig. 5.8

This ratio is of primary importance. It is usually quoted in the reciprocal form, which is given the name of the 'dynamic magnification factor' or 'Q factor'

$$Q = \frac{X}{X_s} = \frac{1}{\sqrt{\left[\left(1 - \frac{\omega^2}{\omega_n^2}\right)^2 + \left(2c\frac{\omega}{\omega_n}\right)^2\right]}} \tag{5.18}$$

The complete solution for the steady state is

$$x = \frac{X_s}{\sqrt{\left[\left(1 - \frac{\omega^2}{\omega_n^2}\right)^2 + \left(2c\frac{\omega}{\omega_n}\right)^2\right]}} \cos\left(\omega t - \tan^{-1}\frac{2c\frac{\omega}{\omega_n}}{1 - \frac{\omega^2}{\omega_n^2}}\right) \tag{5.19}$$

and the complete solution including the transient solution

$$x = A_5 e^{-c\omega_n t} \cos\{\omega_n\sqrt{1 - c^2}\, t + \phi\}$$

$$+ \frac{X_s}{\sqrt{\left[\left(1 - \frac{\omega^2}{\omega_n^2}\right)^2 + \left(2c\frac{\omega}{\omega_n}\right)^2\right]}} \cos\left(\omega t - \tan^{-1}\frac{2c\frac{\omega}{\omega_n}}{1 - \frac{\omega^2}{\omega_n^2}}\right) \tag{5.20}$$

Again we have built up an equation of formidable complexity for the complete solution to the displacement of the mass, and we must not allow this to obscure the essential simplicity of the results in which we are likely to be interested. If indeed a force of the form $F \cos \omega t$ is *suddenly* applied to the mass, and we wish to know the displacement of the mass *at every subsequent instant*, there is no escape: we have to insert the starting conditions (e.g. at $t = 0$, $x = 0$ and $\dot{x} = 0$ if the mass starts from rest) into equation (5.20) and determine the arbitrary constants A_5 and ϕ. To do so once is

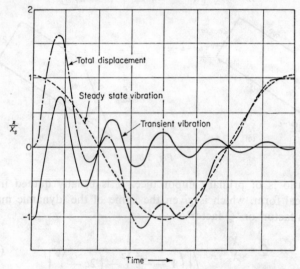

Fig. 5.9 Response of damped spring-mass system to a suddenly applied harmonic force, $F \cdot \cos \cdot \omega t$. $\left(c = 0 \cdot 1, \ \frac{\omega}{\omega_n} = \frac{1}{5} \right)$

instructive: the result (for a few vibrations starting from rest of a system in which $\omega_n > \omega$) is shown in fig. 5.9, and since the complementary function or transient and particular integral or steady state function have been plotted separately before their addition, one can see how they combine to give the total answer, producing early displacements greater than will subsequently occur. A much more important case is the one in which the impressed force is at nearly resonant frequency, since this in effect arises when a lightly damped system is subjected to a small force at a slowly rising frequency. Until resonance is approached the vibration is small, so the system might be considered as starting approximately from rest. This special case will be treated in §§ 5.9 and 5.10.

The general case is one which can rarely, if ever, obtain in mechanical engineering (though it might well do so in electrical circuitry), and as far as the vibration of machines is concerned we can generally ignore the transient and concentrate our attention exclusively on the steady state solution given by fig. 5.8 or equation (5.19).

Here again we may note that the mathematical statement of the result is astonishingly clumsy compared with the vectorial presentation in fig. 5.8. In fact, given the magnitudes of ω_n, ω, and c we need only sketch or draw to scale three lines and we can at once estimate, calculate, or measure the amplitude X of the resulting 'forced vibration' as it is called, and the 'angle of lag' ε.

This case is of such importance that we may well spend more time and space on its consideration.

Suppose that we are given a spring-mass system with a known (let us suppose, as is commonly the case, relatively small) amount of damping, so that the vector $2c\omega/\omega_n$ in fig. 5.8b is short compared with the unit vector. Now consider what happens as the frequency of the exciting force $F\cos\omega t$ is varied. If ω is small compared with ω_n or ω/ω_n is small compared with unity the vector X_s/X will be nearly as large as the unit vector and nearly collinear; in other words, a force with a frequency low compared with the natural frequency of the system will give a vibration amplitude very little greater than the static displacement (since of course if $X_s/X = 1$, $X = X_s = F/\lambda$) and in phase with the force (which is represented by the vector X_s/X). As the frequency of the applied force is further increased, the situation changes, slowly at first, and then as $\omega \rightarrow \omega_n$ rapidly, the amplitude and the angle of lag between displacement and force both increasing until when $\omega = \omega_n$ the angle of lag will be $\pi/2$ and $X_s/X = 2c\omega/\omega_n = 2c$ or $X = X_s/2c$. When the radiancy ω becomes much greater than ω_n, then ω/ω_n is large compared with unity and the vector X_s/X will be nearly as large as ω^2/ω_n^2 and nearly collinear with it. In other words, a force with a frequency large compared with the natural frequency of the system will produce a vibration amplitude which lags behind the force by π radians and so is *opposite* in phase. The vibration amplitude will be given by

$X_s/X = \omega^2/\omega_n^2$ and $X = \dfrac{X_s}{\omega^2/\omega_n^2}$, which is asymptotic to zero at large values of ω/ω_n.

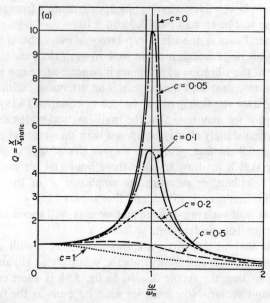

Fig. 5.10a Response of damped spring-mass system to a harmonically varying force

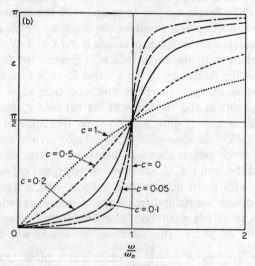

Fig. 5.10b Change of phase of the displacement of the mass of a damped spring-mass system, as the frequency of the harmonic force is varied

Curves of Q and phase lag ε for various values of the damping ratio, c, are shown in figs. 5.10a and 5.10b. The importance of mechanical vibrations arises mainly from the large values of the dynamic magnifier experienced in practice when ω/ω_n has a value near unity, which means that a relatively small harmonic force can produce a very large amplitude of vibration. This phenomenon is normally referred to as 'resonance'. The maximum value of the dynamic magnifier actually occurs at values of ω/ω_n rather less than unity, and this critical value can be found by differentiating equation (5.18) with respect to ω/ω_n:

$$\frac{d(Q)}{d(\omega/\omega_n)} = -\tfrac{1}{2}\left\{\left(1 - \frac{\omega^2}{\omega_n^2}\right)^2 + 4c^2\frac{\omega^2}{\omega_n^2}\right\}^{-\frac{3}{2}}\left\{-4\left(1 - \frac{\omega^2}{\omega_n^2}\right)\frac{\omega}{\omega_n} + 8c^2\frac{\omega}{\omega_n}\right\}$$

so for Q_{max} $\qquad \left\{4\left(1 - \frac{\omega^2}{\omega_n^2}\right)\frac{\omega}{\omega_n} - 8c^2\frac{\omega}{\omega_n}\right\} = 0$

or $\qquad\qquad\qquad \frac{\omega}{\omega_n} = \sqrt{(1 - 2c^2)}$ $\qquad\qquad$ (5.21)

The maximum value of the dynamic magnifier is given by substituting this value of (ω/ω_n) in (5.18):

$$Q_{max} = \left(\frac{X}{X_s}\right)_{max} = \frac{1}{2c\sqrt{(1 - c^2)}} \qquad\qquad (5.22)$$

For small values of c, $Q_{max} \simeq 1/2c$ as at $\omega = \omega_n$, and this factor often reaches values of 50 in engineering practice, and sometimes much more in whirling shafts.

5.4.5 *Response of a damped spring-mass system to a force of magnitude proportional to the square of its radiancy*

In the great majority of cases which arise in relation to machines, for instance when we are dealing with rotating or reciprocating masses, the disturbing force with which we have to deal has an amplitude which is not constant, but is proportional to the square of its radiancy. We may write down such a disturbing force as $R\omega^2 \cos \omega t$, where R is a constant (having the dimensions of mass and length, cf. mass-radius in chapter 3) or perhaps more conveniently as $F_1(\omega/\omega_n)^2 \cos \omega t$, where F_1 has the dimensions of force, and is in fact the magnitude of the disturbing force at a radiancy

$\omega = \omega_n$. The equation of motion will then become (cf. equation (5.9))

$$\ddot{x} + 2c\omega_n\dot{x} + \omega_n{}^2 x = \frac{F_1}{m}\left(\frac{\omega}{\omega_n}\right)^2 \cos \omega t$$

and since for any given value of ω the quantity $F_1/m(\omega/\omega_n)^2$ is a constant, the analysis will follow exactly that in § 5.4.4.

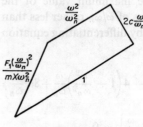

Fig. 5.11

We shall never have a system such as the one we are considering suddenly running at speed: the transient therefore no longer concerns us. Instead, we shall later (§ 5.10) have to consider what happens when ω varies steadily. For the time being, however, we may confine our attention to the steady state solution, for which the vector diagram now becomes as in fig. 5.11. From this vector diagram we can deduce that

$$X = \frac{\dfrac{F_1}{\lambda}\left(\dfrac{\omega}{\omega_n}\right)^2}{\sqrt{\left[\left(1 - \dfrac{\omega^2}{\omega_n{}^2}\right)^2 + \left(2c\,\dfrac{\omega}{\omega_n}\right)^2\right]}} \tag{5.23}$$

When $\omega/\omega_n \ll 1$ the disturbing force is small, and so is the amplitude. When $\omega/\omega_n \gg 1$, then the denominator of this expression reduces to $\omega^2/\omega_n{}^2$ and the amplitude becomes F_1/λ, so we see that at high speeds the amplitude of vibration is asymptotic to a finite value.

The radiancy for the maximum amplitude is given by

$$\frac{dX}{d(\omega/\omega_n)} = 0$$

which gives, when simplified

$$\left(\frac{\omega}{\omega_n}\right)_{X=\max} = \frac{1}{\sqrt{(1 - 2c^2)}}$$

This means that the peak of the resonance curve occurs when ω is slightly greater than ω_n. If we substitute this particular value in equation (5.23) we get

$$(X)_{\max} = \frac{F_1/\lambda}{2c\sqrt{(1 - c^2)}}$$

Fig. 5.12 shows the resonance curve for several values of the critical damping ratio, plotted non-dimensionally.

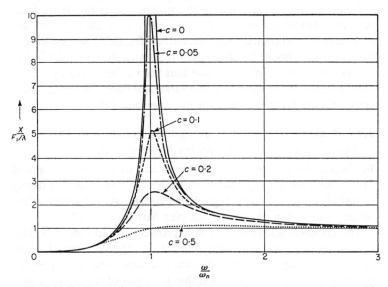

Fig. 5.12 Response of a damped spring-mass system to a harmonic force with an amplitude proportional to radiancy squared

5.4.6 Response of a damped spring-mass system to a complex harmonic force

A complicated piece of machinery may well give rise to a complex harmonic force on its supports, with components at machine frequency and multiples (and occasionally sub-multiples) of that frequency. However, this complex force will repeat itself, and consequently it can be analysed into its Fourier components, so that it can be written in the form

Total force $= F_1 \cos \omega t + F_2 \cos (2\omega t + \alpha_2) + F_3 \cos (3\omega t + \alpha_3) \ldots$

If we now wish to determine the effect of such a force on our standard spring-mass system we can write down the equation as

$$\ddot{x} + 2c\omega_n \dot{x} + \omega_n^2 x = \frac{F_1}{m} \cos \omega t + \frac{F_2}{m} \cos (2\omega t + \alpha_2) \ldots$$

The spring-mass system will respond simultaneously to each component of the exciting force. We can therefore consider each term on the right-hand side in turn and obtain the corresponding solution by drawing a vector diagram. The complete response will then be the summation of all these individual harmonic responses.

If the machine does not run at constant speed, but starts from rest and slowly accelerates, then we can expect each of the harmonics in turn to coincide with the natural frequency so that a series of resonances should be observed. Frequently, however, the higher components are so small in value that their effect is hardly noticeable. Clearly, if we want to avoid large amplitudes we must avoid continuous running at any of the major resonances.

5.4.7 Response of a damped spring-mass system with several degrees of freedom to a complex harmonic force

In chapter 4 we found that most practical structures, as opposed to the idealised single guided mass we have so far been considering in this chapter, had many degrees of freedom. What happens when we apply to such a structure a complex harmonic force?

The complete solution of such a problem would indeed be laborious, though not intrinsically difficult. We should simply have to apply each component force in turn and superpose the results. In practice, however, we can take it that the effect of resonance is such that all other responses pale into insignificance. Thus, our concern will normally be to ensure that no single major force component coincides in frequency with any one of the natural frequencies.

In some cases several complex exciting forces or torques will be applied at different points in the system, such as in a multi-cylinder engine when each cylinder provides a complex exciting torque. This particular problem will be considered in rather more detail in § 5.11.

§ 5.5 Vibration Isolation

The dynamic forces produced by machinery due to unbalanced rotating or reciprocating parts are often very big, and if these machines were mounted solidly in their supporting framework or foundations the vibrations induced might be unacceptable. A typical example is a spin dryer, in which there is frequently a considerable unbalanced rotating mass and the resulting dynamic force is so large that if the dryer tub were mounted in bearings rigidly attached to the framework of the machine it would be necessary to fix the machine rigidly to the floor. A similar situation arises in the case of a car engine which, even if the reciprocating forces are balanced, has a large cyclic torque reaction acting on it, and this might well be unacceptable if it were completely transmitted to the chassis, because of the resulting vibration caused in the body. In practice, these troubles can be overcome or at least reduced by the intelligent use of *flexible mountings*.

For simplicity in this discussion let us again consider first a harmonic force of constant amplitude acting on a damped spring-mass system as shown in fig. 5.13a. Under running conditions the steady-

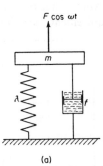

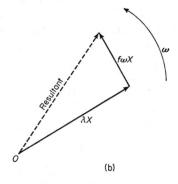

Fig. 5.13a Anti-vibration mounting

Fig. 5.13b Amplitude of resultant force transmitted to foundations

state amplitude of vibration is given by equation (5.18). The forces transmitted to the foundations will consist of the force in the spring and the force transmitted through the dashpot, which are 90° out of phase (see fig. 5.13b). The amplitude of the net force transmitted will be the vectorial sum of these two forces, which gives

$$\sqrt{[(\lambda X)^2 + (f\omega X)^2]}$$

The ratio of the amplitude of the net transmitted force to the amplitude of the harmonic force is

$$\frac{\sqrt{[(\lambda X)^2 + (f\omega X)^2]}}{F} = \frac{X\sqrt{[\lambda^2 + (f\omega)^2]}}{F}$$

If we substitute for X from equation (5.18) this ratio becomes

$$\frac{\text{Transmitted force}}{\text{Applied force}} = \frac{\sqrt{\left(1 + 4c^2\dfrac{\omega^2}{\omega_n^2}\right)}}{\sqrt{\left[\left(1 - \dfrac{\omega^2}{\omega_n^2}\right)^2 + 4c^2\dfrac{\omega^2}{\omega_n^2}\right]}} \qquad (5.24)$$

From inspection it will be seen that when $\omega/\omega_n > \sqrt{2}$ this ratio is less than unity; and it is greater if $\omega/\omega_n < \sqrt{2}$. Curves of this 'transmission ratio' for various values of the damping ratio are given in fig. 5.14.

It is immediately obvious from fig. 5.14 that for the flexible mounting to be beneficial in reducing the net force transmitted to the foundations it is essential that ω/ω_n should be as large as possible, and certainly greater than $\sqrt{2}$. This implies that ω_n, the natural radiancy, should be as small as possible relative to the radiancy of the disturbing force; in other words, that the mounting should be as flexible as possible. This requirement may, however,

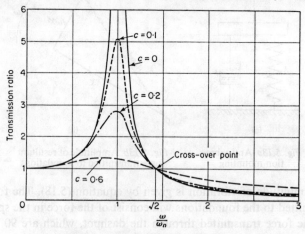

Fig. 5.14 Transmission ratio for a flexible mounting

give rise to other problems; for instance, it is disconcerting for an operator if the machine he is in charge of sways at the lightest touch. Again if the mountings fitted on an engine to reduce the harmonic torques transmitted are excessively flexible, then the rotation of the engine relative to the foundations under the torque reaction may be unacceptably large.

If the mountings are designed so that at the normal running speed of the machine $\omega/\omega_n > \sqrt{2}$, then it is obvious that in the starting and running-up period it will run through the region where $\omega/\omega_n < \sqrt{2}$ and the transmission ratio is greater than unity. This may prove serious if the acceleration through this region is slow, and to prevent excessive values of the transmission ratio and excessive motion of the mass, a considerable amount of damping is desirable.

In some installations metal springs are employed and damping is introduced deliberately by using a dashpot; in other cases 'rubber' mountings are employed and usually these are designed to put the rubber in shear combined with compression. Rubber has very complex mechanical properties, but as a first approximation we can consider it as being equivalent to a linear spring with internal damping. As a consequence, the energy dissipated in damping appears as heat in the rubber, which is acceptable only if the temperature rise is not excessive.

We have so far considered a very much simplified system involving only the longitudinal mode of vibration. In most practical

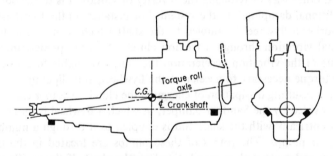

Fig. 5.15 Arrangement of flexible mountings on an automobile engine

systems several of the six modes of vibration possible with a flexibly mounted rigid mass have to be taken into consideration. For instance, in a conventional four-stroke automobile engine with four cylinders there is a harmonic exciting force in the vertical direction due to unbalanced secondary forces, and there are the harmonic torque components of the torque reaction acting about the centre-line of the crankshaft. A Fourier analysis of the torque developed will have a fundamental or lowest frequency component with a frequency equal to the frequency of firing strokes, if we assume that the firing strokes are equally spaced and are identical from cylinder to cylinder (c.f. fig. 1.48). It is this component which is the most difficult to isolate. The flexible mountings for such an engine will have to be designed and positioned to provide effective isolation for these two sources of excitation, and the arrangement of flexible mountings shown in fig. 5.15 is one which can be adopted to give controlled flexibility in the vertical direction and in the torsional sense.

§ 5.6 Dynamic Absorber

If serious vibration is experienced in machinery the obvious first step (after trying to eliminate or reduce the exciting forces) is to see if it can be eliminated by altering the natural frequency of the system, by varying either the stiffnesses or inertias. However, in some cases, such as the torsional vibration of the crankshaft of an internal-combustion engine and the associated drive shaft and load, it may prove impossible to remove the particular vibration causing trouble without introducing some other unacceptable vibration elsewhere in the running-speed range.

Another way of reducing the severity of vibration is to introduce additional damping into the system. For instance, in the Lanchester absorber a flywheel is coupled to the shaft which is subject to torsional vibration through a friction clutch. At some predetermined value of the vibration acceleration the clutch is unable to transmit the torque necessary to accelerate the flywheel, and slipping occurs, with a resulting dissipation of energy in the clutch. Another example is provided by the Sandner damper, in which a flywheel is mounted concentrically with the shaft and is coupled to it through a number of gear pumps. The rotors of these pumps are located in the flywheel and mate with a gear located on the shaft. If the oscillating torque required to accelerate the flywheel to and fro is excessive the oil pressure produced in the pumps is sufficient to overcome relief valves, and energy is dissipated as oil is forced through these valves.

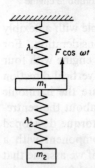

Alternatively, the severity of vibration may be reduced by introducing a non-linear spring into the system, such as the Bibby coupling used for torsional systems.

However, the system with which we are concerned in this section is the dynamic vibration absorber, which, in its undamped form, consists of an additional mass and stiffness added to the

Fig. 5.16 Dynamic vibration absorber

system subjected to vibration. This is shown in a simplified form in fig. 5.16, in which the vibrating system consists of a spring of stiffness λ_1 supporting a mass m_1 subjected to an exciting force of $F \cos \omega t$. The absorber consists of the mass m_2 attached via a spring of stiffness λ_2 to the mass m_1.

For the displaced position the equations of motion for the two masses will be

$$m_1\ddot{x}_1 + \lambda_1 x_1 - \lambda_2(x_2 - x_1) = F\cos\omega t$$
$$m_2\ddot{x}_2 + \lambda_2(x_2 - x_1) = 0$$

Since there is no damping, the solution is of the form $x_1 = X_1\cos\omega t$ and $x_2 = X_2\cos\omega t$. If we substitute $\omega_{n_1}^2$ for $\dfrac{\lambda_1}{m_1}$, and $\omega_{n_2}^2$ for $\dfrac{\lambda_2}{m_2}$, these equations may be rewritten as

$$-\omega^2 X_1 + \omega_{n_1}^2 X_1 - \frac{\lambda_2}{m_1}(X_2 - X_1) = \frac{F}{m_1}$$

and
$$-\omega^2 X_2 + \omega_{n_2}^2(X_2 - X_1) = 0$$

Solving these two equations for X_1 and X_2 gives

$$X_1 = \frac{(\omega_{n_2}^2 - \omega^2)}{\left\{(\omega_{n_1}^2 - \omega^2)(\omega_{n_2}^2 - \omega^2) - \dfrac{\lambda_2}{m_1}\omega^2\right\}} \times \frac{F}{m_1} \qquad (5.25)$$

$$X_2 = \frac{\omega_{n_2}^2}{\left\{(\omega_{n_1}^2 - \omega^2)(\omega_{n_2}^2 - \omega^2) - \dfrac{\lambda_2}{m_1}\omega^2\right\}} \times \frac{F}{m_1} \qquad (5.26)$$

It will be seen that X_1 can be made zero if we design the absorber so that $\omega_{n_2} = \omega^2$, so it is possible to eliminate completely the vibration of mass m_1 at any one particular forcing frequency. Normally the vibration we want to eliminate is that which occurs in the original system at resonance when $\omega = \omega_{n_1}$. In this particular case $\omega = \omega_{n_1} = \omega_{n_2}$ or

$$\frac{\lambda_1}{m_1} = \frac{\lambda_2}{m_2}$$

and the equation for X_2 reduces to

$$X_2 = -\frac{F}{\lambda_2}$$

or
$$\lambda_2 X_2 = -F$$

In retrospect, this result is in a sense obvious, in that if the mass m_1 is to remain stationary, the exciting force must be balanced by an equal and opposite force exerted by the absorber on m_1. It is far from obvious, without going through the algebra, that the absorber will behave in this fashion. It is important to realise that the absorber must be designed to withstand this dynamic force without failure by fatigue or wear; this requirement is not always easy to satisfy.

Although the dynamic absorber has reduced the resonance at $\omega = \omega_{n_1}$ to zero, it must be noted that it has also introduced two fresh frequencies at which resonance will occur. The frequency of these two new resonances can be readily found by considering the conditions which will make X_1 and X_2 infinite, and if we examine

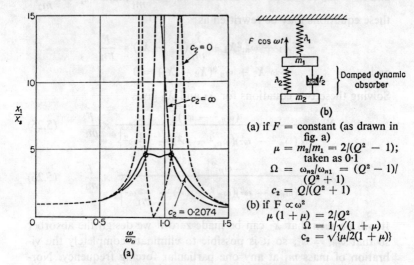

(b)

(a) if F = constant (as drawn in fig. a)
$\mu = m_2/m_1 = 2/(Q^2 - 1)$; taken as 0·1
$\Omega = \omega_{n2}/\omega_{n1} = (Q^2 - 1)/(Q^2 + 1)$
$c_2 = Q/(Q^2 + 1)$

(b) if $F \propto \omega^2$
$\mu(1 + \mu) = 2/Q^2$
$\Omega = 1/\sqrt{(1 + \mu)}$
$c_2 = \sqrt{\{\mu/2(1 + \mu)\}}$

Fig. 5.17 Response of damped vibration absorber

equations (5.25) and (5.26) we see that this would occur if the denominators were zero or

$$(\omega_{n_1}{}^2 - \omega^2)(\omega_{n_2}{}^2 - \omega^2) - \frac{\lambda_2}{m_1}\omega^2 = 0$$

This is a quadratic in ω^2 which gives the two new natural radiancies.

A dynamic absorber can be made effective over the complete frequency spectrum by introducing damping into the absorber system, that is between mass m_1 and m_2 in fig. 5.16. The behaviour of such a system can be considered vectorially.* The analytical solution† also provides the optimum design of such systems, and it is found that it is necessary not only to have a definite value of damping

* R. E. D. Bishop and D. B. Welbourn, 'The Problems of the Dynamic Absorber', *Engineering*, 1952, Vol. 174, p. 796.
† J. P. Den Hartog, *Mechanical Vibrations*, McGraw-Hill, 4th. edn. 1956.

but also to reduce the natural frequency of the absorber relative to the frequency of the exciting force acting on mass m_1. The response curve for the optimum design of damped absorbers has been calculated for the case considered in fig. 5.17, and the curve is also shown in this figure. It will be seen that the severity of the original resonance has been very considerably reduced without introducing other unacceptable resonances, but at the expense of never reducing X_1 to zero.

In the case of torsional vibrations of crankshaft systems the dynamic absorber can take the form of a flywheel attached by a flexible connection to the shaft, but more frequently it takes the form of a 'pendulum absorber'.

There are several variations of this design, but in essence we can consider it as a simple pendulum pivoted to the rotating shaft at a radius r and free to oscillate in the plane perpendicular to the axis of rotation while rotating with the shaft. Fig. 5.18 shows a diagrammatic arrangement of the pendulum which has a length l and a bob

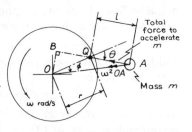

Fig. *5.18* Pendulum absorber

of mass m. If we consider the pendulum displaced by θ from its radial equilibrium position, and assume that gravitational forces are negligible, then the restoring torque about Q will be

$$m\omega^2\,OA \times \text{Perpendicular distance from } Q \text{ to } OA$$
$$= 2m\omega^2 \times \text{Area of the } \triangle\,OQA$$
$$= m\omega^2 \times OQ \times QA \sin\theta$$
$$= m\omega^2 rl \sin\theta$$

If θ is small, then this reduces to

$$m\omega^2 rl\theta$$

and the equation of motion for the pendulum will be

$$ml^2\ddot{\theta} = -m\omega^2 rl\theta$$

since ml^2 is the moment of inertia about Q.

Hence
$$\ddot{\theta} + \omega^2\frac{r}{l}\theta = 0$$

so the natural radiancy of the vibrating pendulum will be

$$\omega_n = \sqrt{\omega^2\frac{r}{l}} = \omega\sqrt{\frac{r}{l}} = \text{const} \times \omega$$

The frequency of the pendulum is not constant, but varies as the rotational speed, and the constant involved can be varied by adjusting r and l. Now major resonances occur when there is coincidence between the frequency of one of the harmonics of the torque diagram and one of the natural frequencies of the system. As the frequencies of any one of the harmonics and that of the pendulum increase with engine speed, the pendulum absorber can eliminate any particular harmonic throughout the running-speed range if the constant $\sqrt{(r/l)}$ has the appropriate value.

The capacity of the pendulum is limited by the maximum amplitude of vibration of the pendulum, which for reasons of linearity of the equation of motion must not exceed 12°–15°, as $\sin \Theta \simeq \Theta$ is not a sufficiently accurate approximation for larger angles. The centripetal force acting on the mass m will be

$$m\omega^2 OA$$

and this gives rise to a tensile component of force in the pendulum arm which will be

$$m\omega^2 OA \cos (\Theta - \Phi)$$

where Θ is the maximum amplitude of the pendulum and Φ is the corresponding value of $LQOA$. This tensile force exerts a moment about O which will be:

$$m\omega^2 OA \cos (\Theta - \Phi) \cdot r \sin \Theta$$
$$= m\omega^2 (l + r \cos \Theta) \cdot r \sin \Theta$$
$$= m\omega^2 r^2 \left(\frac{l}{r} \sin \Theta + \frac{\sin 2\Theta}{2} \right)$$
$$\simeq m\omega^2 r^2 \Theta \left(\frac{l + r}{r} \right)$$
$$= m\omega^2 r^2 \Theta \left(\frac{1 + n^2}{n^2} \right)$$

where $n^2 = r/l$. This will be the maximum vibration torque which the absorber can counteract.

In practice the absorber may take several forms which bear little obvious resemblance to a simple pendulum, but are dynamically equivalent.

§ 5.7 Vibration of a Mass Supported on Foundations Subject to Vibration

Many vibration recording instruments are based on the principle of the seismograph, in which a damped mass is supported on very flexible springs and the relative motion between the mass and the earth is recorded. To obtain a true record the mass must not be affected by the vibration of the earth. Again a car travelling over a rough surface should ideally be mounted so that the motion of its body is unaffected. Dynamically these two situations are similar, as illustrated in fig. 5.19.

Let the mass m be supported on a foundation subjected to a vertical displacement of $A \cos \omega t$. Let the displacement in space of

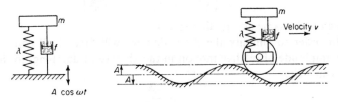

Fig. 5.19 Mass supported from vibrating foundation

mass m at time t be x. The net forces acting on the mass in the displaced position are from the spring $-\lambda(x - A \cos \omega t)$ and from the dashpot $-f \dfrac{d}{dt}(x - A \cos \omega t)$, so the equation of motion will be

$$m\ddot{x} = -\lambda(x - A \cos \omega t) - f \frac{d}{dt}(x - A \cos \omega t)$$

or rearranging and substituting $2c\omega_n$ for $\dfrac{f}{m}$ and $\omega_n{}^2$ for $\dfrac{\lambda}{m}$

$$\ddot{x} + 2c\omega_n \frac{d}{dt}(x - A \cos \omega t) + \omega_n{}^2(x - A \cos \omega t) = 0$$

Let us call the displacement relative to the support $u = (x - A \cos \omega t)$

$$\therefore \ \ddot{x} = \ddot{u} - A\omega^2 \cos \omega t$$

and the equation of motion can be rewritten as

$$\ddot{u} + 2c\omega_n\dot{u} + \omega_n{}^2 u = +A\omega^2 \cos \omega t$$

which can be solved either analytically or vectorially by the methods given in § 5.4.4. The awkward expression for the steady-state solution of this equation is

$$u = \frac{A \dfrac{\omega^2}{\omega_n{}^2}}{\sqrt{\left[\left(1 - \dfrac{\omega^2}{\omega_n{}^2}\right)^2 + 4c^2 \dfrac{\omega^2}{\omega_n{}^2}\right]}} \cos\left\{\omega t - \tan^{-1} \frac{2c \dfrac{\omega}{\omega_n}}{\left(1 - \dfrac{\omega^2}{\omega_n{}^2}\right)}\right\}$$

which is more easily appreciated physically from the vector solution. The motion in space of the mass m is given by

$$x = u + A \cos \omega t$$

which can be found by the vectorial summation of u and $A \cos \omega t$. From this equation the amplitude ratio can be shown to be

$$\frac{X}{A} = \frac{\sqrt{\left(1 + 4c^2 \frac{\omega^2}{\omega_n^2}\right)}}{\sqrt{\left[\left(1 - \frac{\omega^2}{\omega_n^2}\right)^2 + 4c^2 \frac{\omega^2}{\omega_n^2}\right]}}$$

which is the same as equation (5.24).

In order to simplify the discussion we will first of all consider the amplitude of relative motion in the absence of damping, so that we can write

$$U = \frac{A \dfrac{\omega^2}{\omega_n^2}}{\left(1 - \dfrac{\omega^2}{\omega_n^2}\right)}$$

Three particular cases are worthy of special consideration:

(1) If ω/ω_n is very small, $1 - \omega^2/\omega_n^2 \simeq 1$,

so $U = A \dfrac{\omega^2}{\omega_n^2} = \dfrac{1}{\omega_n^2} \times$ maximum acceleration of surface

This forms the basis of an instrument for measuring the maximum vibration acceleration of machines.

(2) $\omega/\omega_n = 1$: For this case the amplitude of relative vibration will be

$$U \to \infty$$

This forms the basis of a very sensitive instrument for indicating the presence of vibration, and it can be employed to measure its frequency. A particular example is depicted in fig. 5.20; it consists

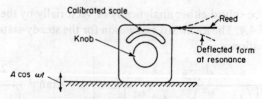

Fig. 5.20 Vibration tachometer

basically of a reed, the free length of which can be varied. If this instrument is clamped to a vibrating surface the reed will be forced into resonance when its natural frequency coincides with that of the vibration. The free length of the reed can be adjusted until this occurs, by rotating the shaft on which it is wound. A scale attached to this shaft indicates the free length and is also calibrated in terms of frequency.

(3) If ω/ω_n is very large, $1 - \omega^2/\omega_n^2 \simeq -\omega^2/\omega_n^2$: The amplitude recorded now becomes

$$U = \frac{A \dfrac{\omega^2}{\omega_n^2}}{-\dfrac{\omega^2}{\omega_n^2}} = -A$$

This means that the mass m is remaining very nearly stationary, so that the amplitude of the relative motion between the mass and the vibrating support is equal to the amplitude A. This, then, is the

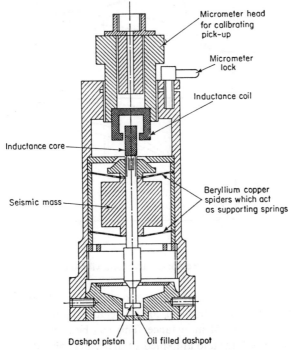

Fig. 5.21 An inductive vibrometer
(By courtesy of Southern Instruments Ltd)

situation which must be achieved in vibration-measuring instruments, including the seismograph. In one mechanical type of vibrograph the relative motion is recorded as a scratch on a strip of celluloid which is fed through the machine at a known speed, thus providing a displacement time record. A mechanical linkage in the instrument gives some magnification, and the celluloid strip can be projected optically to give additional magnification. Another example, this time of an electrical transducer, is shown in fig. 5.21, in which an inductance core is attached to a seismic mass supported by beryllium–copper spiders which act as springs.

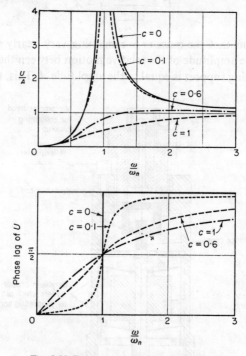

Fig. 5.22 Response curve for a vibrometer

When there is relative motion due to vibration the coil attached to the body of the instrument moves relative to the core, so that there is a change of inductance. This can be converted to a proportionate voltage and amplified so that vibrations of the order of a micron can be measured with ease.

Although the response of an undamped vibrograph will be

nearly linear only if the natural frequency of the instrument is low compared with the frequency of the vibration to be measured, it can be used for frequencies not very much greater than its natural frequency by using a correction factor. Alternatively, the approximately linear range can be extended towards $\omega/\omega_n = 1$ by introducing damping. Fig. 5.22 shows typical response curves, from which it will be seen that damping of about 0·6 times the critical value is a good compromise. Solving example 5.14 will be found to be instructive.

In the case of a car suspension, even if we neglect the flexibility of the tyres, the theory we have outlined is, of course, far too unsophisticated, since we are concerned not with one degree of freedom of the car body but six, of which the most important are pitching, rolling, and vertical motion. Each can be treated independently, but in principle we can see that to obtain a smooth ride we should provide 'soft' springs to reduce ω_n to as low a value as possible, and provide damping of the order 0·6 times critical damping.

§ 5.8 Whirling of Shafts

If the speed of a shaft or rotor is slowly increased from zero it is found that at some speed its deflection suddenly increases, and it rotates in a bent state about the axis of the bearings. This is known as the whirling speed. Even with a very carefully balanced rotor the deflection may well be sufficient to cause the shaft to fail. If before the elastic limit is exceeded the speed is raised further the deflection will usually decrease and a region of stability will be reached where the deflection becomes very small. However, in a system which has distributed masses and stiffnesses further whirling speeds will be found if the speed is raised even more.

The nature of the above phenomenon has been known for a very long time.* Much effort has been devoted to the understanding of the stability of the system above (and even at) the whirling speed, including investigations of the sources and magnitudes of the damping forces, as well as the causes of a more minor whirl which occurs at about half the normal whirling speed, the effect of skew stiffness of shafts and so on.

Let us first consider a simply supported shaft with a centrally

* W. J. M. Rankine, 'Centrifugal Whirling of Shafts', *Engineer*, 1869, Vol. 27, p. 249.

located disk of mass m, which has a centre of gravity displaced a distance e from the centre of the shaft. When the shaft is rotating at an angular speed ω we shall assume that the displacement of the mass is r, and that it is collinear with the eccentricity e. The inward restoring force due to the stiffness λ of the shaft provides the centripetal force, so

$$\lambda r = m\omega^2(r + e)$$

Now for transverse vibrations of such a shaft (§ 4.11.3) we know that $\omega_n^2 = \lambda/m$

so

$$\omega_n^2 r = \omega^2(r + e)$$

$$\therefore \ r = e\,\frac{\omega^2}{\omega_n^2 - \omega^2}$$

$$= e\,\frac{\dfrac{\omega^2}{\omega_n^2}}{1 - \dfrac{\omega^2}{\omega_n^2}} \tag{5.27}$$

Thus, when the rotational speed coincides with the natural radiancy of transverse vibration of the shaft the deflection becomes infinite however small the value of e. This is the whirling speed.

Similarly, we could consider a continuous shaft and for the deflected state calculate the value of the distributed restoring force, and hence obtain a fourth-order differential equation identical to that deduced in § 4.11.5. Again the displacement would become infinite if the rotational speed coincided with any one of the natural radiancies of transverse vibration calculated.

Though this elementary treatment gives the correct whirling speeds of shafts, and is often used to predict the equilibrium configuration at any speed, it is in fact completely inadequate for this purpose, and a more sophisticated treatment is necessary to obtain a real understanding of the behaviour of shafts.

Fig. 5.23a shows a vertical shaft carrying a single central mass, and fig. 5.23b gives a plan view of the disk and shaft, in which the centre of rotation is at O and the centre of the deflected shaft is at A. It is assumed that the centre of gravity is at B, which is a distance e from the centre of the shaft, though in practice the eccentricity would normally be very small. For the general position we also assume that A, B, and O are not collinear. The rectangular coordinates of the centre of gravity of the disk will be $x + e \cos \omega t$

and $y + e \sin \omega t$. The disk is rotating and travelling tangentially. It will therefore be subjected to an aerodynamic force similar to that on a rotating golf ball or tennis ball (the 'Magnus effect') which will act radially inwards, and therefore in the same direction as the elastic restoring force, but will be proportional to the velocity. Whether this is the *main* source of such a force or not (and we know

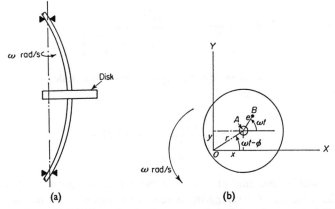

Fig. 5.23 Whirling of a simply supported shaft carrying a central disk

too little about the numerical values to be sure), it provides us with a reason for inserting in the equation of motion a 'damping term' with constant $= f$.

The equations of motion for the x and y displacements will be:

$$m\frac{d^2}{dt^2}(x + e \cos \omega t) = -\lambda x - f\frac{dx}{dt}$$

$$m\frac{d^2}{dt^2}(y + e \sin \omega t) = -\lambda y - f\frac{dy}{dt}$$

where λ is the stiffness of the shaft. These equations give

$$\ddot{x} + 2c\omega_n\dot{x} + \omega_n^2 x = e\omega^2 \cos \omega t$$
$$\ddot{y} + 2c\omega_n\dot{y} + \omega_n^2 y = e\omega^2 \sin \omega t \qquad (5.28)$$

We have already considered the steady-state solution of equations of this form, so we can write down the solutions straight away

$$x = \frac{e\dfrac{\omega^2}{\omega_n^2}}{\sqrt{\left[\left(1 - \dfrac{\omega^2}{\omega_n^2}\right)^2 + 4c^2\dfrac{\omega^2}{\omega_n^2}\right]}} \cos\left\{\omega t - \tan^{-1}\dfrac{2c\dfrac{\omega}{\omega_n}}{\left(1 - \dfrac{\omega^2}{\omega_n^2}\right)}\right\}$$

$$y = \frac{e\dfrac{\omega^2}{\omega_n{}^2}}{\sqrt{\left[\left(1 - \dfrac{\omega^2}{\omega_n{}^2}\right)^2 + 4c^2\dfrac{\omega^2}{\omega_n{}^2}\right]}} \sin\left\{\omega t - \tan^{-1}\frac{2c\dfrac{\omega}{\omega_n}}{\left(1 - \dfrac{\omega^2}{\omega_n{}^2}\right)}\right\}$$

These are the co-ordinates of a point moving in a circle centred at O and having a radius

$$r = \frac{e\dfrac{\omega^2}{\omega_n{}^2}}{\sqrt{\left[\left(1 - \dfrac{\omega^2}{\omega_n{}^2}\right)^2 + 4c^2\dfrac{\omega^2}{\omega_n{}^2}\right]}}$$

In the absence of damping the expression for the deflection of the central mass becomes

$$r = \frac{e\dfrac{\omega^2}{\omega_n{}^2}}{\left(1 - \dfrac{\omega^2}{\omega_n{}^2}\right)} \text{ as in (5.27)}$$

and it will be seen that at $\omega/\omega_n = 1$ the deflection becomes infinite. In the presence of damping the maximum radius of whirl will occur at a speed of $\omega_n/\sqrt{(1 - 2c^2)}$, exactly as calculated for vibrations in § 5.4.5.

It will, of course, be seen at once that the equations of motion (5.28) are identical in form with that for vibration in § 5.4.5, and indeed whirling can be regarded as the summation of two equal simultaneous transverse vibrations in rectangular planes. Inevitably therefore the whirling speed of any shaft coincides identically with the radiancy for transverse vibration.

In the presence of damping the phase angle ϕ, that is the angle between the radius and the eccentricity AB (fig. 5.23b), will be

$$\tan^{-1}\frac{2c\dfrac{\omega}{\omega_n}}{\left(1 - \dfrac{\omega^2}{\omega_n{}^2}\right)}$$

At low frequencies when $\omega/\omega_n \to 0$ this angle will approach zero, so that the points OAB are collinear, with the centre of gravity radially out from A. With large values of ω/ω_n the phase angle approaches 180°, which means that the points OBA are collinear, with the centre of gravity now radially in from the shaft centre A. It will also be noted that at high values of ω/ω_n the deflection of the shaft, even with damping present, approaches a value of e, since $1 - \dfrac{\omega^2}{\omega_n{}^2} \simeq \dfrac{\omega^2}{\omega_n{}^2}$ and $4c^2\,\dfrac{\omega^2}{\omega_n{}^2}$ is negligible in comparison with $\left(1 - \dfrac{\omega^2}{\omega_n{}^2}\right)^2$. This means that at very high speeds the disk rotates about its own centre of gravity. This fact was long ago recognised by de Laval in the design of the impulse turbine named after him, in which the rotor is mounted on a very thin shaft so that the whirling speed is very low. As a consequence, the turbine is accelerating fairly fast as it passes through the whirling speed, and it runs at a very much higher speed, so that the deflection of the shaft is negligibly small. To prevent excessive

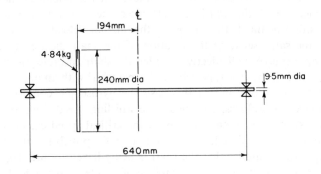

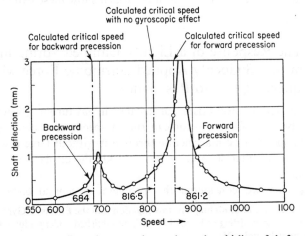

Fig. 5.24 Effect of gyroscopic couple on the whirling of shafts

bending of the shaft as the rotor accelerates through the whirling speed, special guard rings are provided which restrict the deflection within safe limits.

It will be noted that when a rotor is rotating at speeds near the whirling speed the shaft is rotating in the deflected state and it will be subjected to a constant bending stress with no cyclically varying stress. In consequence, there is no damping arising from hysteresis of the shaft material, and the main sources of damping are probably the fluid surrounding the rotor and the bearings. However, these damping forces tend to be very small, with the result that the deflection at the whirling speed may be unacceptably large, even though the eccentricity of the centre of gravity is very small.

If the disk or disks are not situated at the points of maximum deflection of the whirling shaft their axes must in the deflected position be tilted. If, as is usual, the angular velocity of the whirl is in the same sense as the rotation of the shaft the resulting gyroscopic couples will effectively stiffen the shaft and raise the whirling speed. It is, however, possible to have a whirl with an opposite sense to the rotation, and in this case the gyroscopic couples tend to reduce the stiffness, and hence the whirling speed. Two resonance peaks, one below and one above the critical speed calculated neglecting gyroscopic effects, are consequently possible, but the lower one, which implies that the shaft is being flexed to and fro, will be more heavily damped and therefore of smaller magnitude. Fig. 5.24 illustrates some experimental results* which show these two speeds.

§ 5.9 The Response of a Lightly Damped Spring-mass System to a Suddenly Applied Harmonic Force with a Frequency Close to Resonance

This situation in effect arises, for instance in gas turbines, where an extremely well-balanced rotor can run at speeds very close to the whirling speed or even accelerate through it without large displacements. It is of interest to know how quickly the displacement does build up.

In § 5.4.4 we considered this problem in general terms and

* P. R. Lever and P. P. Mehta, 'An Investigation into the Effects of Gyroscopic Moments on the Whirling of Loaded Shafts', unpublished undergraduate thesis, University of Bristol, 1951.

stated that this special case would be discussed separately. Equation (5.20) gave the complete solution. For convenience it is repeated

$$x = A_5 e^{-c\omega_n t} \cos\{\omega_n \sqrt{1 - c^2}\, t + \phi\}$$

$$+ \frac{X_s}{\sqrt{\left[\left(1 - \dfrac{\omega^2}{\omega_n^2}\right)^2 + \left(2c\, \dfrac{\omega}{\omega_n}\right)^2\right]}} \cos\left(\omega t - \tan^{-1} \frac{2c\, \dfrac{\omega}{\omega_n}}{1 - \dfrac{\omega^2}{\omega_n^2}}\right) \quad (5.20)$$

where $X_s = F/\lambda$; and F, for the whirling problem considered in § 5.8, would be $m\omega^2 e$.

It considerably reduces the algebra involved if we assume that the forcing radiancy is exactly equal to that of the natural damped vibration, i.e. $\omega = \omega_n \sqrt{(1 - c^2)}$, as the frequency of the first and second terms in the above equation are then identical. The exact value we choose is of no importance, since we are considering small values of critical damping (i.e. $c \ll 1$). The complete expression for x immediately reduces to

$$x = A_5 e^{-c\omega_n t} \cos\{\omega t + \phi\} + \frac{X_s \cos\{\omega t - \beta\}}{\sqrt{(4c^2 - 3c^4)}} \quad (5.29)$$

where

$$\beta = \tan^{-1} \frac{2\sqrt{(1 - c^2)}}{c} \simeq \frac{2}{c}$$

$$\therefore \cos\beta \simeq \frac{c}{2}, \quad \sin\beta \simeq 1$$

As c is small $\sqrt{(4c^2 - 3c^4)} \simeq 2c$

If the applied harmonic force is applied suddenly when $t = 0$, then at $t = 0$, $x = 0$, and $\dot{x} = 0$.

$$(x)_{t=0} = 0 = A_5 \cos\phi + \frac{X_s \cos(-\beta)}{2c}$$

Hence

$$A_5 = -\frac{X_s}{4\cos\phi}$$

$$(\dot{x})_{t=0} = 0 = -A_5(c\omega_n \cos\phi + \omega \sin\phi) - \frac{X_s \omega \sin(-\beta)}{2c}$$

Whence

$$\tan\phi \simeq -\frac{2}{c} \text{ and } \phi \simeq -\frac{\pi}{2}$$

Substituting in equation (5.29),

$$x \simeq -\frac{X_s}{2c} e^{-c\omega_n t} \cos\left(\omega t - \frac{\pi}{2}\right) + \frac{X_s}{2c} \cos\left(\omega t - \frac{\pi}{2}\right)$$

$$\simeq \frac{X_s}{2c}(1 - e^{-c\omega_n t}) \sin\omega t \quad (5.30)$$

We can therefore see that the amplitude will build up rapidly at first and approach asymptotically the maximum value of $X_s/2c$. Moreover, if we differentiate the amplitude with respect to time we obtain

$$\frac{dX}{dt} = \frac{X_s}{2} \cdot \omega_n e^{-c\omega_n t} \tag{5.31}$$

which shows that in the early stages the rate of growth of amplitude is independent of the damping, and that the addition per cycle is $\pi \cdot X_s$. For a whirling shaft the damping is very small; if we were to assume a value of $c = 0 \cdot 001$, then the number of revolutions for the displacement to build up to 90 per cent of its final value would be about 366.

§ 5.10 Response of Spring-mass System to a Force of Steadily Increasing Frequency

A slightly different case which is of importance is the behaviour of a system in which the frequency of the applied force varies with

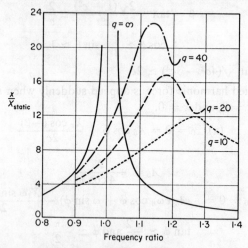

Fig. 5.25 Response of spring-mass system to a harmonic force with a frequency varying linearly with time

time and passes through the resonant frequency: for instance, a rotor which is accelerated through the whirling speed. The exact solution is very difficult: even a simplified solution is very tedious

and too difficult for inclusion, but the results are of interest.* Fig.
5.25 shows how the amplitude varies with speed as it passes through
resonance with different values of uniform angular acceleration. In
this case the damping ratio has been taken as zero, and the value of
q attached to each curve is the number of revolutions (or cycles)
taken from starting to reach the resonant speed.

§ 5.11 Sources of Vibration Excitation

So far we have considered excitation of vibration by simplified
forcing functions, but we have not discussed the sources of excitation
in practice. From the vast range of possibilities, let us choose one
example as an illustration.

Perhaps the importance and destructiveness of mechanical vibrations was first realised in the development of high-speed reciprocating engines for marine propulsion. It was soon discovered that

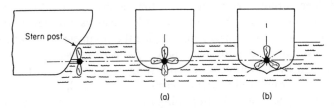

Fig. 5.26 Hydrodynamic interference between a ship's propeller and
stern-post

serious torsional vibration of the propeller shaft was possible at
certain critical speeds of rotation, and that as a consequence of
such vibrations, fatigue failure occurred at weak points in the system,
such as key-ways in the shafting. In some cases these failures were
accentuated by the presence of a corrosive environment in the form
of salt water, which causes fatigue failure to occur at much lower
stress levels.

These torsional vibrations can be excited either by the non-
uniform torque produced by the engine or by the periodically
varying torque resisting the rotation of the propeller. On first

* S. Hother-Lushington and D. C. Johnson, 'The Acceleration of a Single-degree
of Freedom System Through its Resonant Frequency'. *J. roy. Aero. Soc.*, 1958,
Vol. 62, pp. 752–7.

consideration it might appear surprising that there is a variation in the torque resisting the rotation of the propeller. If, however, we look at fig. 5.26a we may appreciate that whenever a blade passes the stern post there will be a variation in the hydrodynamic forces acting on the blades due to the flow of water round the hull. With an even-bladed propeller the largest variation in the force will be when one of the blades is passing the stern post above the centre line of rotation; simultaneously another blade passing the stern post below the centre line will make a small contribution to the variation in force. In consequence, there is an exciting torque at blade frequency (i.e. Number of blades × Rotational speed). Applying the same reasoning to an uneven number of blades, it can be seen that there will be an exciting torque at both blade frequency and twice blade frequency (see fig. 5.26b).

The torque developed by a reciprocating engine is made up of the torque arising from the gas or steam pressure forces on the piston and the torque caused by inertia of the piston and connecting-rod. The variation of torque with time is cyclic; in a four-stroke single-acting engine the complete cycle occupies two revolutions, and in a two-stroke single-acting engine the complete cycle occupies one revolution. The torque diagram *for each cylinder* can be represented by a Fourier series consisting of a constant term and a series of harmonically varying terms having 1, 2, 3, 4, etc., repetitions per cycle. Fig. 5.27 (repeating fig. 1.47) shows the torque diagram for a single-cylinder four-stroke engine taking into account the inertia forces, and the first five terms of the Fourier series are shown underneath. If we add up sufficient terms of the Fourier series for a particular angular position of the crank it will, of course, be equal to the engine torque in that position. In the case of the four-stroke cycle the radiancy of the successive harmonic terms of the Fourier series are $\frac{1}{2}$, 1, 1$\frac{1}{2}$, 2, 2$\frac{1}{2}$, etc., times the engine speed. Any one of these harmonic terms can cause resonance if its frequency, within the running speed range of the engine, coincides with one of the natural frequencies of torsional vibration.

Some of these harmonic exciting torques developed by individual cylinders in a multi-cylinder engine tend to cancel one another, and the speeds at which resonance is excited by these harmonics are known as 'minor critical speeds'. Other harmonics, however, tend to add up or assist one another, and the speeds at which resonance occurs due to these harmonics are known as 'major critical speeds'.

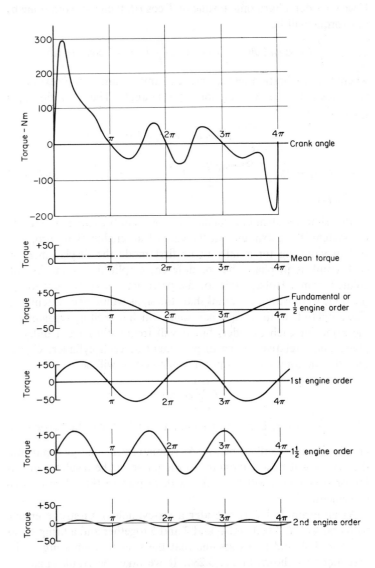

Fig. 5.27 Fourier analysis of torque diagram for one cylinder of four-stroke engine

If we consider a harmonic torque of $T \cos \omega t$, then the work done by this torque will be

$$\int T \cos \omega t \,.\, d\theta = \int T \cos \omega t \,.\, \frac{d\theta}{dt}\, dt = \int T \cos \omega t \,.\, \dot{\theta} dt$$

where θ, the vibration displacement at any instant, can be expressed as $\Theta \cos (\omega t - \beta)$, β being some phase angle. The work done per cycle will be

$$-\int_{t=0}^{t=2\pi/\omega} T \cos \omega t \,.\, \Theta\omega \sin (\omega t - \beta) dt$$

$$= -T\Theta\omega \cos \beta \,.\, \int_{t=0}^{t=2\pi/\omega} \frac{\sin 2\omega t}{2}\, dt + T\Theta\omega \sin \beta \,.\, \int_{t=0}^{t=2\pi/\omega} \cos^2 \omega t \,.\, dt$$

$$= \pi T\Theta \; \sin \beta$$

With a simple system in resonance β is 90° and the energy input into the system is a maximum. In the case of an engine–shaft–propeller system vibrating torsionally in its first mode the displacement at each crank is in phase, but of different amplitude, as can be calculated from a Holzer table for the particular mode of vibration (if the node were within the crankshaft the signs would need attention). The amplitude of the torque harmonic which is exciting resonance is the same for each cylinder, but it will frequently have a different phase angle relative to the displacement at each cylinder. Consequently, the total energy input into the system will be

$$\pi T \sum_{i=1}^{i=n} \Theta_i \sin \beta_i$$

where $\sum_{i=1}^{i=n} \Theta_i \sin \beta_i$ is a vector summation which can be carried out graphically as described below. At resonance the energy input into the system will be a maximum, so the phase of the resultant of this vector summation will be at 90° with respect to the vibration displacement.

As an example let us consider a six-cylinder four-stroke engine with cranks 1 and 6, 3 and 4, and 2 and 5 together and a firing order 1–3–5–6–4–2, and let us assume that the system is vibrating in the first mode as shown in fig. 5.28a. If we take the crank angle for crank 1 to be 0°, then the crank angles for the firing of successive cylinders starting with 1 will be 0°, 600°, 120°, 480°, 240°, and 360°. We do not, of course, know the phase angle between the vibration displacement and any harmonic torque component acting at any

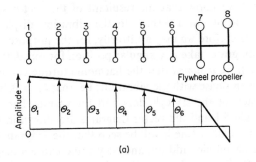

Fig. 5.28a Elastic line for first mode of vibrations of six-cylinder engine and propeller

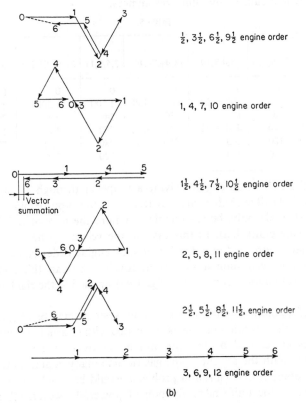

$\frac{1}{2}$, $3\frac{1}{2}$, $6\frac{1}{2}$, $9\frac{1}{2}$ engine order

1, 4, 7, 10 engine order

$1\frac{1}{2}$, $4\frac{1}{2}$, $7\frac{1}{2}$, $10\frac{1}{2}$ engine order

2, 5, 8, 11 engine order

$2\frac{1}{2}$, $5\frac{1}{2}$, $8\frac{1}{2}$, $11\frac{1}{2}$, engine order

3, 6, 9, 12 engine order

(b)

Fig. 5.28b Vector diagrams for various harmonics

cylinder, but we know that the resultant of the vector summation $\sum \Theta_i \sin \beta_i$ is at 90° with respect to the vibration displacement at resonance. Therefore we may arbitrarily, for the purpose of drawing the vector diagram, take any angular position for our datum. In this example we have *assumed* that the harmonic torque component for cylinder 1 is in phase with the vibration; i.e. that $\beta_1 = 0$. The values of β_i for each of the six cylinders and for various harmonics are given in Table 5.1 and the vector diagrams are shown in fig. 5.28b. For this particular engine it will be seen that the 3rd, 6th, 9th, 12th, etc., engine harmonics add up, and they could cause a major resonance if the frequency of any of these harmonics coincided with the natural frequency anywhere in the running-speed range of the engine. The vector diagrams for the other harmonics tend to cancel out, so these would cause only minor resonances.

Table 5.1

CYL. NO.	$\frac{1}{2},3\frac{1}{2},6\frac{1}{2},9\frac{1}{2}$	1,4,7,10	$1\frac{1}{2},4\frac{1}{2},7\frac{1}{2},10\frac{1}{2}$	2,5,8,11	$2\frac{1}{2},5\frac{1}{2},8\frac{1}{2},11\frac{1}{2}$	3,6,9,12
1	0	0	0	0	0	0
2	300	600 = 240	900 = 180	1 200	1 500	1 800
3	60	120	180	240	300	360
4	240	480 = 120	720 = 0	960	1 200	1 440
5	120	240	360 etc.	480	600	720
6	180	360 = 0	540	720	900	1 080

In the case of an engine driving a propeller through a long and relatively flexible shaft it may, at least as a first approximation for the first mode only, be reasonable to ignore the lack of rigidity in the engine crankshaft. In this case all the vectors in fig. 5.28b are of equal length, and only the 3–6–9, etc., engine order components give a non-zero summation. Alternatively, one can think of the torque summation curve (as in fig. 1.49) applied to the rigid engine inertia.

Brief qualitative notes on a few other systems may be of interest. Many of the vibration problems associated with internal-combustion engines are caused by the intermittent nature of the power production. With the advent of the gas turbine many engineers tended to the opinion that vibration problems would be much less serious because of the uniform production of power. However, vibration problems in gas turbines have proved to be even more serious than in reciprocating engines, and the causes of these vibrations are frequently obscure and difficult to analyse. Indeed, it has, so far, proved to be impossibly difficult even to calculate the natural frequencies of any but simplest modes of vibration of the blades used in axial compressors and turbines.

Vibration of blading is caused mainly by:

(*a*) fixed vane wake excitation
(*b*) classical torsion–flexural flutter
(*c*) stalled flutter

(*b*) and (*c*) are self-excited vibrations which are discussed in the next section. (*a*) is the most readily understood source of blade vibration excitation. In many engines there are fixed vanes at entry to the compressor to support the forward bearing and to carry lubricating oil to the rotor. The air flow past these vanes creates wakes of low air velocity compared with the rest of the annulus. Consequently, as the rotor blades pass through these zones of reduced velocity they will feel a reduction of aerodynamic force. The frequency of these pulses will be equal to the number of vanes times the speed of rotation. If this frequency coincides with one of the natural frequencies of the blade within the running-speed range of the engine resonance will occur. The amplitude will be limited by aerodynamic damping, hysteresis damping and frictional damping at the blade-root fixing. A similar interference occurs between stator and rotor blades. In any such situation, if the highest common factor between the number of blades in stator and rotor is large, then the input of vibrational energy to the rotor as a whole may be dangerously large.

Obviously it is essential in a gas turbine that no serious disk vibrations should occur in the running-speed range. Disks can vibrate in many ways. Fig. 4.28 shows the sand patterns associated with two particular modes. Some of these vibrations can be excited by irregularities in the flow pattern from the combustion chambers, but usually the exciting force is too small to cause serious amplitudes. However, failure in the combustion due, for instance, to a blocked burner can give rise to serious vibrations with resulting failure of the disk.*

* See A. C. Lovesay, 'The Art of Developing Aero-engines', *J. roy. Aero. Soc.*, 1959, Vol. 63, p. 429.

§ 5.12 Self-excited Vibrations

So far we have considered the vibrations of simple systems excited by an externally applied cyclic force, which has a magnitude and frequency independent of the vibration it excites. The system vibrates with the frequency of the external force, and it is only if this frequency is close to the natural frequency that resonance occurs. With a self-excited vibration there is no externally applied alternating force which is independent of the motion, but the force sustaining the vibration is a consequence of the vibratory motion itself.

It is perhaps not sufficiently well recognised that many of the vibration problems met with in practice are self-excited vibrations. We have already noted one of the most elementary examples of self-excitation, which is afforded by passing a violin bow at a constant velocity over a violin string. At first the string is deflected by the bow until the restoring force in the string is sufficient to overcome the static frictional force. Slipping between the bow and string can then occur, and as the kinetic coefficient of friction is less than the static value, the string continues to return towards its equilibrium position until the restoring force in the string is insufficient to overcome the kinetic frictional force. The slipping between the bow and string then ceases and the events described above are repeated. Similar types of 'slip–stick' vibration are common in engineering practice.

Although we have assumed two different coefficients of friction, a very similar end result obtains if we assume a coefficient of friction falling with speed. It is very easy to write down the equation of motion for this situation and show that a term similar to the damping term but opposite in sign appears in the equation. Such a condition is often referred to as one of 'negative damping'. In the corresponding solution the exponential term has a positive index, which implies that the vibration builds up until the conditions postulated no longer apply.

Two of the most prolific sources of self-excitation are hydrodynamic and aerodynamic forces, for instance oil whip in rotating shafts caused by hydrodynamic forces in film-lubricated bearings* and aircraft wing flutter due to aerodynamic forces.

---

* See B. L. Newkirk and H. D. Taylor, 'Shaft Whipping due to Oil Action in Journal Bearings', *General Electric Review*, 1925, also B. L. Newkirk, 'Instability of Oil Films and more Stable Bearings', General Discussion on Lubrication and Lubricants, Group 1, *Inst. Mech. Eng.*, 1937.

In order to understand the rudiments of the phenomenon of flutter, let us consider a simple cantilever wing rigidly built in at the root and without ailerons, mounted in a wind tunnel at a small angle of incidence. At low air velocities any vibration of the wing which occurs dies away quickly. However, at some critical air velocity an oscillation of the wing is maintained at a steady amplitude, and if

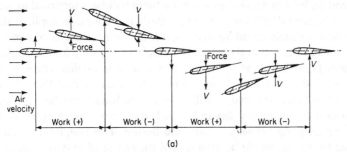

(a)

Vibrations in phase: net work input zero

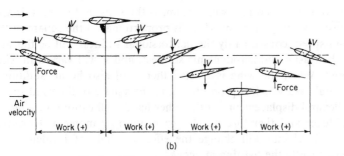

(b)

Vibrations 90° out of phase: net work input positive

Fig. 5.29 Vibrations of aerofoil in windstream

the wind speed is further increased the vibration grows until the amplitude may be sufficient to cause failure. Under these conditions the wing is said to 'flutter'. The essential facts in this problem are that the vibration of the wing must involve both flexural and torsional components, and there must also be a phase difference between these two components of motion.

If we consider the torsional and flexural motions in phase with one another, as shown in fig. 5.29a, then it will be seen that the net work done by the airstream on the structure is zero. In the extreme case, if there is a phase difference of 90° as shown in fig. 5.29b, there is a continuous input of work into the structure, and the vibration amplitude will increase until the work input is balanced by the energy lost in damping, or the structure fails. In the intermediate case there will be a net work input.

In still air there will in general be coupling between the torsional and flexural modes due to elastic and inertial causes. Normally if a bending load is applied to the wing there will be a torsional as well as a flexural displacement, and similarly if a couple is applied: this indicates elastic coupling (see § 4.13). However, there is one point, known as the 'flexural centre', for which if a bending load is applied there is no twist, or for which if a couple is applied there is no flexural displacement. If the wing is deflected so that there is no twist and then released, it will normally be found that the resulting motion is not purely flexural, which indicates that there must be some coupling between torsion and flexure. This coupling, which is due to inertia, would be zero only if the centre of gravity coincided with the flexural centre.

We can now consider the effect of the aerodynamic forces when the wing is placed in a windstream. If the wing is twisted this will cause a twisting moment which is proportional to the square of the wind speed, consequently the net torsional stiffness will be the sum of the elastic and aerodynamic stiffness, and it will vary with wind speed. When the wing is twisted there will also be an additional flexural load which is proportional to the square of the wind speed. A flexural displacement, on the other hand, will cause no change in incidence and therefore no change in twisting moment. However, flexural velocity will change the effective angle of incidence and consequently the twisting moment.

Enough has been written to indicate the complexity of the flutter problem and also to show that a large change in the amplitude ratio and phase difference of the flexural and torsional motions could be expected as the air velocity is varied. Normally the effect of these changes is to alter the net dissipation of energy in a vibration at low speed, with resulting stability, to a positive input of energy above a certain critical wind speed, with a resulting instability or flutter vibration. A relatively simple analytical treatment of this problem is given by Duncan.*

We have stated that classical flutter can occur only when there

* W. J. Duncan, 'The Fundamentals of Flutter', Aeronautical Research Committee, *Reports and Memoranda No. 2417*, November 1948.

are flexural and torsional components of the motion, but at large angles of incidence instability can arise when only flexural motion is involved. This phenomenon is known as 'stall flutter'. If we consider an aeroplane wing (or aerofoil blade) vibrating in a purely flexural mode at a small angle of incidence, then during the downward motion the effective angle of incidence is increased so that the flexural load is increased, and vice versa. The net result is that the change in lift is always a force resisting the velocity of the wing, i.e. a damping force, so vibrations will die away. However, if the effective angle of incidence exceeds the stalling angle the flow breaks away from the upper surface of the wing and the slope of the graph of lift against angle of incidence is reversed. As a consequence, the change in the lift is now a force acting in the same direction as the velocity, so that energy is being fed into the system and the amplitude of vibration will increase exactly as in the case of 'slip–stick' vibrations. Stall flutter is a very serious problem in axial compressors of jet engines, as though the angle of incidence under normal operating conditions is much less than that required to cause stalling, this may not be the case under other operating conditions.

These few examples give some idea of the sources and complexities of self-excited vibrations. The student is referred to a paper by J. P. Den Hartog, 'Vibrations: A survey of industrial applications', *The Chartered Mechanical Engineer*, 1958, Vol. 5, which gives many examples of such problems.

Reference Books

1. LORD RAYLEIGH. *The Theory of Sound*, 2nd edn. Macmillan, 1894.
2. S. TIMOSHENKO. *Vibration Problems in Engineering*, 3rd edn. Van Nostrand, 1955.
3. J. P. DEN HARTOG. *Mechanical Vibrations*, 4th edn. McGraw-Hill, 1956.
4. A. STODOLA. *Steam and Gas Turbines*, trans. L. C. Loewenstein. New York, Smith, 1945.
5. L. S. JACOBSEN and R. S. AYRE. *Engineering Vibrations*. McGraw-Hill, 1958.
6. W. KER WILSON. *Practical Solution of Torsional Vibration Problems*, 3rd edn. Chapman & Hall, 2 vols, 1956, 1963.
7. E. J. NESTORIDES. *Handbook of Torsional Vibration*. Cambridge University Press, 1958.

Examples

5.1 A mass of 50 kg is attached to a rigid body by a spring of rate 35 kN/m. Determine:

(a) the undamped natural frequency;

(b) the viscous damping required to give critical damping;

(c) the damping ratio if the viscous damping coefficient is 1·75 kNs/m.

5.2 A system consists of a spring of stiffness 3·5 kN/m and a mass of 5 kg. When it is vibrating freely the amplitude of the mass decays 1 per cent per cycle. Calculate:

(a) the frequency of vibration;

(b) the damping constant;

(c) the critical damping ratio.

If this system is subjected to a harmonic force, plot the non-dimensional response curve up to a value of $\omega/\omega_n = 2$, and calculate the magnitude of the harmonic force to give an amplitude of vibration at resonance of 10 mm.

5.3 An instrument of mass m is supported by rubber springing in a packing case which is dropped vertically on the floor from a height of 1 metre. If the spring limits the movement of the mass to 100 mm, calculate the maximum acceleration to which the mass is subjected. The rubber spring can be assumed to provide a restoring force which is a linear function of displacement and a damping force which is proportional to velocity, and which gives a critical damping ratio of 0·2.

It may be assumed that the box does not rebound.

5.4 An instrument for measuring fluctuating pressures consists essentially of a spring-loaded piston with some damping deliberately introduced to damp out transient vibrations. The piston mass is 0·05 kg, its cross-sectional area 320 mm², the spring has a stiffness of 140 kN/m, and the damping may be assumed to be half the critical damping.

If a steady pressure of 14 bar is suddenly applied, find the time for the amplitude of the transient vibration to become less than 5 per cent of the steady state displacement.

5.5 A pressure-measuring device consists of a light piston of 9 mm diameter and of mass 2·5 g supported by a liquid spring which deflects 1 mm when a pressure of 1·2 kbar is applied to the piston. Viscous damping occurs between the piston and cylinder, and the damping constant is calculated to be 1·5 Ns/m.

A pressure of 3 kbar is applied to this measuring device in 3 milliseconds and thereafter remains constant. Calculate as a percentage of the steady state displacement the amplitude of the transient vibration when the maximum pressure has been reached. It may be assumed that the pressure waveform during the pressure rise is $P \sin \omega t$ where $0 < \omega t < \pi/2$.

5.6 A single-cylinder engine is mounted at the centre of two 4·5 m long beams, which are simply supported at each end. Calculate the amplitude of vibration at the running speed of the engine, which is 500 rev/min, assuming that the secondary force can be neglected.

Mass of reciprocating parts 2·5 kg
Crank radius 60 mm
Mass of whole engine and bedplate 200 kg
2nd moment of area of each beam 2×10^{-5} m^4
Young's modulus for steel 2·07 Mbar

5.7 A four-cylinder in-line engine on a bedplate is mounted on flexible supports which deflect 0·6 mm under the combined weight of 150 kgf. The mass of the reciprocating parts per cylinder is 1 kg, the crank radius is 60 mm, and the connecting-rod length is 275 mm. Find the amplitude of vibration when the engine is running at 1 000 rev/min, assuming that the crankshaft arrangement is that normally associated with a four-cylinder four-stroke engine and the line of the out-of-balance force coincides with the centre of gravity.

5.8 A six-cylinder engine is perfectly balanced for reciprocating and rotating masses, but it is supported on rubber mountings designed to reduce the vibration torque transmitted to the foundations. The full torque developed by the engine consists mainly of a steady torque with a superimposed 3rd engine order component, which has an amplitude of 950 Nm.

To determine the damping in the rubber mountings the engine is disturbed torsionally, when mounted in position, and the amplitudes of successive vibrations are recorded. From these tests it is found that the ratio of successive amplitudes is $\frac{1}{4}$ and the natural damped frequency is 400 c/min. The moment of inertia of the engine about its axis of oscillation is 15 kg m².

Find the maximum torque transmitted to the foundations at an engine speed of 250 rev/min.

5.9 A single-cylinder petrol engine is mounted on a spring-supported base-plate. The following particulars apply to the engine and mountings:

Combined stiffness of spring supports 15×10^3 kgf/m

Mass of engine with base plate 190 kg

Engine speed 750 rev/min

Reciprocating mass 1·4 kg

Stroke 150 mm

A damping force is provided by a dashpot, which is fitted between the floor and the base-plate, and the damping is adjusted to keep the amplitude of vibration down to 0·50 mm. Ignoring the secondary force, determine the damping required and find the force transmitted to the floor.

5.10 A machine of mass 550 kg is flexibly supported on rubber mountings which provide a force proportional to displacement of 210 kN/m together with a viscous damping force. The machine gives an exciting force of the form $R\omega^2 \cos \omega t$, where R is a constant. At very high rotational speeds the measured amplitude of vibration is 0·25 mm and the maximum amplitude recorded as the speed is slowly increased from zero is 2 mm. Determine the value of R and the damping ratio.

5.11 A machine which produces a harmonic force of constant amplitude is mounted on vibration isolators which provide a total stiffness of λ and a force which gives a damping ratio of c. Determine the frequency at which the force transmitted to the foundations is a maximum, and the amplitude of this force.

5.12 An engine installation has a '1½ engine order torsional resonance' at an engine speed of 3 000 rev/min. To reduce the magnitude of this vibration a simple pendulum absorber is fitted. The pivot of the pendulum is at a radius of 50 mm and absorber mass is 2 kg. Find the length of the pendulum and the maximum vibration torque it can absorb if the amplitude of swing of the pendulum is limited to 12°.

5.13 An undamped vibrometer is used to measure translational vibrations. It consists of a seismic mass of 90 g suspended on a light spring. If the instrument is to measure vibrations with a minimum frequency of 20 Hz to an accuracy of 2 per cent, find the stiffness of the spring.

5.14 Find the expression for the frequency ratio at which a damped vibrometer gives the maximum ratio of recorded amplitude of vibration to the amplitude of the actual vibration. If the maximum *positive* error recorded by the instrument is not to exceed 2 per cent, find the damping ratio necessary. What is the minimum frequency ratio at which the instrument could be used if an error of ± 2 per cent could be tolerated?

5.15 A damped vibrometer is calibrated by clamping it to a vibrating table whose frequency and amplitude can be varied. When the table is vibrating with an amplitude of 2·5 mm the amplitude recorded by the instrument at 25 Hz is 2·44 mm and at 50 Hz 2·59 mm. Determine the damping ratio and the natural undamped frequency.

5.16 A two-wheeled trailer has springs fitted which deflect 100 mm under the weight of the sprung mass of 500 kg. Dashpots are provided between the road wheels and body such that the damping is 0·4 of the critical damping. The trailer is pulled along a road with an approximate sine-wave contour; the distance between successive peaks is 14 m, and the depth between peaks and valleys is 50 mm. Assuming that the wheels do not leave the ground and that the tyres are stiff compared with the springs, find the variation in the force in the springs if the trailer is pulled at 65 km/hr.

5.17 A 25 mm diameter shaft is supported at each end by plain bearings. The free length of the shaft is 300 mm, and it supports a disk at its centre which has a mass of 16 kg. If it is assumed that the mass of the shaft can be neglected, find the whirling speed of the shaft it is considered that:

(a) the shaft is directionally constrained at each end;

(b) the shaft is freely supported at each end.

Young's modulus for the shaft material may be taken as 2·07 Mbar.

5.18 The propeller shaft of a car is in the form of a steel tube 50 mm outside diameter and 44 mm inside diameter, and it is carried between two universal joints 2 m apart. Find the whirling speed.

It is proposed to stiffen the shaft by:

(a) using a solid shaft 50 mm diameter; *or*

(b) adding a central self-aligning bearing.

Find, and comment on, the effect on the whirling speed of these alterations. Density 7 850 kg/m³ and $E = 2 \cdot 07$ Mbar.

5.19 A five-cylinder four-stroke engine is coupled to a load. In the first mode of vibration the node is far removed from the engine, and it may be assumed that the amplitude of torsional vibration is the same at all cranks. In a higher mode of vibration one of the nodes is at the centre of the crankshaft, and it may be assumed that the amplitudes of torsional vibration are the same at cranks 1 and 5, and 2 and 4. If the firing order is 1–2–4–5–3, find the engine orders for each mode of vibration which could give rise to serious vibrations.

5.20 A machine which weighs 3 020 kgf and runs at 600 rev/min produces at this speed an unbalanced vertical harmonic force of maximum value ±50 kgf. A number of these machines have to be installed on an upper floor of a factory. This floor may be considered as carrying slabs of concrete supported by steel joists, each slab carrying one machine. The slab weighs 8 940 kgf, and the central deflection of the floor due to its weight is 3·25 mm. A spring mounting for the machine is available, which provides a natural frequency of vibration of 177 c/min.

Show graphically how the amplitude of vibration of the floor will vary as the machine speed is slowly increased from zero to 800 rev/min, both with and without the mounting.

Write a short report advising as to the course to be adopted, and making any suggestions you can think of for improving the position, if necessary.

6. Automatic Control

§ 6.1 Introduction

As civilisation develops, machines replace human beings for many purposes, not only because they relieve men from laborious or boring tasks but also because they are limited neither to the power developed by men nor by the speed at which they can operate. As the machines themselves become faster and more complex, the task of controlling them, once the prerogative of man, is in turn and for very similar reasons handed over to subsidiary instruments and mechanisms. This is no recent development. In the early steam engines built by Newcomen the valve gear was operated by hand: when as a boy Humphrey Potter was given this task he rigged up strings to do it while he slept, so achieving a measure of automatic control of the engine. Not only did he save himself from boredom: he paved the way to designing engines to run at speeds at which human control of the valve gear would be ludicrously inadequate. When Watt adapted the fly-ball governor, originally invented by Thomas Mead to control the clearance between upper and nether millstones, to limit the speed of a steam engine, he took another big step in the direction of automatic control. If today we watch the pilot of an aircraft sitting with his hands in his lap while the craft pursues its determined course or even completes a blind landing, or walk through a chemical plant erected at a cost of millions of pounds and see the few men who are nominally in charge of it reading newspapers or playing cards while the process is in full operation, we are witnessing the present state of the art. Clearly automatic control is a subject of vital importance to the modern engineer, whether he regards himself primarily as a chemical, electrical, mechanical, or aeronautical specialist. The techniques involve mainly electrical and mechanical engineering: thus, for the early stages in a controller electrical devices will often provide the best solution, while for the final operating stages mechanical, pneumatic, or hydraulic devices are likely to be more acceptable because they give a better power-to-weight ratio. We shall here be discussing basic principles without bothering too much about whether the practical devices will be electrical or mechanical, and take in illustration whichever type best suits our purposes.

Control systems may be completely automatic or may include human operators. They may be designed to serve a number of functions between which the boundaries are not necessarily well defined; indeed, no common terminology has yet been agreed, though in B.S. 1523 an attempt is being made at standardisation. Thus, a device which controls a 'process' (for instance, a change in pressure, temperature, voltage, or speed) may be known as a *process controller*, or if it keeps such a quantity at a constant value it may be called a *regulator*. A device which controls the position or velocity or acceleration of a member may be called a *kinetic control* (or in the more usual case, where only position is of importance, a *remote position control*), and if it includes a power amplifier a *servo-mechanism*. However, all these controllers are fundamentally similar, and can be treated theoretically in basically the same way.

When we were studying mechanisms we found that from a complex machine such as a car engine we could extract (cf. § 1.1) a particular mechanism and represent its elements diagrammatically in an extremely simple way, so that we could elucidate its kinematics or even its dynamics without getting confused by all the detail which is of such vital importance in practical design. In a very similar way we shall study control devices in a simplified form, using for the purpose what are called '*block diagrams*', in which each 'block', represented by a labelled rectangle, and thought of as a 'black box' with a defined function, is connected to other blocks by lines representing 'flow', where the flow can be one of material fluid or immaterial information. Thus, a single block might represent an engine, a meter, a relay, or a reservoir, and so on; if an engine we may be wholly unconcerned to know anything about it in detail, merely that if supplied with steam or fuel it will supply power; if a flowmeter that it will supply a signal corresponding to the flow through it, and so on. This does not imply that detail is not as important in the practical design of a control system as in the practical design of a car engine, but simply that it is not our immediate concern. The choice we make in dividing the system into blocks is, however, an arbitrary one. Thus, for some purposes we might be content to represent a hydraulic relay such as that shown in fig. 6.4 by a single block, while for other purposes we might find it preferable to represent the piston valve and the actuating piston by separate blocks linked by a line representing the flow of information in the form of a flow of oil.

§ 6.2 Unmonitored and Monitored Control Systems

Systems of automatic control may be divided into two broad groups, usually described by the terms *unmonitored* or *open-loop* and *monitored* or *closed-loop*. In the first case control is achieved essentially by previous design, calibration, and perhaps trial and error: in the second the results are continually monitored and the answer, in the form of the current 'deviation' or 'error', 'fed back' to the controlling devices, so closing the 'information loop' and allowing correcting action to be taken.

As an example of an unmonitored system we may consider the current means of supplying to a petrol engine the appropriate fuel–air mixture. The commonest device for this purpose is a carburettor. In the early days these were relatively crude, and a considerable measure of control was provided for the operator, who could strengthen or weaken the mixture according to his impressions of the engine's requirements. As they developed, carburettors became more and more automatic, providing an approximately constant mixture strength under normal running conditions, with perhaps an overriding control to strengthen the mixture when the engine was cold. Additional refinements might provide modifications in mixture to cope with variations in power demand; or in the case of aircraft engines with variations in ambient pressure; or to save the operator the trouble of remembering to alter the setting when his engine had warmed up; and so on. Clearly such devices can become very complicated, because they depend essentially upon providing predetermined means of coping with each extraneous variable. Whether they are satisfactory or not depends on the skill of the inventors, designers, and makers, and the degree of complication for which the user is willing to pay; when they are installed their performance is normally subject only to the crudest kind of check in relation to insufficient power output or excessive fuel consumption. Fig. 6.1 shows a diagrammatic sketch of, and a possible block diagram for, an elementary carburettor.

We do not need to consider this situation for very long to realise that the basic difficulty in improving it is that we have no easy means of measuring the variable we are trying to control. If we had a device which measured the strength of the mixture as it issued from

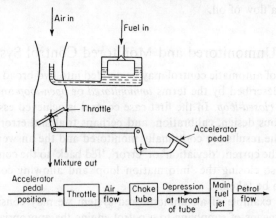

Fig. 6.1 Design of elementary carburettor showing flow of material fluids.
Block diagram of a carburettor to show flow of information

the carburettor we might feed back this information to the control; if we could even analyse the exhaust gas as it left the engine and deduce the error in the mixture strength we might use this information to close the loop – though the inevitable delay or 'lag' in obtaining the information would make control more difficult. In the absence of a means of measurement we are forced to accept the unmonitored system. The problems involved in designing such automatic devices are not our concern: we shall confine our attention to cases in which the variable we are trying to control can be measured.

To clarify our ideas, let us now consider a typical specific problem, that of controlling a steam turbine so that it runs at a constant speed. We may assume that the turbine will be supplied with steam through an adjustable valve and that it will be coupled to a load; moreover, we have seen that we must be able to measure the quantity we are trying to control, so we must have a suitable tachometer, which might well, at least diagrammatically, take the form of one of the governors we discussed in § 0.12 These units can thus be illustrated as in fig. 6.2a, or even more simply as in the full lines in the block diagram, fig. 6.2b.

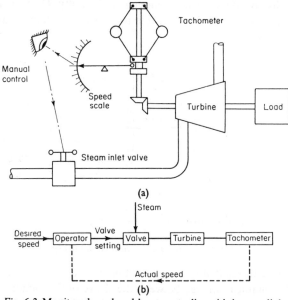

Fig. 6.2 Monitored or closed loop controller with human link

Perhaps initially we might think of instructing a man to control the turbine. Basically, we should require him to watch the tachometer, compare the actual turbine speed with the intended speed, and if a 'deviation' or 'error' existed to adjust the control valve. This, then, is an example of a monitored system using a 'human operator', and it is illustrated by the addition in dotted lines to fig. 6.2b. At first sight it seems to provide a complete answer to the problem, because whatever variations occur in the load or steam conditions, or whatever the response of the turbine to a movement of the valve, the man can apparently take appropriate correcting action. However, it does not take long to realise that the solution is far from perfect. The major long-term variations in speed may be eliminated, but the man may find himself quite unable to cope with sudden changes in conditions, since his rate of response is very seriously restricted; moreover, in an attempt to do so he is very likely to 'over-correct', and move the valve too far. If this continues to happen the actual speed will fluctuate about the desired value instead of remaining at it. He is bound to get bored or tired, so his performance will deteriorate, and although he has the ability to think for himself and allow for unforeseen troubles, this ability may be disadvantageous if his judgement is faulty. He costs a lot of money. Clearly it may be desirable to replace the man by an automatic element.

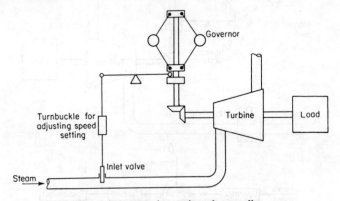

Fig. 6.3 Automatic monitored controller

Our next thought might well be that we could easily arrange a simple mechanical linkage between the sleeve of the tachometer and the valve as sketched in fig. 6.3, the controlled speed being initially adjusted to the desired value by the turnbuckle shown. If, however, operation of the steam valve requires a considerable force this system is most unlikely to be adequate. Unfortunately, as we can easily deduce even from the elementary discussion in § 0.12, the result of requiring a sensitive governor to operate such a valve is almost certain to be hopeless instability: the valve will not move at all until the speed is grossly in error, and it will then move far too far. This is indeed a common situation in automatic control, that the 'error signal' is too weak to be used directly to effect a correction, and must first be amplified, using for the purpose an external source of power.

We might next think of the device sketched in fig. 6.4, in which the governor sleeve is coupled not to the steam valve but to a piston valve which controls the supply of oil to a 'working' piston which in turn operates the steam valve. Since the hydraulic forces on this piston valve are balanced (this in fact is not quite true when oil is flowing through the valve, but it is near enough for our present purposes), a very small force will suffice to displace it from its neutral position, and the reaction on the governor sleeve may be ignored. If, then, the turbine speed rises the sleeve will rise, oil will be admitted

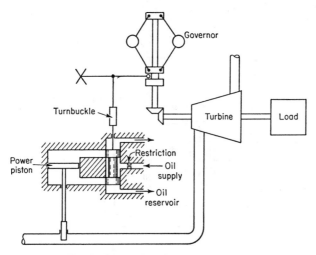

Fig. 6.4 Integrating automatic controller

above the working piston, and the steam valve will move towards the closed position. A moment's consideration is enough to show that we shall have to restrict severely the flow of oil, so that the closing of the valve is effected very slowly indeed; otherwise we shall have the valve shutting completely every time the turbine speed rises slightly above the correct value, and vice versa, giving intolerable hunting. This means that the system will be incapable of dealing effectively with sudden major changes of load, so by itself it would be quite unsatisfactory. Before discarding it out of hand, however, we might note that it has the virtue that in the long term it selects a steam-valve position which is entirely unaffected by any variable other than turbine speed, whereas the simple governor system described above provides a greater valve opening only by allowing a continuing, if small, error in speed. In this type of controller the movement of the piston valve is some function of the error in speed, and thus to a first approximation the flow of oil through the piston valve, and therefore the displacement of the working piston, are proportional to the time integral of the function of the error. Consequently, this type of controller is said to have an 'integral action'.

Instead of this integral controller we could employ a hydraulic

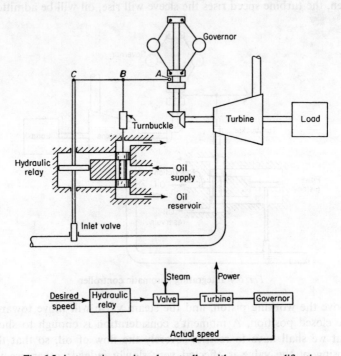

Fig. 6.5 Automatic monitored controller with power amplification

relay designed to increase the available force to operate the steam valve but giving an output displacement which under static conditions is equal to or linearly related to the input displacement. Such a relay is shown in fig. 6.5. If the turbine speed increases the governor sleeve rises, and the motion is communicated to the end *A* of a link *ABC*, of which the other end *C* is temporarily fixed by being attached to the 'working' piston in the closed oil-filled cylinder. The intermediate point *B* therefore rises, carrying with it the 'balanced' piston valve, which as mentioned above requires very little force to move it. Oil under pressure is therefore admitted to the upper side of the working piston which moves downwards, so (*a*) reducing the flow of steam and (*b*) returning the piston valve to its original neutral position, whereupon the operation may come to an end, with the turbine running at a slightly higher speed than before and the steam valve slightly less fully open.

But it is not difficult to see that this sort of reasoning may be altogether too ingenuous. There are in it many tacit assumptions which may not be justifiable. For instance, if the load is suddenly thrown off the turbine the speed will start to rise fairly rapidly and the governor balls will move outwards, acquiring momentum in this direction; clearly the governor will tend to overshoot and oscillate about, instead of stopping at, the new equilibrium position. Again, the hydraulic relay will move in a direction to close the steam valve, but unless the piston-valve ports are made with literally zero overlap there will be some 'lag' in this operation, since the first movement of the piston valve will not be sufficient to open the ports. During this lag or delay period the turbine speed will rise too far, the governor sleeve will rise too far. and the steam valve, when it does begin to move, may close too far. The turbine will now begin to slow down, but for exactly the same reasons the steam valve will be opened up too late and perhaps too much, and the cycle will repeat itself, perhaps with diminishing intensity, perhaps even with increasing intensity. Thus the addition of a hydraulic relay may not be sufficient to stop 'hunting', but may even accentuate this behaviour. Whether our system will be stable or not depends on many factors, of which a few have been indicated – the rate of response of the turbine to load and valve opening, the sensitivity of the governor and the amount of damping in its movement, and the presence of any 'imperfections' such as overlap of the piston-valve ports. Clearly, in designing governor systems we have problems which are not covered by the elementary treatment in § 0.12.

§ 6.3 Continuous and Discontinuous Controllers

In this example – the turbine-speed regulator system – we made the initial assumption that a continuously variable controlling element, in the form of a steam valve, was available to us. In other cases we may not have this facility. Thus, for instance, we might wish to control the temperature of a furnace either at a constant value (another regulator situation) or in accordance with a pre-determined time schedule (a process-controller situation). If the furnace were, say, gas- or oil-fired, continuously variable control would probably be available, but if electrically heated we might have to use a device which altered the heat supply in finite steps – in the extreme case by switching the furnace on or off. If such a discontinuous device is to control at all, one of two alternative conditions

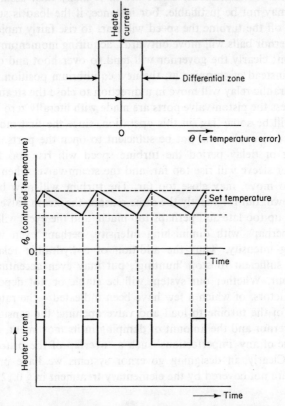

Fig. 6.6 Behaviour of an on–off temperature controller

available at a given instant must provide more heat than is needed, and the other less. Thus, if the temperature is below the desired value some thermometer (e.g. a thermocouple) will give an output which can be compared with the desired value and the higher heating rate selected; or perhaps a 'thermostat' incorporating an expanding element and a buckling device will hold the switch in the position to produce this result. The temperature will rise. At some instant the thermocouple will indicate that the desired value has been reached and the heat input will be reduced; or the thermostat will operate the switch; but always there will be lags in the system because the temperature-measuring device is not in intimate contact with the heating element, and because of this the furnace temperature will inevitably oscillate between two values above and below the correct one, as shown in fig. 6.6. The range and frequency of this oscillation will depend on the response of the temperature-measuring device, the thermal inertia of the windings, the magnitude of the 'step' controlled by the on–off switch, and so on.

It is usually fairly easy to estimate the qualitative behaviour of a discontinuous controller, and, perhaps after building a prototype, to adjust in the light of common sense the available variables to make it give the most satisfactory performance of which it is capable. A more sophisticated approach in general terms is frustrated, or at least made extremely difficult, by the fact that a discontinuity is essentially rather intractable mathematically. We shall give no further consideration to discontinuous systems.

§ 6.4 Transducers

We have seen that the ability to measure the quantity to be controlled is an essential element in any monitored system. The device which carries out this function normally converts one physical quantity, such as temperature, into another more easily usable quantity, such as displacement of a pointer on a scale or voltage. Such a device is called a transducer. If, for instance, we wanted to control temperature we might be able to measure it by allowing a man to feel the element which was being heated. This might be good enough for a domestic hot-water system, but is hardly suitable for scientific purposes. The provision of a mercury-in-glass thermometer would improve the situation as regards accuracy, but the result would still be in a form which required either a human operator (or a complicated photo-electric device) to use it. A bi-metallic element provides the result in the form of a linear displacement, and this can be used very simply indeed, as suggested in the previous paragraph, to operate an on–off switch, or as in the case of an ordinary domestic gas oven, to provide continuous control of the gas supply. Alternatively, we might use a pressure-type thermometer in which a fluid-filled bulb is heated and the resulting pressure is made to operate a bellows or a Bourdon tube, to give a displacement. Any of these displacements could be converted to rotational motion, which might be used to operate a rheostat or potentiometer, so producing a change in current or voltage; alternatively, we might produce such signals more directly by using a thermocouple or a resistance thermometer and Wheatstone bridge, and so on. Clearly the variants are endless; and the same situation holds for most of the quantities we are likely to want to measure. Usually we find that we have a wide choice from the cheap but very crude to the very sophisticated but expensive, and it is sensible to choose carefully. It might perhaps be noted in passing that for a reasonably sophisticated control system it often pays handsome dividends to convert the signal into an electrical quantity because of the relative ease of its manipulation, in particular because we can readily integrate and differentiate such quantities (see § 6.8).

§ 6.5 Transfer Functions

After these excursions into the qualitative aspects of automatic control we must get down to the quantitative behaviour of the various

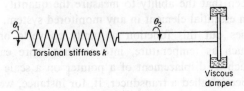

Fig. 6.7

elements which may be included in a system. It will be remembered that in the introduction it was pointed out that to avoid confusing the issues with irrelevant detail it was customary and helpful to illustrate a system by means of a 'block diagram'. We are therefore vitally concerned with the relationship between the output from and input to each unit or block, and this relationship or ratio is termed the 'transfer function' for the block. From a knowledge of the transfer functions of the individual blocks we have then to build up the equation of motion of the system. Let us take a few examples.

Consider first the mechanical system shown in fig. 6.7, which consists of a massless torsional spring of stiffness k (which might in practice be a flexible shaft of negligible mass) coupled to a viscous damper which provides a resisting torque proportional to the angular velocity. If at any instant the left- and right-hand ends of this spring have angular displacements of θ_1 and θ_2 respectively, the twist in the shaft will be $(\theta_1 - \theta_2)$, so the torque applied to the damper will be $k(\theta_1 - \theta_2)$, and the resisting torque will be $f \cdot \dot{\theta}_2$, where f is the damping constant. Equating these torques, we can write down

$$k(\theta_1 - \theta_2) = f\dot{\theta}_2$$

or preferably

$$(\theta_1 - \theta_2) = (f/k)D\theta_2{}^* \qquad (6.1)$$

D is known as an 'operator', which indicates that the operation of differentiating with respect to time has to be performed.

We can now write down the transfer function for the system in the most general terms as

$$\frac{\theta_2}{\theta_1} = \frac{1}{1 + \dfrac{f}{k}D} \qquad (6.2)$$

or substituting T for f/k

$$\frac{\theta_2}{\theta_1} = \frac{1}{1 + TD} \qquad (6.3)$$

Now let us suppose that at some instant before which $\theta_1 = \theta_2 = 0$ the left-hand end of the spring is suddenly given a displacement Θ_1, then for $t > 0$

$$(1 + TD)\theta_2 = \Theta_1$$

This is a first-order equation, and its solution consists of a transient and steady state solution. For the transient solution we write

$$(1 + TD)\theta_2 = 0$$

and by substituting $\theta_2 = Ae^{\gamma t}$ find that $\gamma = -1/T$. The steady state solution will be of the same form as the right-hand side of the equation, so the total solution will be

$$\theta_2 = \Theta_1 + Ae^{-t/T}$$

If we satisfy the starting condition that $\theta_2 = 0$ at $t = 0$, then A is $-\Theta_1$ and the complete solution is

$$\theta_2 = \Theta_1(1 - e^{-t/T})$$

* See, for instance, L. R. Ford, *Differential Equations*, 2nd edn. McGraw-Hill, 1955; or H. T. H. Piaggio, *Differential Equations*, G. Bell, 1942.

This solution is sketched in fig. 6.8. It is clear that θ_2 will approach Θ_1 exponentially and never reach this value. By differentiating we see that

$$\dot{\theta}_2 = \frac{\Theta_1}{T} e^{-t/T}$$

so that initially

$$\dot{\theta}_2 = \frac{\Theta_1}{T}$$

If $\dot{\theta}_2$ were to retain this value, θ_2 would reach the value Θ_1 in a time T. T is known as the 'time constant' for the device.

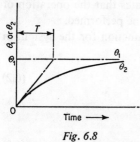

Fig. 6.8

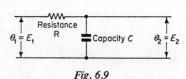

Fig. 6.9

Many other systems give a transfer function of this type. Consider, for instance, the resistance–capacity 'network' sketched in fig. 6.9. Let the electromotive force at the left- and right-hand ends at any instant be θ_1 and θ_2.* The current i through the resistance R will be $(\theta_1 - \theta_2)/R$, so the e.m.f. in the condenser will build up at a rate $\dot{\theta}_2 = i/C = (\theta_1 - \theta_2)/RC$. We can therefore write down

$$(\theta_1 - \theta_2) = RC \cdot D\theta_2$$

and the transfer function as

$$\frac{\theta_2}{\theta_1} = \frac{1}{1 + RC.D} = (\text{say}) \frac{1}{1 + TD}$$

by analogy with equation (6.3). $T \, (=RC)$ is the time constant of the electrical network. We see that the behaviour of the two systems is exactly analogous.

* We should note the unusual notation for e.m.f. In control theory the use of θ_1 and θ_2 for the input and output signals of an element is customary, *whatever* the physical quantity concerned. Similarly, θ_i and θ_o are used for the input and output signals of the complete system, whatever the quantities, and θ for the difference between the desired and actual values of θ_o.

Again in § 6.2 we have mentioned a hydraulic relay which is illustrated in isolation in fig. 6.10. To obtain the transfer function for this system we shall 'linearise' the problem by assuming that the flow through the piston valve is proportional to valve opening,

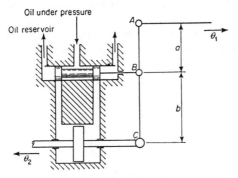

Fig. 6.10 Hydraulic relay

which is only approximately true for small displacements, and we shall ignore the mass of the piston, which could be added to that of the element being driven by the relay.

If A has been displaced to the right by the amount θ_1 and the working piston has moved to the left by an amount θ_2, then the displacement of the piston valve will be

$$\frac{b}{a+b}\theta_1 - \frac{a}{a+b}\theta_2$$

Assuming that the oilflow is q times the opening of the piston valve and A is the cross-sectional area of the working piston, then

$$D\theta_2 = \frac{q}{A}\left\{\frac{b}{a+b}\theta_1 - \frac{a}{a+b}\theta_2\right\}$$

so the transfer function $\dfrac{\theta_2}{\theta_1} = \dfrac{b/a}{(1+TD)}$

where $T = \dfrac{A}{aq}(a+b)$

If we consider a simple bimetal strip thermometer in which a pointer is rotated by the deformation of the strip, then we could deduce a similar transfer function, on the assumption that the rate of heat transfer to the bimetal strip was a linear function of the temperature difference between the applied and recorded temperatures.

Similarly, many other very different components can be shown to have the same type of transfer function. It is this fact that apparently very different elements give a similar transfer function which is one of the main conveniences of the use of such functions. The response of the components we have so far considered is referred to as an '*exponential lag*'.

So far we have neglected the mass of the moving parts of the components we have considered, but in fig. 6.11 we show a system similar

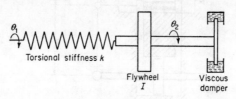

Torsional stiffness *k*

Flywheel
I

Viscous
damper

Fig. 6.11

to that of fig. 6.7, except that a flywheel of moment of inertia I has been added to the viscous damping unit. Considering the torque transmitted to the flywheel and damper, we can write down

$$k(\theta_1 - \theta_2) = fD\theta_2 + ID^2\theta_2$$

so the transfer function $\theta_2/\theta_1 = \dfrac{k}{k + fD + ID^2}$

$$= \dfrac{k/I}{\dfrac{k}{I} + \dfrac{f}{I}D + D^2}$$

but from § 5.3 we can note that $k/I = \omega_n^2$ and $f/I = 2c\omega_n$, where ω_n is the natural radiancy of the flywheel oscillating torsionally on the spring and c is the critical damping ratio. Consequently, we can write down the transfer function as

$$\frac{\omega_n^2}{\omega_n^2 + 2c\omega_n D + D^2} = \frac{1}{1 + 2cT_1 D + T_1^2 D^2} \qquad (6.4)$$

where $T_1 = 1/\omega_n$

The electrical circuit in fig. 6.12 is analogous to the mechanical system of fig. 6.11. The difference in e.m.f. will be

$$\theta_1 - \theta_2 = LDi + Ri$$

or $\qquad\qquad\qquad\qquad i = \dfrac{\theta_1 - \theta_2}{LD + R}$

and the build-up of e.m.f. across the condenser will be

$$D\theta_2 = \frac{i}{C}$$

or $\qquad \theta_1 - \theta_2 = (LD + R)i = (LD + R)CD\theta_2$

so the transfer function $\quad \dfrac{\theta_2}{\theta_1} = \dfrac{1}{1 + RCD + LCD^2}$

which is similar in form to equation (6.4). Such electrical analogues

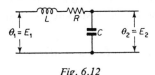

Fig. 6.12

of mechanical systems can be and are used to simulate mechanical vibration problems* or some of the elements of a control system.

As an example of a rather more complicated component let us consider the spring-controlled governor shown in fig. 0.13, which is redrawn for convenience in fig. 6.13. In this case the transfer function relates the change in the vertical displacement of the spring-

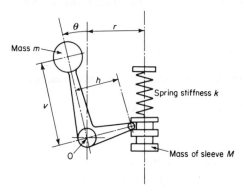

Fig. 6.13 Hartnell governor

loaded governor sleeve of mass M (which could, for example, be connected to the piston valve of a hydraulic relay such as that shown in fig. 6.10) to the change in angular speed. We shall assume that when the arms holding the balls are vertical the compression of the spring, which has a stiffness k, is X_0, and the equilibrium speed ω_e.

* See H. F. Olsen, *Dynamical Analysis*, Van Nostrand, 1958.

We shall further assume that there is some viscous damping acting at the governor sleeve, with a damping constant f.

If due to a small change, ω, in the speed the small displacement of the governor sleeve at some subsequent time is x the governor arms will be displaced by a small angle θ, shown in an exaggerated form in fig. 6.13, where

$$\sin \theta = \frac{x}{h} \simeq \theta \tag{6.5}$$

Considering moments about the bell crank pivot at O, ignoring the moment due to gravity, and assuming there are two balls, we can write down

$$-k(X_0 + x)h = 2mv^2 D^2\theta + MhD^2x + hfDx - 2vm(\omega_e + \omega)^2(r + v\theta)$$

↓	↓	↓	↓	↓
Moment due to spring force	Inertia torque of the balls	Inertia torque of governor sleeve mass	Damping torque	Moment to provide centripetal force

Substituting for θ from equation (6.5), this becomes

$$-k(X_0 + x)h$$
$$= \frac{2mv^2}{h}D^2x + MhD^2x + hfDx - 2mv(\omega_e + \omega)^2\left(r + \frac{vx}{h}\right)$$

Neglecting second-order small terms, this reduces to

$$\left(\frac{2mv^2}{h} + Mh\right)D^2x + hfDx + kh(X_0 + x)$$
$$= 2mv\left\{\omega_e^2 r + \omega_e^2 \frac{vx}{h} + 2\omega_e\omega r\right\}$$

Under steady conditions, however,

$$2mvr\omega_e^2 = kX_0h$$

so the equation of motion can be simplified to

$$\left(\frac{2mv^2}{h} + Mh\right)D^2x + hfDx + \left(kh - \frac{2mv^2}{h}\omega_e^2\right)x = 4mv\omega_e r\omega$$

or, dividing through by $\left(\frac{2mv^2}{h} + Mh\right)$, can be put in the form

$$D^2x + 2c\omega_n Dx + \omega_n^2 x = \frac{4mvh\omega_e\omega r}{(2mv^2 + Mh^2)}$$

So the transfer function $\dfrac{\theta_2}{\theta_1} = \dfrac{x}{\omega} = \dfrac{4mvh\omega_e r/(2mv^2 + Mh^2)}{\omega_n^2 + 2c\omega_n D + D^2}$

where the natural radiancy $\omega_n = \sqrt{\dfrac{kh^2 - 2mv^2\omega_e^2}{Mh^2 + 2mv^2}}$

This is the natural radiancy of the governor when the balls are disturbed from their equilibrium position and c is the critical damping ratio.

Basically these last three examples are of elements which have a second-order equation of motion and, consequently, if the damping is not excessive their response will be an oscillation dying away with time (see §§ 5.3 and 5.4). The response of such elements is known as a *'complex exponential lag'*.

§ 6.6 Open- and Closed-loop Transfer Functions

We now proceed to consider the behaviour of a number of elements in sequence. We shall assume that the output of each element is unaffected by the elements to which it is connected. This is often, but not always, the case. For instance, we have already noted that in the hydraulic relay shown in fig. 6.4 the forces on the piston valve are not really balanced, since the flow of fluid through the valve gives rise to an unbalanced force, and this means that there is some interaction between the working piston and the valve.

Let us take first an unmonitored control system such as that depicted in fig. 6.14a, in which a number of elements 1, 2, 3, ... n are

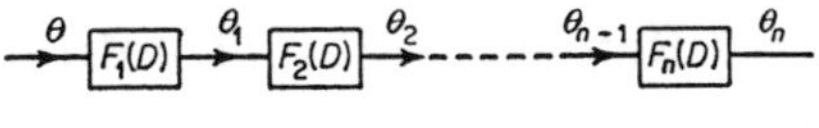

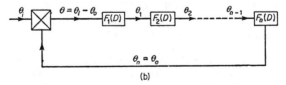

(a)

Fig. 6.14a Unmonitored or open-loop controller

(b)

Fig. 6.14b Monitored or closed-loop controller

arranged in sequence. With the standard notation we have adopted, the transfer functions of the individual elements are

$$\frac{\theta_1}{\theta} = F_1(D), \quad \frac{\theta_2}{\theta_1} = F_2(D), \ldots \frac{\theta_n}{\theta_{n-1}} = F_n(D)$$

and multiplying these transfer functions together, we find the overall transfer function, which is referred to as the 'loop transfer function' or 'open-loop transfer function',

$$\frac{\theta_n}{\theta} = F_1(D) . F_2(D) . F_3(D) \ldots F_n(D)$$

$$= \text{(say)} \, K . G(D) \tag{6.6}$$

K being the total loop gain or amplification, which is independent of D, and $G(D)$ a function of the operator D.

In a monitored control system, however, the output from the last element, which is the value of the controlled variable $\theta_o \, (= \theta_n)$ is compared with the desired value of the controlled variable θ_i, and the deviation or error $\theta \, (= \theta_i - \theta_o)$* is fed into the first element, as depicted in fig. 6.14b. So, substituting into equation (6.6), we can write

$$\frac{\theta_n}{\theta} = \frac{\theta_o}{\theta_i - \theta_o} = K . G(D)$$

or rearranging this equation, we can obtain the 'closed-loop transfer function', which relates the value of the controlled variable to the desired value

$$\frac{\theta_o}{\theta_i} = \frac{K . G(D)}{1 + K . G(D)} \tag{6.7}$$

§ 6.7 Remote-position Controller

We can now consider a complete system, and as a relatively simple example we shall choose a remote-position controller. There are many practical examples of such a system. One of the earliest is the 'steam steering gear' for ships. A rather more recent application is the control of an anti-aircraft gun or rocket launcher from information obtained by a radar station, and this is a convenient example

* The sign of θ is a matter of convention. Although B.S. 1523 has recently defined θ as $(\theta_o - \theta_i)$, which is in accord with practice in instrument technology, nevertheless it seems unlikely that this will obtain wide acceptance, and we shall use the convention above.

to think of, in that it illustrates understandably some of the situations with which a system such as a remote-position controller has to cope. Thus, for instance, if the radar suddenly picks up a target in a direction different from that in which the gun is pointing there will be an idealised requirement for the gun to swing round as rapidly as possible and stop instantaneously when it reaches the position demanded. Since the gun possesses inertia, it is obvious that we shall not be able to satisfy this requirement exactly, but we may be able to design the system to approximate closely to it, perhaps by including a damping device. But the target may now move steadily across the field of view. How do we design the system so that the gun follows without intolerable error, and will this design conflict with the earlier one? Or if the target takes evasive action, demanding, say, a sinusoidal motion of the gun, will the gun follow the demands?

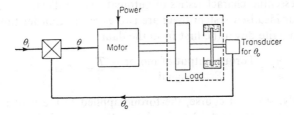

Fig. 6.15 Remote-position controller

There are essential similarities between these situations and those dealt with in chapter 5 (§ 5.4), in which various types of disturbing force were applied to elastically controlled inertias and the motions of the latter determined. For instance, the sudden demand for a change in position corresponds to a suddenly applied force, and so on. As we proceed with the calculations we shall find that we can save time and effort by abstracting from that paragraph solutions to the equations of motion we set up. When we were discussing vibrations we usually concentrated our attention on the steady-state solution and often ignored the transient, but in the problems we are now discussing the latter may well be the dominant factor.

Fig. 6.15 shows in bare outline the sort of system we are now considering. The desired position signal (let us suppose angular in nature) θ_i is fed into the first unit of the system, where it is compared with the actual position signal θ_o, the difference $\theta = \theta_i - \theta_o$ being passed on to the unit labelled 'motor'. This motor could, of course, take many forms, electric or hydraulic, for instance; all we need to know (or to assume) is that it produces an output in the form of a torque T proportional to θ, say $K\theta$. This torque is then applied to the third unit, entitled 'load', which is the member whose position is to be controlled. For the sake of definitive clarity, this load is shown not only as a block but also as a combination of a flywheel and rotational viscous damper, to indicate that these, inertia and viscous resistance to motion, are the properties we assume it to have. The fourth unit is a device for measuring the actual position θ_o of the load. Of course, in a practical system we might not achieve these idealised conditions: for instance, the motor torque might not be directly proportional to θ; the resistance to motion might not be truly viscous; there might be 'lags' in measuring θ_o and comparing it with θ_i and so on. Such complications have to be left for later study, but the simplified system demonstrates some of the essential characteristics of automatic controllers.

In our idealised system there are two essential transfer functions. We can write down the first from the data

$$\frac{\theta_1}{\theta} = \frac{\text{Torque output of motor}}{\text{Error signal}} = \frac{T}{\theta} = K = \frac{T}{\theta_i - \theta_o}$$

Secondly, since, of course, the torque applied to the load must be equal to that required (*a*) to accelerate the flywheel, $ID^2\theta_o$, and (*b*) to turn the damper at the current speed, $fD\theta_o$, we can write

$$T = ID^2\theta_o + fD\theta_o$$

so the transfer function

$$\frac{\theta_o}{\theta_1} = \frac{\text{Angular position of output shaft}}{\text{Torque applied to load}} = \frac{\theta_o}{T} = \frac{1}{ID^2 + fD}$$

Multiplying these two equations together (or from equation (6.6)), we can write

$$\frac{\theta_o}{\theta_i - \theta_o} = \frac{\theta_o}{T} \cdot \frac{T}{\theta_i - \theta_o} = \frac{K}{ID^2 + fD}$$

whence $(ID^2 + fD + K)\theta_o = K\theta_i$ (6.8)

and it is, of course, clear that in this straightforward case the use of transfer functions was of minimal value.

It should not take long to realise that we have met equations very like this before, but the use of the operator D notation may confuse the situation. If we rewrite (6.8) in the fluxional notation we have used previously, and take the opportunity to divide out by I, we get

$$\ddot{\theta}_o + \frac{f}{I}\dot{\theta}_o + \frac{K}{I}\theta_o = \frac{K}{I}\theta_i$$

and now we should certainly recognise K/I, an (angular) stiffness divided by a (moment of) inertia, as analogous to the quotient k/I for a moment of inertia I positioned by a shaft of stiffness k, and therefore equal to ω_n^2, where ω_n is the natural radiancy. Similarly, we should recognise f/I as analogous to the damping in such a system: we can therefore replace it by $2c\omega_n$, whereupon the equation becomes

$$\ddot{\theta}_o + 2c\omega_n \cdot \dot{\theta}_o + \omega_n^2\theta_o = \omega_n^2 \cdot \theta_i = \text{(say) const.} f(t) \quad (6.8a)$$

since θ_i, the input, is an independent variable which may be any arbitrary function of time and θ_o is the dependent variable.

Now equation (5.9) was

$$\ddot{x} + 2c\omega_n\dot{x} + \omega_n^2x = \frac{1}{m}f(t) = \text{const.} f(t)$$

and we surely need not elaborate the analogy further.

Having established the analogy between the equation (6.8a) and the forced damped vibration equation (5.9), there is no need to repeat the analysis given in detail in § 5.4 for various forcing functions. For instance, if we apply a sudden displacement to the input (that is $\theta_i = 0$ for $t < 0$ and $\theta_i = \Theta_i$ for $t > 0$), then for $c < 1$ (i.e. sub-critical damping) we can from the analogy with a suddenly applied force considered in § 5.4.1 write down the solution

$$\frac{\theta_o}{\Theta_i} = 1 - \frac{e^{-c\omega_n t}}{\sqrt{(1 - c^2)}} \cos\left\{\omega_n\sqrt{1 - c^2}\, t - \tan^{-1}\frac{c}{\sqrt{(1 - c^2)}}\right\}$$

where θ_o/Θ_i replaces the ratio x/X_s. The form of the response for $c = 0\cdot1$ will be exactly that depicted in fig. 5.5, and the percentage overshoot will be identical. Similarly, response curves can be derived for a critically damped or for an over-critically damped system. From fig. 5.3, however, we can easily deduce that the quickest response with *no* overshoot is given by critical damping, but for $c \simeq$ (say) $0\cdot6$ the response is even faster and the overshoot is very small. The speed of response is dependent on ω_n, so that for a fast response we want to keep K as large as possible and the inertia I as small as possible. This is equally true for the more realistic case where the change in θ_i is extremely rapid but not instantaneous, as in § 5.4.2.

Similarly, the response of this remote-position controller to a suddenly applied constant-velocity input (that is $\theta_i = 0$ for $t = 0$ and $\theta_i = \omega_i t$ for $t > 0$, where ω_i is the input velocity) is exactly analogous to the case of steadily increasing force applied to a spring-mass system which is discussed in § 5.4.3, and the form of the response would be the same as that shown in fig. 5.7. After the transient vibration has died away the solution for θ_0 will be

$$\theta_o = \omega_i t - 2c \frac{\omega_i}{\omega_n}$$

But at time t the input will have rotated by

$$\theta_i = \omega_i t$$

and consequently there will be a constant error between the input and output of

$$\theta = \theta_i - \theta_o = 2c \frac{\omega_i}{\omega_n}$$

The physical reason for this position error is that when the transient has died away the output shaft is rotating at the same constant speed as the input, and there is a viscous drag torque acting on the output of magnitude of $f\omega_i$. This torque must be provided by the motor, which implies a constant error signal of

$$\frac{f\omega_i}{K} = 2c \frac{\omega_i}{\omega_n}$$

The error can be reduced by keeping c small, but this would make the error due to the transient response much worse. From this point of view it is desirable to make the natural radiancy ω_n as large as possible in relation to ω_i.

For a sinusoidal input $\theta_i = \Theta_i \cos \omega t$ the equation of motion becomes

$$\ddot{\theta}_o + 2c\omega_n\dot{\theta}_o + \omega_n^2\theta_o = \omega_n^2 \Theta_i \cos \omega t$$

which is similar to that deduced in § 5.4.4. For a system with less than critical damping the solution will be identical to equation (5.20), except that x will be replaced by θ_o and X_s by Θ_1. When the transient solution has died away the dynamic magnifier or *gain* (i.e. Θ_o/Θ_1) as the frequency ω is varied is shown in fig. 5.10a and the phase lag of the output relative to the input is shown in 5.10b.

If it is necessary for the output to follow closely the sinusoidal input, then from fig. 5.10 it can be seen that the natural frequency of the system should again be as high as possible. Alternatively, the damping should be of the order of 0·6 of critical damping, which would give a gain of nearly unity up to values of ω/ω_n approaching unity, but the error in phase would be considerable. The harmonic response of elements of control systems and complete control systems is of vital importance in the analysis of systems in relation to stability (see § 6.9).

If a steady torque τ is applied to the output shaft, then the equation of motion becomes

$$\ddot{\theta}_o + 2c\omega_n\dot{\theta}_o + \omega_n^2\theta_o = \pm\frac{\tau}{I}$$

and the steady state solution will be

$$\pm\frac{\tau}{I\omega_n^2} = \pm\frac{\tau}{K}$$

This implies that there will be a steady state error of this magnitude in the position of the output shaft, which is obvious because the torque motor is providing a steady torque of $\pm\tau$, and consequently there must be an error signal into the motor of $\pm\tau/K$. This kind of error is similar to that experienced by the simple governor system depicted in fig. 6.3a, which gives a reduction in governed speed as the load on the engine is increased. The drop in the governed speed can, however, be reduced by increasing the sensitivity of the governor, or in the case being considered by increasing K. Alternatively, we could use an integral action as shown for the governor system in fig. 6.4. It should be noted that both these modifications to the controller increase the possibility of instability.

§ 6.8 The Effect on the Performance of a Controller of Modifying the Error Signal

When we were discussing the ability of a man to use his intelligence in controlling a process we might have noted that one of the ways in which he could do so is to pay attention not only to the instantaneous value of the deviation but also to the way in which it is altering. If, for instance, a large deviation had occurred and he had taken correcting action, only to see the indicator swinging more and more rapidly towards the desired setting, he would almost certainly start reversing the control long before this setting had been reached, because experience would have taught him that overshoot was otherwise inevitable. This gives us a clear hint that if we can design our automatic controller to take proper account of the *rate of change* of the deviation as well as its magnitude we may achieve a better result. And if the rate of change (or derivative) of the deviation helps us, would not higher derivatives, and perhaps also the time integral of the deviation, be of value? We must investigate these possibilities. The procedure is usually described as modifying the error signal.

In many controllers the input signal and the feedback from the output are in the form of voltages, and it is easy to modify these voltages or their differences by the use of simple electrical networks. For instance, let us consider the network shown in fig. 6.16a, in which the impedance of R is large compared with that of C, so that

$$E_1 = Ri + \frac{1}{C}\int i\,dt \simeq Ri$$

and
$$E_2 = \frac{1}{C}\int i\,dt = \frac{1}{RC}\int E_1\,dt$$

From this it will be seen that the output voltage across the condenser is proportional to the time integral of the input voltage E_1. A slightly more complicated circuit is shown in fig. 6.16b. In this case

$$E_1 = (R_1 + R_2)i + \frac{1}{C}\int i\,dt$$

$$= (R_1 + R_2)i + \frac{i}{CD}$$

or
$$i = \frac{E_1 CD}{(R_1 + R_2)CD + 1}$$

and
$$E_2 = R_2 i + \frac{1}{C}\int i\,dt = R_2 i + \frac{i}{CD}$$

$$= \frac{(R_2 CD + 1)}{CD}i$$

$$= \frac{R_2 CD + 1}{(R_1 + R_2)CD + 1}E_1$$

$$= \gamma \frac{(TD + 1)}{(TD + \gamma)}E_1$$

where $\gamma = \dfrac{R_2}{R_1 + R_2}$ and $T = R_2 C$.

If R_2 is small compared with $R_1 + R_2$ so that γ is small, then this equation reduces to

$$E_2 = \gamma\left(1 + \frac{1}{TD}\right)E_1$$

which is proportional to the input voltage plus the time integral of the input voltage multiplied by a constant.

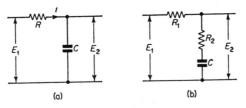

Fig. 6.16 Integrating networks

Yet a further example is provided by the network shown in fig. 6.17a in which

$$i = C\frac{d}{dt}(E_1 - E_2)$$

and $$E_2 = Ri = RC\frac{d}{dt}(E_1 - E_2) = TD(E_1 - E_2)$$

where $T = RC$

so $$E_2 = \frac{TD}{(1 + TD)}E_1$$

In this case the output voltage is proportional to the differential of the input voltage coupled with an exponential delay, i.e. $1/(1 + TD)$.

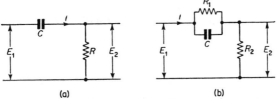

Fig. 6.17 Phase advance networks

If T is very small then the exponential delay is negligible and we effectively have a differentiating network.

When we come to consider the harmonic response of such elements we shall find that this network produces an output signal which leads the input signal in phase: it is often referred to as a *phase advance network*.

Finally, in fig. 6.17b we have another example of a phase advance network for which

$$E_2 = iR_2$$

$$= \frac{E_1 R_2}{R_2 + \dfrac{R_1}{1 + CR_1 D}}$$

$$= \gamma \frac{1 + TD}{1 + \gamma TD} E_1$$

where $\gamma = \dfrac{R_2}{R_1 + R_2}$ and $T = CR_1$. In this case if γ is small, then the output voltage is proportional to the input voltage plus the differential of the input multiplied by a constant T.

These examples serve to show how easy it is to modify the error signal if it is in the form of a voltage. Though equivalent pneumatic and hydraulic circuits have been designed, it is not so easy to alter time constants, to achieve accuracy of the various parameters, or to obtain linearity.

Having demonstrated the feasibility of modifying the error voltage, let us now consider the effect of various modifications to the input signal to the motor on the performance of the controller considered in the previous paragraph.

6.8.1 First derivative of error compensation

In this case the error is fed to a modifying network as indicated in the block diagram of fig. 6.18, which gives an output of

$$\theta + K_1 \dot{\theta}$$

where K_1 is a constant. The equation of motion will now be

$$I\ddot{\theta}_o + f\dot{\theta}_o = K(\theta + K_1 \dot{\theta})$$

or rearranging this equation we get

$$I\ddot{\theta}_o + (f + K_1 K)\dot{\theta}_o + K\theta_o = K\theta_i + K_1 K\dot{\theta}_i \qquad (6.9)$$

From this it will be seen that the natural frequency is unchanged but the critical damping ratio has been raised from f/f_c to

$$\frac{f + K_1 K}{f_c} = \frac{f + K_1 K}{2\sqrt{KI}}$$

As a result, the transient response is improved without the need for an excessive damping torque acting on the load.

With a stepped velocity input of $\theta_i = \omega_i t$ when the transient vibration has died away $\ddot{\theta}_o = 0$ and $\dot{\theta}_o = \omega_i$, so from equation (6.9) the steady-state solution will be

$$(\theta_o)_{SS} = \omega_i t - \frac{f\omega_i}{K}$$

$$= \omega_i t - 2c'\frac{\omega_i}{\omega_n}$$

where $c' = \frac{f}{f_c}$ not $\frac{f + K_1 K}{f_c}$

Thus, we have a means of improving the transient response by increasing the magnitude of the derivative term (i.e. by raising K_1) with-

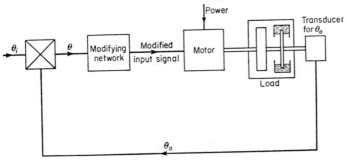

Fig. 6.18

out raising the steady-state error due to a stepped velocity input. Indeed, we can now keep f as small as possible.

6.8.2 Second derivative of error compensation

In this case the input into the torque motor is for practical reasons, which will soon be apparent, given the form $\theta - K_2\ddot{\theta}$ and the equation of motion becomes

$$(I - K_2 K)\ddot{\theta}_o + f\dot{\theta}_o + K\theta_o = -KK_2\ddot{\theta}_1 + K\theta_i$$

Because $(-K_2)$ has been made negative the undamped natural radiancy is raised from $\sqrt{\frac{K}{I}}$ to $\sqrt{\frac{K}{I - K_2 K}}$ and the critical damping ratio from $\frac{f}{2\sqrt{KI}}$ to $\frac{f}{2\sqrt{[K(I - K_2 K)]}}$, while the steady-state error due to a stepped velocity input or an output torque is unaffected.

The advantages are an improved response rate due to the increased natural radiancy and an improved transient response due to the increase in the critical damping ratio. Alternatively, as the critical damping ratio has been raised, it may be possible to keep f to a minimum without harming the transient response, so that the steady-state error for a stepped velocity input is a minimum.

6.8.3 First derivative of output compensation

This requires a second feedback loop and a differentiator to give a signal proportional to $\dot{\theta}_o$ so that the signal from the mixing unit

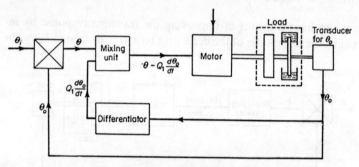

Fig. 6.19 First derivative of output compensation

shown in fig. 6.19 is preferably $(\theta - Q_1\dot{\theta}_o)$, which gives an equation of motion

$$I\ddot{\theta}_o + (f + Q_1 K)\dot{\theta}_o + K\theta_o = K\theta_i$$

The natural frequency is unchanged, but the critical damping ratio is raised from $\dfrac{f}{2\sqrt{KI}}$ to $\dfrac{f + KQ_1}{2\sqrt{KI}}$, and the steady-state error arising from a stepped velocity input will also be increased. However, damping is achieved without the necessity of fitting a mechanical damper on the output shaft dissipating energy which would have to be supplied by the torque motor.

6.8.4 Second derivative of output compensation

In this case the signal into the torque motor is preferably $\theta + Q_2\ddot{\theta}_o$ and the equation of motion becomes

$$(I - KQ_2)\ddot{\theta}_o + f\dot{\theta}_o + K\theta_o = K\theta_i$$

The natural radiancy is increased to $\sqrt{\dfrac{K}{I - KQ_2}}$ and the critical damping ratio is raised to $\dfrac{f}{2\sqrt{[K(I - KQ_2)]}}$.

Both the speed of response and the transient response will be improved. The damping constant f can be kept small, which will reduce the steady-state error for a stepped velocity input, while the transient can in effect be heavily damped.

6.8.5 First derivative of input compensation

This requires a feed forward as shown in fig. 6.20, coupled with a differentiating unit, so the output of the mixing unit is $\theta + S_1\dot{\theta}_i$ and the equation of motion becomes

$$I\ddot{\theta}_o + f\dot{\theta}_o + K\theta_o = (K\theta_i + KS_1\dot{\theta}_i)$$

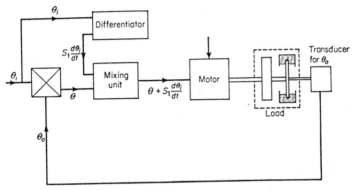

Fig. 6.20 First derivative input compensation

From this equation it will be seen that the transient response is unaffected. For a velocity input $\theta_i = \omega_i t$ the steady-state solution is

$$(\theta_o)_{SS} = \omega_i t - \left(\frac{f}{K} - S_1\right)\omega_i \text{ or } \theta = \left(\frac{f}{K} - S_1\right)\omega_i$$

Consequently, the steady-state error can be made zero if $f/K = S_1$. Except for this, the first derivative input compensation has no effect.

6.8.6 Integral control

So far we have seen that with one exception there will be an error due to a steady velocity input, and in all cases an error due to a torque applied to the output shaft. This can be overcome, however, by adding an integral-of-error compensation term to the signal being fed to the torque motor. A practical example of this has already been given in the governor system shown in fig. 6.4, in which the error in the speed opens a piston valve by a proportionate amount so that oil can pass through into the working cylinder. As a result, the working piston, which controls the flow of steam into the turbine, moves a distance which is proportional to the time integral of the piston valve opening, and consequently to the time integral of the error.

For an integral control the signal out of the shaping element is

$$\theta + K_3 \int_0^t \theta \, dt$$

and the equation of motion is most conveniently written in the form

$$I\ddot{\theta} + f\dot{\theta} + K\theta + KK_3 \int_0^t \theta \, dt = I\ddot{\theta}_i + f\dot{\theta}_i \pm \tau$$

which is a third-order equation, and as we shall see in the next paragraph, this is more likely to lead to instability than a second-order equation.

For a stepped velocity input, when $t > 0$ $\dot{\theta}_i = \omega_i$ (or $\theta_i = \omega_i t$), and $\ddot{\theta}_i = 0$; and for $t \gg 0$ $\ddot{\theta} = 0$, $\dot{\theta} = 0$.

Therefore $\qquad K\theta_{ss} + KK_3 \int_0^t \theta \, dt = f\omega_i \pm \tau$

The right-hand side is by inspection finite, but if θ_{ss} were finite then the term $KK_3 \int_0^t \theta \, dt$ would tend to infinity since θ is the sum of the steady state and transient solutions. Consequently, θ_{ss}, the steady-state error, must be zero and therefore

$$KK_3 \int_0^t \theta \, dt = f\omega_i \pm \tau$$

Integral control eliminates steady-state errors due to a stepped velocity input and that caused by an output torque. The time required to eliminate these errors will depend on the coefficient KK_3.

§ 6.9 Stability

We have already noted that instability can arise in monitored control systems, in the sense that when the system is disturbed the resulting oscillations do not die away but instead increase in amplitude until limited by the power input into the system. We have also seen that the equation of motion of a simple control system is a second-order linear differential equation, but that this can be raised to a third-order equation if integral action is introduced. However, the remote-position controller which we considered was a very idealised system, and in practice there would be lags in measuring, transmitting, and modifying information. Such lags have the effect of raising the order of the equation of motion, and the conditions for stability are then more difficult to investigate and ensure.

The stability of a control system is determined by the transient solution, since the steady-state solution is (by definition) of no importance as far as stability is concerned. For a second-order equation we are consequently involved in the solution of

$$\ddot{x} + b_1\dot{x} + b_2x = 0$$

We have already considered the solution of this form of equation in § 5.3, and found that when $(b_1/2)^2 < b_2$ it is of the form

$$x = Ae^{-\frac{b_1}{2}t} \cos\left\{t\sqrt{\left[b_2 - \left(\frac{b_1}{2}\right)^2\right]} + \phi\right\}$$

This oscillation will die away exponentially only if b_1 is positive; if it were negative (i.e. if negative damping were to occur in the system), then the amplitude of oscillation would increase exponentially.

Though we cannot ignore the possibility of instability arising from negative damping, instability is more usually associated with higher-order equations. For instance, let us consider a third-order equation

$$\dddot{x} + b_1\ddot{x} + b_2\dot{x} + b_3x = 0$$

Then the transient solution will be of the form $x = Ae^{\gamma t}$ and substituting

$$\gamma^3 + b_1\gamma^2 + b_2\gamma + b_3 = 0$$

For stability none of the three roots of this equation must contain positive real parts, as this would imply that one of the terms of the transient solution would be increasing exponentially with time. Routh showed* that the conditions to be satisfied are

$$b_1b_2 > b_3$$

and $b_1 > 0, b_2 > 0,$ and $b_3 > 0$

---

* For instance, see T. von Kármán and T. V. Biot, *Mathematical Methods in Engineering*, McGraw-Hill, 1940.

For a fourth-order equation such as

$$\frac{d^4x}{dt^4} + b_1 \frac{d^3x}{dt^3} + b_2\ddot{x} + b_3\dot{x} + b_4x = 0$$

the conditions to be satisfied are

$$b_1b_2b_3 > b_3^2 + b_1^2b_4$$

and $\qquad b_1 > 0, b_2 > 0, b_3 > 0, b_4 > 0$

Hurwitz has generalised the conditions for stability for third- and higher-order equations, but as this approach is of little value for examining the stability of control systems, we shall not go into any further detail. The main disadvantage is that even though this kind of analysis tells us whether a system is stable or not, it does not give us any indication of the margin against instability. It is possible, for instance, for a system to be very near the point of instability so that the oscillation due to a disturbance dies away extremely slowly, or else to be excessively stable so that the system only slowly returns to its equilibrium position after a disturbance. If the control system is near to instability, then slight differences in performance as a consequence of errors in manufacture or slow changes in the characteristics of individual components with time, such as the ageing of electronic valves or variations in friction due to wear, may cause the system to become unstable. In addition, the Hurwitz type of analysis gives us no real indication of what changes are desirable to improve the performance or make the system stable: we should be faced with changing the various parameters at our disposal one by one and then recalculating the transient response. Even if satisfactory parameters governing stability were established, the probability that the system could have been further improved would remain.

Because of these limitations of the classical approach to stability, graphical methods for determining the margin of stability and transient response have been developed, particularly by Nyquist and Bode, and these are the subject of the next two paragraphs.

§ 6.10 Graphical Methods of Investigating Stability and Overall Response

The first of these methods is based on the *steady-state* response of a system to a sinusoidal input whose radiancy is varied from 0 to ∞. This approach leads to graphical methods which enable the margin of stability to be determined, and allow optimisation of the design to be achieved. Perhaps more importantly, however, the method lends itself to experiment, since the frequency response of individual elements in the control system can be established by applying an appropriate harmonic signal of varying frequency to the input of the element and measuring the amplitude and phase of the output. Alternatively, if we cut the loop to make the controller an un-monitored controller, as suggested in fig. 6.21, we can inject a known

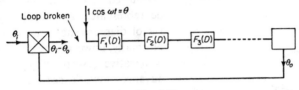

Fig. 6.21

harmonic signal in place of the error signal and determine the overall frequency response by measuring the amplitude and phase of the output.

The harmonic signal of unit amplitude injected into the system depicted in fig. 6.21 can be replaced by a vector $e^{j\omega t}$ (see § 0.14) and D (or d/dt) will then be $j\omega$. Consequently, the open-loop transfer function (see § 6.6) will be

$$\frac{\theta_o}{\theta_i - \theta_o} = \frac{\theta_o}{\theta} = F_1(j\omega) \, . \, F_2(j\omega) \, . \, F_3(j\omega) \, . \, . \, . \, .$$

$$= KG(j\omega)$$

The transfer function for each of the elements of the controller for a harmonic input will be

$$\frac{\theta_1}{\theta} = A_1 e^{j\phi_1}$$

$$\frac{\theta_2}{\theta_1} = A_2 e^{j\phi_2}$$

$$\frac{\theta_3}{\theta_2} = A_3 e^{j\phi_3} \, . \, . \, . \, \text{etc.}$$

where A_1 is the amplitude ratio and ϕ_1 the phase shift (usually negative) across the first element and so on. Consequently, the overall frequency response will be

$$\frac{\theta_o}{\theta} = (A_1 \, . \, A_2 \, . \, A_3 \, . \, . \, .)e^{j(\phi_1 + \phi_2 + \phi_3 \, . \, . \, .)} = KG(j\omega)$$

Thus, we see that for the unmonitored system the output for a harmonic input of unit amplitude will be a harmonic quantity of the same frequency and having an amplitude equal to the product $A_1 . A_2 . A_3 \ldots$ and with a phase change equal to the sum $\phi_1 + \phi_2 + \phi_3 \ldots$

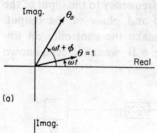

This relationship between the harmonic input and output of the unmonitored system can in principle be represented graphically by the vector diagram shown in fig. 6.22a. Normally, however, the phase angle is negative (except with a phase-advance network: see § 6.8), so the vector diagram is more likely to appear as in fig. 6.22b.

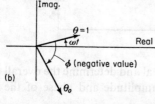

Having found the open-loop relationship between θ_o and θ, we can immediately deduce the closed-loop relationship between θ_o and θ_i because we know that $\theta = \theta_i - \theta_o$, so that $\theta_i = \theta + \theta_o$, this equation having, of course, to be interpreted graphically, as in fig. 6.22c. Alternatively, it could have been found from equation (6.7), which could be rewritten as

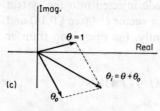

Fig. 6.22 Vector diagrams for unmonitored controller shown in fig. 6.21

$$\frac{\theta_o}{\theta_i} = \frac{KG(j\omega)}{1 + KG(j\omega)}$$

Similarly, if we had taken an input signal in the form $\theta_i = 1 \cos \omega t$, then we could have put down

$$\frac{\theta_o}{\theta_i} = Me^{j\lambda} = Y_o(j\omega) = \frac{KG(j\omega)}{1 + KG(j\omega)}$$

where $Y_o(D)$ is the overall transfer function relating the input and output of a monitored control system, and $Y_o(j\omega)$ the overall harmonic response. M is the overall dynamic magnification or amplitude ratio and λ the overall phase change.

The amplitude and phase angle of θ_o will both vary with ω, so if we draw the vector for θ always in the position where $\omega t = 0$ (the co-ordinates of the tip of the vector being therefore $+1$, 0), then as ω is varied from 0 to ∞ the end of the vector representing θ_o will describe a locus such as that illustrated in fig. 6.23. For the particular case when $\omega = \omega_1$, the output vector will, let us assume, be in the position OC. If now we were to close the loop the input signal θ_i required to give an error signal θ of unit amplitude would be represented by the vector OB obtained by the vector summation of OA and OC. Alternatively, the magnitude of the input vector could equally well be found by joining C to the point E, which has co-ordinates -1, 0, as the triangles OAB and EOC are congruent.

Fig. 6.23 is known as a '*Nyquist diagram*'. It is evident that if the

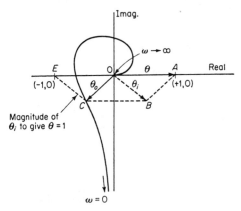

Fig. 6.23 Overall frequency response of an unmonitored controller

locus of θ_o passes through the point having co-ordinates $(-1, 0)$, then at the frequency which corresponds with this particular point on the locus the input signal required to maintain an error signal of unit amplitude from the closed loop would be zero. This implies that if the system were to be disturbed by some random input, then any harmonic component of the input having this frequency would not die away. In fact, the system is on the verge of instability, and it can be shown, though this is beyond the scope of this book, that if one traces the locus from $\omega = 0$ to $\omega = \infty$ the system will be stable if the point -1, 0 lies to the left as in fig. 6.24c, and unstable if the point is to the right as in fig. 6.24b. The system shown in fig. 6.24d is described as a conditionally stable system because, although (by the rule just given) it is clearly stable, it will become unstable if the gain K of the system is either increased or decreased to such an extent that the locus passes above the point $-1, 0$. If the gain is decreased far enough another region of stability will be reached, as the whole of the locus will be to the right of the point $-1, 0$.

The margin of stability can be judged from the Nyquist diagram

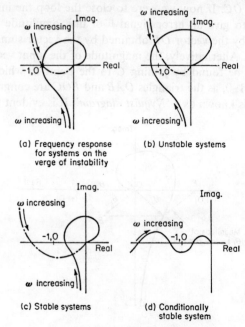

(a) Frequency response for systems on the verge of instability

(b) Unstable systems

(c) Stable systems

(d) Conditionally stable system

Fig. 6.24 Nyquist stability criterion

by the closeness of approach of the curve to the point $-1, 0$. One method is to draw a circle centred on the point $-1, 0$ just to touch the curve as shown in fig. 6.25a. It is found from experience that the radius of this circle should exceed 0·6 to give sufficient stability. More usually, however, the closeness of approach is indicated by two quantities, the 'gain margin' and the 'phase margin', which are shown in fig. 6.25b. The gain margin is the distance of the response curve from the point $-1, 0$ when ϕ is 180°, and it should have a value in excess of 0·6. The phase margin is the angle between the negative real axis and the vector θ_0 when its amplitude (i.e. $|\theta_0|$) is unity, and again from experience this should exceed 30°. If these values of the gain and phase margin are achieved, then the dynamic magnification at resonance of the system will not be large.

Having established the value of the Nyquist diagram, we must now consider how this diagram may be constructed from the frequency responses of the elements which go to make up the control system. In particular, we shall consider those elements for which we have found the transfer functions in §§ 6.5 and 6.8.

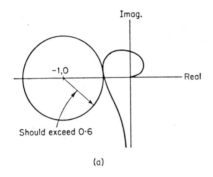

(a)

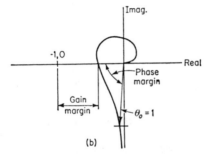

(b)

Fig. 6.25 Margin of stability

First let us consider an element which gives an exponential lag: this, as we have seen, is represented by the transfer function

$$\frac{\theta_2}{\theta_1} = \frac{1}{1 + TD}$$

which for a harmonic input becomes

$$\frac{\theta_2}{\theta_1} = \frac{1}{1 + j\omega T}$$

or $\theta_2(1 + j\omega T) = \theta_1$

This equation can be represented by fig. 6.26a, where θ_1 has a real component θ_2 and an imaginary or unreal component $\theta_2\omega T$. If the unit input vector θ_1 is drawn in the position $\omega t = 0$ (since it corresponds to 0 in the overall system), then we get the diagram shown in fig. 6.26b. As ω is increased from 0 to ∞ the tip of the vector θ_2 describes a semicircle as shown.

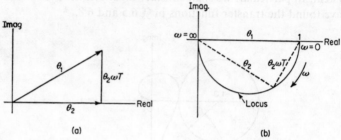

(a)   (b)

Fig. 6.26 Harmonic response for element with transfer function $\dfrac{\theta_2}{\theta_1} = \dfrac{1}{1 + TD}$

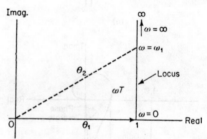

Fig. 6.27 Harmonic response for element with a transfer function $\dfrac{\theta_2}{\theta_1} = 1 + TD$

Phase-advance networks have been mentioned in § 6.8, and they can give rise to a transfer function of the type

$$\frac{\theta_2}{\theta_1} = 1 + TD$$

which for a harmonic input becomes

$$\frac{\theta_2}{\theta_1} = 1 + j\omega T$$

In an exactly similar fashion, if we draw θ_1 in the position $\omega t = 0$ the result is represented graphically in fig. 6.27.

Several examples of elements which have a complex exponential lag have been considered in § 6.5; they have a transfer function

$$\frac{\theta_2}{\theta_1} = \frac{\omega_n^2}{\omega_n^2 + 2c\omega_n D + D^2}$$

or

$$\frac{1}{1 + 2cT_1 D + T_1^2 D^2}$$

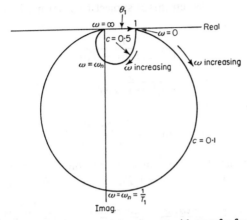

Fig. 6.28 Harmonic response for element with transfer function
$$\frac{\theta_2}{\theta_1} = \frac{1}{1 + 2cT_1D + T_1^2D^2}$$

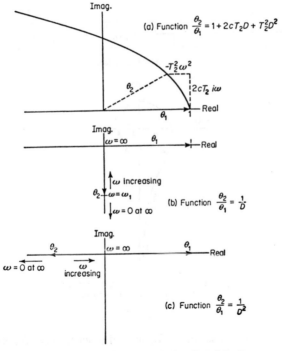

Fig. 6.29 Harmonic response of various functions

where $T_1 = 1/\omega_n$. When this is subject to a harmonic input of unit amplitude then

$$\frac{\theta_2}{\theta_1} = \frac{1}{1 + 2cT_1 j\omega - T_1^2\omega^2}$$

or

$$|\theta_2| = \frac{|\theta_1|}{\sqrt{[(1 - \omega^2 T_1^2)^2 + 4c^2\omega^2 T_1^2]}}$$

and the phase angle is

$$\phi = \tan^{-1} - \left\{\frac{2c\omega T_1}{1 - \omega^2 T_1^2}\right\}$$

which is a phase lag. This harmonic response is shown graphically in fig. 6.28. When $\omega = \omega_n = 1/T_1$, then $\phi = -90°$ and $\theta_2 = 1/2c$.

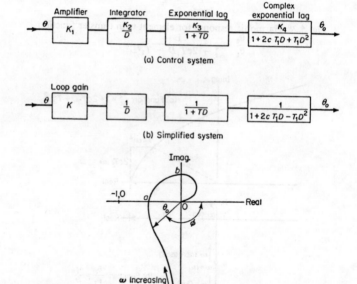

(a) Control system

(b) Simplified system

(c) Frequency responce or Nyquist diagram
for system shown in (b) above

Fig. 6.30

Harmonic response curves for several other transfer functions are shown in fig. 6.29, and the response for an unknown element can be found experimentally by applying a suitable harmonic input to the element and measuring the amplitude and phase of the output.

Now returning to the complete control system we can readily build up the complete open-loop transfer function for a harmonic error signal from the harmonic response of each element. For example consider the system shown in fig. 6.30a. This can first of all be simplified by collecting together the frequency independent terms for each element and representing these by a block having a simple transfer function K, as shown in fig. 6.30b. K, which is equal to $K_1K_2K_3K_4$ is, of course, the total loop gain (see § 6.6). The harmonic loop transfer function will be

$$KG(j\omega) = \frac{K}{(j\omega)(1 + Tj\omega)(1 + 2cT_1j\omega - T_1^2\omega^2)}$$

which can readily be computed from the individual responses of each of the terms in the denominator by adding the phase angles for the individual elements at a given radiancy and multiplying the moduli. Thus at $\omega = 0$ the phase angle will be $0 - 90 + 0 + 0 = -90°$ and the modulus $A = K \times \infty \times 1 \times 1 = \infty$, and at $\omega \longrightarrow \infty$ the phase angle will be $0 - 90 - 90 - 180 = -360°$, while the modulus $= K \times 0 \times 0 \times 0 = 0$. Intermediate values for radiancies between 0 and ∞ can be calculated in a similar fashion, and a curve shown in fig. 6.30b can be drawn. The values of the intercepts *oa* and *ob* will, of course, depend on the gain K.

§ 6.11 Bode Diagram or Logarithmic Plot

In the previous paragraph we developed methods of examining the stability of a control system from a knowledge of its open-loop frequency response, by drawing the locus of the output vector on what is in essence an Argand diagram. Alternatively, the open-loop frequency response can be conveniently presented by plotting the phase angle and the logarithm of the modulus against the logarithm of the frequency or frequency ratio. The advantages of this are that it permits simple straight-line approximations of the functions in which we are mainly interested, and the multiplication or division of the moduli of these functions is achieved simply by addition or subtraction.

Appropriately enough in relation to 'harmonic' quantities, the units we shall use are derived from the theory of sound. Thus, an '*octave*', as is well known, indicates a *doubling* of the frequency, e.g. from 1 to 2, or 100 to 200 c/s (or for the notes A immediately below and above middle C from 220 to 440 c/s in the internationally agreed standard musical pitch). The musical reasons for its adoption are too obvious to need elaboration: it might be remarked that the human ear is (variably) sensitive to vibrations over a range of perhaps 20 to 20 000 c/s, a ratio of about 2^{10} or 10 octaves. Sometimes the preferred unit is a '*decade*', which indicates a factor of 10 in the frequency, e.g. from 100 to 1 000 c/s. The human ear is even more remarkable in the range of acoustic energy it can accommodate: in the middle frequency range it can tolerate a sound of intensity a million times as great as the least sound it can just hear. The *smallest* increase in intensity which it can detect is not an arithmetic addition of sound but a multiplication of intensity by a factor of about $1\frac{1}{4}$. Again a logarithmic scale is clearly appropriate, and the basic unit chosen is a 'bel' (the word being derived from the name of Alexander Graham Bell, 1847–1922, who invented the telephone), which means an intensity or power *ratio* of 10. This is rather too big a ratio for convenience, and the unit in common use is a 'decibel' (db), which, of course, implies a power ratio of $^{10}\sqrt{10}$ or roughly 1·259 – just about the minimum audible change. In general, a change of power from P_1 to P_2 means a change of $\log_{10}\dfrac{P_2}{P_1}$ bels or $10\log_{10}\dfrac{P_2}{P_1}$ decibels; thus, for instance, a change of 3 db means a power ratio of $10^{0\cdot3} \simeq 2$. Unfortunately there is another source of confusion. If an e.m.f. E_1 feeding into a constant impedance R changes to E_2, the *power* changes from $E_1{}^2/R$ to $E_2{}^2/R$; thus this change is one of

$$10\log_{10}\frac{E_2{}^2}{E_1{}^2}\ \text{db}$$

or

$$20\log_{10}\frac{E_2}{E_1}\ \text{db}$$

This last form is the one in which we usually meet the term decibel in control engineering.

In general, the open-loop transfer function will be made up of the transfer functions considered in §§ 6.5 and 6.8, and consequently, for a harmonic input signal it can be expressed in the form

$$KG(j\omega) = \frac{K}{(j\omega)^n}\frac{(1 + T_a j\omega)\dots\{1 + aT_1 j\omega + T_1{}^2(j\omega)^2\}}{(1 + T_b j\omega)\dots\{1 + bT_2 j\omega + T_2{}^2(j\omega)^2\}}$$

If we consider the modulus and phase of this function separately, then, taking logs, we can write down

$\log |KG(j\omega)|$

$= \log K + \log |1 + T_a j\omega| \ldots + \log |1 + aT_1 j\omega + T_1^2(j\omega)^2|$
$- n \log |j\omega| - \log |1 + T_b j\omega| \ldots - \log |1 + bT_2 j\omega + T_2^2(j\omega)^2|$

and the phase angle

$$KG(j\omega) = \phi = \tan^{-1} \omega T_a \ldots + \tan^{-1} \frac{\omega a T_1}{1 - \omega^2 T_1^2}$$

$$- n \times \frac{\pi}{2} - \tan^{-1} \omega T_b \ldots - \tan^{-1} \frac{\omega b T_2}{1 - \omega^2 T_2^2}$$

Normally ω is in rad/s and the amplitude or modulus $|KG(j\omega)|$ is in decibels, so that

$$20 \log_{10} |KG(j\omega)| = 20 \log_{10} |K| + 20 \log_{10} |1 + T_a j\omega| \ldots \text{etc.}$$

Usually we plot separate diagrams showing the logarithmic amplitude in decibels and the phase in degrees as ordinates, each against a logarithmic scale of ω.

It is helpful to be familiar with the log plots for various common transfer functions. First, let us consider the function $1/j\omega$, which, as we saw in § 6.10, has a phase angle of $-90°$ for all values of ω and a logarithmic amplitude in decibels of

$$20 \log_{10} \left|\frac{1}{j\omega}\right| = -20 \log_{10} |j\omega| = -20 \log_{10} \omega$$

as the modulus of $j\omega$ (i.e. $|j\omega|$) is ω. For $\omega = 0\cdot1$ the amplitude in decibels will be $-20 \log_{10} 0\cdot1 = +20$, and similarly for other values of ω. The amplitude in decibels and the phase angle in degrees are shown plotted against ω on a log scale in fig. 6.31. The slope of the amplitude curve is -20 db/decade or $-6\cdot02$ db/octave. For $(1/j\omega)^2$ the amplitude will be $-40 \log_{10} \omega$ db, the phase angle will be $-180°$ and the slope of the amplitude curve will be -40 db/decade or $-12\cdot04$ db/octave.

For the function $1/(1 + T_b j\omega)$ the logarithmic amplitude will be

$$-20 \log_{10} |1 + T_b j\omega| \text{ or } -20 \log_{10} \sqrt{(1 + T_b^2 \omega^2)}$$

and the phase angle $-\tan^{-1} T_b \omega$. For this case it is convenient to plot the logarithmic amplitude against ωT_b. At small values of ωT_b the logarithmic amplitude approaches zero and the phase angle

0, and when $T_b\omega \gg 1$ then $\sqrt{(1 + T_b^2\omega^2)} \simeq T_b\omega$ and the amplitude becomes $-20 \log_{10} T_b\omega$ and the phase angle approaches $-90°$. At low frequencies the slope of the amplitude curve is zero, but at high frequencies the slope becomes -20 db/decade or $-6\cdot02$

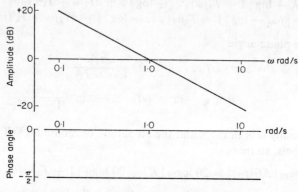

Fig. 6.31 Log. plot of function $\dfrac{1}{j\omega}$

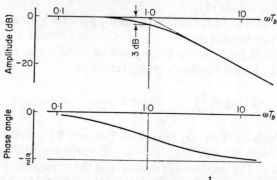

Fig. 6.32 Log. plot of function $\dfrac{1}{1 + T_b j\omega}$

db/octave. The shape of the amplitude curve is shown in fig. 6.32. If the slope of the final curve is extended backwards it will intersect the horizontal axis through the origin at a radiancy of $1/T_b$. For a radiancy of $1/T_b$ the actual amplitude in decibels will be

$$-20 \log_{10} \sqrt{\left(1 + \frac{T_b^2}{T_b^2}\right)} = -3 \text{ db}$$

Similarly, for the function $1 + T_a j\omega$ we can draw the amplitude and phase-angle curves shown in fig. 6.33.

Lastly let us consider the function

$$\frac{1}{1 + 2cT_2j\omega + T_2^2(j\omega)^2}$$

for which the logarithmic amplitude will be

$$-20 \log_{10} | 1 + 2cT_2j\omega + T_2^2(j\omega)^2 |$$
$$= -20 \log_{10} \sqrt{[(1 - T_2^2\omega^2)^2 + 4c^2T_2^2\omega^2]}$$

and the phase angle is

$$-\tan^{-1} \frac{2cT_2\omega}{1 - T_2^2\omega^2}$$

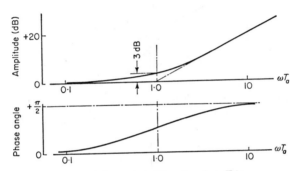

Fig. 6.33 Log. plot of function $1 + T_a j\omega$

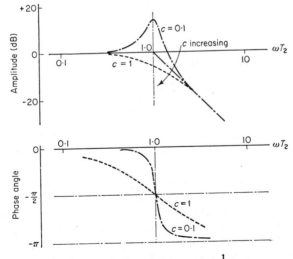

Fig. 6.34 Log. plot of function $\dfrac{1}{1 + 2cT_2j\omega + T_2^2(j\omega)^2}$

Again it is convenient to plot the logarithmic amplitude against ωT_2. When ωT_2 is small the logarithmic amplitude is nearly zero, as is the slope of the curve; and the phase angle approaches zero. At large values of ωT_2 the amplitude becomes $-20 \log_{10} T_2^2 \omega^2$, so its slope is -12.04 db/octave, and the phase angle approaches $-180°$. If the slope of the final curve of amplitude is extended backwards it will intersect the horizontal axis through the origin at a

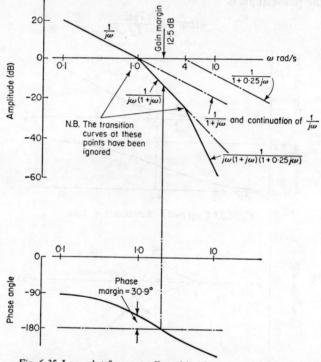

Fig. 6.35 Log. plot for controller with a loop transfer function
$$\frac{1}{j\omega(1 + j\omega)(1 + 0.25j\omega)}$$

radiancy $\omega = 1/T_2$, and the value of the phase angle at that point will be exactly $-90°$. At a radiancy of $1/T_2$ the exact value of the amplitude curve will depend on the constant c, which in a mechanical system represents the critical damping ratio. The true shape of the curves in this region are indicated in fig. 6.34. Similarly, the inverse curves to those shown in fig. 6.34 can be drawn for the function $1 + 2cT_1 j\omega + T_1^2(j\omega)^2$.

The main advantage of these curves is that one can make simple straight-line approximations to the logarithmic amplitude curves, which facilitates the rapid construction of logarithmic amplitude curves for complete control systems by the simple addition of the curves for each function. Similarly, phase-shift curves can be readily constructed. If greater accuracy is desirable, then templates or standard charts can be used. Having constructed the complete logarithmic amplitude and phase shift curves, we can deduce the gain and phase margins, but more importantly if these margins are not acceptable this curve enables the designer to appreciate more readily which of the components is responsible.

In fig. 6.35 we have plotted approximately the logarithmic amplitude curve for a controller and also the phase-angle curve. The logarithm of the gain margin is the vertical distance below the horizontal axis when the phase shift is $-180°$. As stated in § 6.10, an acceptable value of this is

$$20 \log_{10} (1 - 0·6) = -8 \text{ db}$$

or less. In the example the gain margin is $-12·5$ db, which is ample. The phase margin is the phase angle when the amplitude or modulus of $KG(j\omega)$ is unity or $\log_{10} | KG(i\omega) |$ is zero, minus $-180°$. As stated in § 6.10, this phase margin should exceed $30°$ for stable operation. In the example shown in fig. 6.35 the phase margin is $-140·1 + 180 = 30·9°$, which is just sufficient.

Reference Books

1. A. PORTER. *Introduction to Servomechanisms.* Methuen, 1952.
2. R. H. MACMILLAN. *An Introduction to the Theory of Control in Mechanical Engineering.* Cambridge University Press, 1951.
3. J. C. WEST. *Servo-mechanisms.* English Universities Press, 1953.
4. J. C. WEST. *Analytical Techniques for Non-linear Control Systems.* English Universities Press, 1960.
5. W. R. EVANS. *Control-system Dynamics.* McGraw-Hill, 1955.
6. H. LAUER, R. LESNICK, and L. E. MATSON. *Servomechanism Fundamentals.* McGraw-Hill, 1947.
7. H. CHESTNUT and R. W. MAYER. *Servomechanisms and Regulating System Designs.* Wiley, 1959. Vols. 1 and 2.
8. L. A. STOCKDALE. *Servomechanisms.* Pitman, 1962.
9. D. P. ECKMAN. *Automatic Process Control.* Wiley, 1958.
10. H. R. MARTIN. *Introduction to Feedback Systems.* McGraw-Hill, 1968.

Examples

6.1 Find the transfer functions of the systems depicted in fig. 6.36.

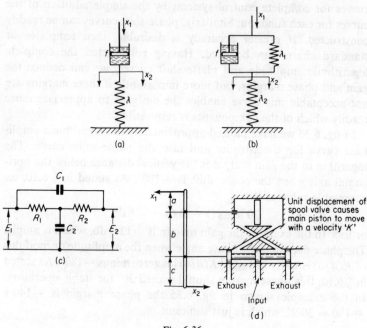

Fig. 6.36

6.2 With a remote position controller the load on the output has a moment of inertia of $0 \cdot 2$ kg cm^2 and the damping is $0 \cdot 6$ of the critical damping. The driving motor gives a torque of 2×10^{-4} Nm/rad error.

If the output shaft is subject to a disturbance in the form of a sudden displacement from its equilibrium position, determine the maximum value of the overshoot of the output shaft during its recovery to its equilibrium position and the time after this disturbance when the maximum is reached.

6.3 A simple linear servomechanism consists of a proportional controller which supplies a torque K times the error and drives a rotational load of moment of inertia I and with viscous damping f. Derive an expression giving the output position as a function of time for a step displacement input if the damping is less than critical.

For a particular servomechanism I is $1 \cdot 25$ kg m² and it is found that an external torque of 3 kgf m applied to the output shaft gives a steady-state error of $0 \cdot 2$ rad and that for a constant velocity input of 1 rad/s the steady state error is $0 \cdot 1$ rad. If a step displacement input of 20° is applied to the system when it is at rest, how long is it before the error is always less than 1°?

6.4 A remote position controller controls the angular displacement of a shaft loaded by an inertia of 4 kg m² and viscous damping. When the controller gives a torque proportional to the misalignment the damping ratio is found to be $0 \cdot 6$ and the steady-state velocity lag is 5° for an input speed of 30 rev/min.

The specification requires that the minimum value of the damping ratio and the steady state velocity lag shall be $0 \cdot 75$ and $2 \cdot 5$° respectively. This is to be met by a reduction in the viscous damping and the use of first derivative of error compensation. Find the reduction in the viscous drag and the amount of first derivative of error compensation necessary.

6.5 In a remote position controller the load has an inertia of $0 \cdot 217$ kg cm² and the viscous damping is 10^{-4} Nm s/rad; the controller gives a torque of K times the error and the damping ratio of the system is $0 \cdot 4$.

The transient response of the system is satisfactory, but it has four times the acceptable error for a velocity input. This error is brought within the required limits and the transient response maintained by

(a) increasing the value of K and adding first derivative of error compensation
(b) first derivative of input compensation

Draw the block diagrams of the modified systems and calculate the amount of compensation required in each case.

6.6 An automatic radar tracker has a motor with a rotor of inertia $2 \cdot 72$ kg mm² which turns an aerial with a moment of inertia of $0 \cdot 272$ kg m² through a 100:1 step-down box. A transducer detects the angular position of the aerial and the signal passes into a receiver which gives 1 V/degree error. The voltage is amplified and applied to the motor, which gives a torque proportional to the applied voltage, the torque being $0 \cdot 055$ Nm at 100 V. The viscous friction measured at the motor shaft is 70 μNm s/rad and the damping ratio is 0·25.

If the wind exerts a static torque of 0·40 Nm on the aerial what are

    (a) the error in degrees relative to a fixed target

    (b) the motor voltage under these conditions?

6.7 The level of liquid in a reservoir is controlled by means of a float and hydraulic relay, as shown in fig. 6.37. The hydraulic relay

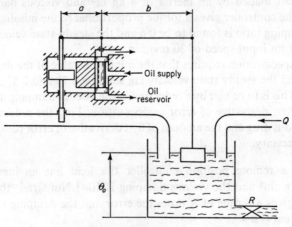

Fig. 6.37

is supplied with oil under pressure, and the flow of oil into the working cylinder, which has a cross-sectional area A, will be k_1 times the displacement of the piston valve from its equilibrium position. Similarly, the change in flow into the tank is k_2 times the displacement of the working piston.

The tank has a cross-sectional area C and an outflow which is limited by a restrictor which gives a resistance which is proportional to flow (i.e. Head above restrictor = Flow × Constant R). In addition, the conditions in the tank are disturbed by another inflow of Q.

Determine the transfer function for the controller and draw a block diagram for the whole system.

    Given that

$$R = 500 \text{ s/m}^2$$
$$k_2 = 1·0 \text{ m}^2/\text{s}$$
$$b = 5a$$

determine the change in level of fluid in the tank if the inflow is raised by 0·06 m³/s.

6.8 A controller has an open-loop transfer function

$$\frac{\theta_o}{\theta} = \frac{K}{D(1 + T_1D)(1 + T_2D)}$$

For this controller find the equation for the gain margin, and the value of K that makes the system unstable. Check your answer by using the Routh stability criterion (see § 6.8).

HINT: When the phase between θ_o and θ is 180° the imaginary part of the transfer function is zero.

6.9 If the open-loop transfer function of a controller is

$$\frac{\theta_o}{\theta} = \frac{K}{D(1 + TD)}.$$

determine the value of the forcing radiancy ω to give this function a value of unity, and the phase margin of the controller.

6.10 Prove that a line of constant dynamic magnification M (see § 6.10) on the Nyquist diagram is a circle with a radius $\dfrac{M}{M^2 - 1}$ which is centred at the point with the co-ordinates $-\dfrac{M^2}{M^2 - 1}, 0$.

The open-loop frequency response of a controller with a second-order characteristic equation is found experimentally. During the test the gain constant is fixed at unity. A Nyquist diagram is constructed and the co-ordinates of several points on the diagram are given below

ω rad/s	Real component	Imaginary component
0·8	−0·61	−0·76
0·9	−0·55	−0·61
1·0	−0·50	−0·50
1·1	−0·45	−0·41
1·2	−0·41	−0·34
1·5	−0·31	−0·21

Using M circles, determine the value of the gain constant if the maximum dynamic magnification is to be limited to 1·414 and determine the frequency at which this occurs.

HINT: Consider the triangle *OEC* in fig. 6.23, and note that $|\theta|$ is unity, so that *EC* is equal to $\left|\dfrac{\theta_i}{\theta_o}\right|$ and *OC* to $\left|\dfrac{\theta_o}{\theta}\right|$.

Consequently, $\dfrac{OC}{EC} = \left|\dfrac{\theta_o}{\theta_i}\right| = M.$

6.11 A simple position control has an open-loop transfer function

$$\frac{\theta_o}{\theta} = \frac{K}{D(D+1)}$$

Plot the Nyquist diagram for $K = 1$ and using this determine the maximum value of K if the dynamic magnifier M is to be limited to 1.3. Find the frequency which gives this maximum and check your results analytically.

HINT: Before doing this question it is necessary to understand question 6.10.

6.12 Fig. 6.38 shows the block diagram of a controller stabilised by a feedback from the output, consisting of an accelerometer whose

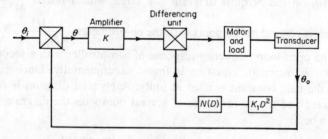

Fig. 6.38

output is modified by a network $N(D)$. If the load is a pure inertia with no viscous damping, investigate the stability if

$$(a)\ N(D) = 1$$
$$(b)\ N(D) = \frac{1}{1 + TD}$$
$$(c)\ N(D) = 1 + TD$$

6.13 Fig. 6.39 shows a block diagram for a controller. Construct a Bode diagram, and from this determine the maximum value of K

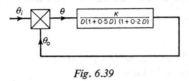

Fig. 6.39

that can be used without instabilities arising. Your graph can be restricted to the radiancy range 0·1 to 10 rad/s.

6.14 A controller has an open-loop transfer function

$$\frac{\theta_o}{\theta} = \frac{2}{D(1 + D)(1 + 0·25D)}$$

Plot a Bode diagram for values of ω between 0·1 and 10 rad/s and determine the gain and phase margins. Also determine the steady state error for a velocity input of 1 rad/s.

7. Friction and Lubrication

The action of friction is divided into parts of which one is simple and all the others are compound.

Simple [friction] is when the object is dragged along a plain smooth surface without anything intervening; this alone is the form that creates fire when it is powerful . . . as is seen with water-wheels when the water between the sharpened iron and this wheel is taken away.

Friction produces double the amount of effort if the weight be doubled.

The friction made by the same weight will be of equal resistance at the beginning of its movement although the contact may be of different breadths or lengths.

The others [forms of friction] are compound and are divided into two parts; and the first is when any greasiness of any thin substance is interposed between the bodies which rub together; and the second is when other friction is interposed between this as would be the friction of the poles of the wheels. The first of these is also divided into two parts, namely the greasiness which is interposed in the aforesaid second form of friction and the balls and things like these.

Everything whatsoever however thin it be which is interposed in the middle between objects that rub together lightens the difficulty of this friction.

Observe the friction of great weights, which make rubbing movements, how I have shown that the greater the wheel that is interposed the easier this movement becomes; and so also conversely the less easy in proportion as the intervening thing is thinner as would be any thin greasy substance; and so increasing tiny grains such as millet make it better and easier, and even more balls of wood or rollers, that is wheels shaped like cylinders, and as these rollers become greater so the movements become easier.

That thing which is entirely consumed by the long movement of its friction will have part of it consumed at the beginning of this movement.

*The greatness of the contact made by compact bodies in their friction will
have so much more permanence as it is of greater bulk; and so also conversely it will be so much less enduring as it is of less size.*

LEONARDO DA VINCI (1452–1519)

§ 7.1 Introduction

The quotations at the top of this chapter are culled from the 'notebooks' of Leonardo da Vinci as translated by E. MacCurdy.* It is
true that some of the terminology is a little archaic, but it would be
very easy to modify the wording (which is, after all, only a translation) to make it sound less old-fashioned. Who can doubt that
Leonardo was, in this instance as in so many others, some four
hundred years ahead of his time? He defines 'simple' friction, and
propounds the Amontons–Coulomb Laws as taught today to nearly
every schoolboy; to the effect that friction between unlubricated
bodies is proportional to the normal force between them and
independent of their area. He points out that friction between such
surfaces can produce a very large rise in temperature – can in fact
'create fire'. He outlines the influence of *any* 'thin substance' interposed between the surfaces, and makes particular reference to a
'greasy substance'. He points out that wear inevitably accompanies
friction. He makes a plea for adequate bearing area to provide long
life. He foreshadows ball and roller bearings. He seems to have failed
only to discover the possibilities of hydrodynamic lubrication, and to
be a little less than up-to-date with his explanation of why materials
behave as they do when rubbed together. In order to show a significant advance on Leonardo we shall have to concentrate on these
topics.

§ 7.2 Dry Friction

Every well-brought-up boy makes some attempt in his school
physics laboratory to verify or disprove the Amontons–Coulomb
'laws of friction' by measuring the force required to pull a 'slider'
along a reasonably smooth surface, and most eventually succeed in
showing that these 'laws' are a fair approximation to the truth. The
frictional force is roughly proportional to the normal reaction

* *The Notebooks of Leonardo da Vinci.* Jonathan Cape, 1938.

between the surfaces; it is nearly independent of the area of contact; the force required to initiate sliding is usually greater than that required just to cause sliding to continue. But most boys have a little trouble in achieving these answers, finding first that their results are very variable, and dependent on the care with which they have cleaned the surfaces and avoided subsequent contamination. The experimental value of the coefficient of friction μ they obtain will probably be in the region 0·15–0·3 for most metals. Amontons's original value was $\frac{1}{3}$. It must occur to the more intelligent that a more effective cleaning than they have been able to use might give a different answer, and this is indeed the case. A thorough de-greasing treatment, for instance, will give, for dissimilar common metals, values nearer unity, and for similar pure metals, values which usually exceed unity. But if contamination is still further reduced by 'outgassing' the metals by heating them in a high vacuum μ will rise to very much higher values, of the order 4–8, and with still better cleaning techniques, to 100 or even ∞, in the sense that the rubbing materials may weld together.

Results such as these clearly make nonsense of the 'explanations' of friction which could reasonably in 1699 satisfy Amontons, who assumed that the 'asperities' or irregularities on the surfaces tended to interlock, so that the load had to be 'lifted' in order to allow sliding to occur. Although this explanation gives rise to some rather awkward supplementary questions, such as why the slider requires to be *continually* lifted against the load, it was not seriously challenged for about two hundred years. Coulomb in 1781 considered molecular adhesion as a possible cause of friction, but rejected it. Ewing in 1892 advanced molecular forces as the basic cause of friction, and similar theories were propounded by Tomlinson (1929) and Hardy (1936). The overwhelming weight of modern opinion, however, accepts, at least in principle, the work of Bowden and his collaborators as providing the most satisfactory explanations for frictional phenomena, and it is to this work in the main which we shall now give brief attention. Needless to say, only the barest outline can be given in the space available: reference to more detailed treatments will be found at the end of the chapter.

What really happens when two smooth surfaces are pressed together? We shall have to start by considering what we mean by 'smooth'. In engineering terms we might think of a ground, scraped, honed, or lapped surface. The very best of these finishes, over a small area, deviate from perfection by tiny amounts, usually measured in small fractions, say from $\frac{1}{10}$ to $\frac{1}{100}$, of a micron (or μm). But even 0·01 micron compared with the magnitudes we now have to consider is rather a large distance: the distance over which two molecules can exert an appreciable force on each other, and might therefore be regarded as influencing or 'touching' each other, is smaller again by a factor of the order 50. So if we bring two such clean surfaces of similar metals very gently together they will first 'touch' at a few, perhaps only three, places or 'points' where projections happen to coincide. The area of these 'points' will be tiny, so even if the normal force between the surfaces is extremely small the resulting compressive stresses in the materials will be large and may easily exceed the elastic limit. Under increasing force increasing elastic and plastic distortions will certainly occur, and their amount will to a first approximation be defined by the requirement that the *true* area of contact must be able to support the load: in other words, the true area will be roughly proportional to the load.

By 'touching', we must mean that some atoms in one surface have moved so close to those in the other surface that the repulsive forces between them exceed the attractive forces: they are therefore actually in at least as intimate contact as neighbouring atoms in the parent metal. How, then, do we distinguish between one piece of metal and the other? We are driven to the thought that over these areas of true contact 'welding' may, or rather must, occur.

This reasoning is not unsupported by experimental evidence. The true area of contact between surfaces of various materials has been measured by a number of methods, in particular by the electrical resistance between them, and it has been shown that whether the surfaces are in the form of crossed cylinders, where even the apparent area is initially small, or in the form of 'flat' surfaces, where it is apparently very much larger, over a wide range of force normal to the surface – from a few to several thousand newtons – this true area of contact is very nearly directly proportional to the force. Moreover, in particular cases, for example, when a clean steel ball is pressed on to the clean surface of a soft metal with a relatively small force a similar force is needed to pull it off again, and 'pick-up' of the soft metal on the steel surface shows that 'welding' has indeed occurred even between these dissimilar materials. If two steel surfaces are pressed together they do not so obviously 'adhere' to each other, because as the force is removed the release of the elastic distortion breaks down the tiny 'welds' one by one, and few are left intact by the time that the normal force is zero.

What will happen if two clean surfaces are pressed together in this fashion, and one is caused to slide over the other? Clearly the 'welds', or 'junctions', as they are usually called, will have to shear, so, initially at least, the force required to cause this sliding will be proportional to the total area of the junctions, which we have agreed will be proportional to the normal force between the surfaces. We have thus a logical basis for the experimental fact that the force to cause sliding to start is proportional to the normal force and independent of the apparent area of the surface. But, particularly if we are discussing friction between similar materials, there is no obvious reason why the junctions should rupture along the original surfaces, and if not, the originally 'smooth' surface will become progressively rougher: it is not really surprising, after all, that the coefficient of friction in these circumstances should rise to values so high that the term really becomes almost meaningless, and that 'seizing' should occur.

Why, then, does this catastrophic type of occurrence not happen every time we make one surface slide over another? We have been discussing so far surfaces which are *clean*, and it has been pointed out that this sort of cleanliness is achieved only by very special techniques, such as heating previously well-cleaned members in a high vacuum to drive off the adsorbed gases. Under such conditions seizing does occur very readily indeed. But the admission of even a trace of oxygen to the vacuum chamber is enough to effect an immense change, the coefficient of friction dropping to a value of the order unity. If we bring our surfaces into the open air they will soon be covered 'with a layer of oxide, and probably a physically adsorbed gas and water layer as well . . . a trace of contaminating grease which may be only one molecule thick and may arise from fatty films migrating from the fingers even if the cleaned side is not handled'.* Under ordinary laboratory or workshop conditions, then, the phenomena associated with clean metal surfaces will never be observed.

In general, the effect of oxide films may be regarded as that of protecting the underlying metals from intimate contact and so from seizure, but this is a complicated situation, and in detail the effect

* F. P. Bowden and D. Tabor, *The Friction and Lubrication of Solids*. Oxford, Clarendon Press, 1950, p. 77.

depends on the properties of the metals *and* their oxides. Chromium, for instance, is a hard metal which forms a strongly adherent oxide. Such junctions as are formed are therefore limited to the oxide, and these junctions tend to shear in the actual surface of contact, so the tendency is for wear, both of the chromium and the surface against which it is rubbing, to be relatively small. Aluminium, on the other hand, is a soft metal which forms a very hard and brittle oxide. Even under very small stresses, this film may break up and embed itself in the parent metal, so forming a very effective lap. Copper is intermediate in its behaviour. At light loads the oxide forms an effective protective – the thicker the film, the better – but if the loads are increased the film will be ruptured and metal-to-metal contact will occur, probably with catastrophic effects. Steel is not, of course, an element, but its behaviour is important to engineers. The usual oxide of iron (Fe_2O_3) is much harder than iron but softer than really hard steel: soft iron on soft iron forms a most unsatisfactory combination for sliding surfaces, but case-hardened or nitrided surfaces are very satisfactory indeed: the reason begins to be comprehensible.

§ 7.3 Boundary Lubrication

The effect on friction of a *trace* of lubricant, say oil or grease, is so well known as to be proverbial, but we are still far from knowing or understanding much about its exact mode of action. This type of lubrication is known as boundary lubrication, and in carrying out experiments we have to choose conditions of high pressure (approximating to the yield stress of the materials) and low sliding speed to avoid confusion with an entirely different type of lubrication which we shall discuss later.

Engineers have known for many years that some animal or vegetable oils were more effective than mineral oils in reducing friction. Experiments carried out with homologous series of paraffins, alcohols, and fatty acids* showed that the coefficient of friction

---

* A paraffin is a hydrocarbon which may be pictured in a simplified form as

$$H-\underset{\underset{H}{|}}{\overset{\overset{H}{|}}{C}}-\underset{\underset{H}{|}}{\overset{\overset{H}{|}}{C}}-\underset{\underset{H}{|}}{\overset{\overset{H}{|}}{C}}-\underset{\underset{H}{|}}{\overset{\overset{H}{|}}{C}}-\underset{\underset{H}{|}}{\overset{\overset{H}{|}}{C}}----\underset{\underset{H}{|}}{\overset{\overset{H}{|}}{C}}-\underset{\underset{H}{|}}{\overset{\overset{H}{|}}{C}}-\underset{\underset{H}{|}}{\overset{\overset{H}{|}}{C}}-H$$

though the chain of carbon atoms may be branched.   [*footnote continued over* depended on the molecular weight of the lubricant used and was largely independent of the surface finish, and that with each of these types of lubricant the coefficient fell with increasing molecular weight, reaching a minimum value of about 0·1. The first two types of lubricant tended when liquid to give a 'stick–slip' type of sliding, suggesting that the lubrication was not fully effective and that serious wear or seizure was imminent, but a smooth motion when the temperature was so low that they were solid; whereas the fatty acids always gave a smooth motion. (It was also observed that the amount of surface damage which occurred could be correlated with the friction.) Further, it has been shown that the addition of a small amount of a fatty acid to a paraffinic type of oil such as a mineral oil is enough greatly to improve its lubricating pro-perties, the coefficient of friction being reduced by a factor of about 5 by 0·01 per cent, and about 10 by 1 per cent of the acid; and from more elaborate experiments it has been shown that the start-ling effect of such a trace of fatty acid is confined to the reduc-tion of friction between materials with which it reacts to form a metallic soap, and that this lubricating effect disappears at the temperature at which the metallic soap melts. From these and many other experiments it has been deduced that lubricants of this type are much more effective when they are in the solid rather

---

In an alcohol one of the hydrogen atoms is replaced by an OH group, giving

$$H-\underset{\underset{H}{|}}{\overset{\overset{H}{|}}{C}}-\underset{\underset{H}{|}}{\overset{\overset{H}{|}}{C}}----\underset{\underset{H}{|}}{\overset{\overset{H}{|}}{C}}-\underset{\underset{H}{|}}{\overset{\overset{H}{|}}{C}}-OH$$

while in a fatty acid the ending is COOH, thus

$$H-\underset{\underset{H}{|}}{\overset{\overset{H}{|}}{C}}-\underset{\underset{H}{|}}{\overset{\overset{H}{|}}{C}}----\underset{\underset{H}{|}}{\overset{\overset{H}{|}}{C}}-C\overset{\diagup O}{\diagdown OH}$$

The paraffins (as the name implies) are 'without affinity' or non-reactive, while the alcohols tend to 'attach' themselves to metal surfaces and the fatty acids to 'react' with certain metals, the metal radical replacing the H in the OH group to form what is described as a 'metallic soap', a substance which has a melting point higher than the corresponding fatty acid.

It is worth noting that the metals with which the fatty acids react in this way include many of those commonly used as bearing materials – copper, zinc, lead, cadmium, and so on – but not with many other metals, such as platinum, nickel, aluminium, and silver.

than the liquid state, and that the superiority of the fatty acid type of lubricant is due to the formation of a film of a metallic soap which is solid at the operating temperature. Again, solids such as calcium oleate or stearic acid, or even soft metals, such as indium or cadmium, can be astonishingly effective lubricants under these conditions of high pressure and low speed. We may picture this condition of boundary lubrication as the interposition between two relatively hard and strong surfaces of a layer of material which tends to prevent the formation of welds or junctions between them, and in which the shearing which must occur when one surface is caused to slide over the other requires a relatively weak tangential force.

Under extreme conditions of load and rubbing speed, such as are experienced in the hypoid gears often used in car driving axles, the boundary lubrication provided by metallic soaps may break down, because these soaps melt at the high surface temperatures reached. The consequent large increase in the coefficient of friction causes more heat to be evolved, and clearly this gives rise to a disastrous spiral. In such cases it is usual to add to the oil certain chemically active materials, usually called 'extreme pressure additives', though the term is perhaps misleading. Their characteristic is that at high *temperatures* they tend to react with the metal surfaces to form a contaminant film of chloride, sulphide, or phosphide. Thus, at the very points where excessive surface temperatures are attained this film is formed and it replaces the metallic soap as a boundary lubricant. Clearly if the materials are so selected that the soap-melting temperature and chemical-reaction temperature are similar the 'oil' will be able to provide boundary lubrication under much more difficult conditions than will an untreated oil. Care must, however, be exercised in selecting such additives, since a compound which was reactive at a temperature lower than that at which its effect was required might lead to excessive corrosion.

Interesting as the condition of boundary lubrication may be, and important as it is in cases where no other type of lubrication is available to ease sliding motion, we have to remember that the coefficient of friction obtainable is still far too high to be tolerated in most engineering practice. To recapitulate briefly and crudely, very thoroughly cleaned metals will weld together on contact and seize catastrophically under sliding motion. Chemically clean metals will produce coefficients of friction of the order 1–0·5, and usually seize at the slightest provocation. Metals covered with a thin film of paraffinic (or *pure* mineral) oil will give a coefficient of friction of perhaps 0·5–0·1, exhibit stick–slip motion, and give rapid wear. Metals covered with a thin film of oil containing traces of fatty acids, such as most animal or vegetable oils, or strong metals covered with a film of a soft metal, will give a coefficient of perhaps half this magnitude, and correspondingly less wear, but still could not be contemplated for use in, say, high-speed engine bearings. How, then, can such bearings be made to work?

§ 7.4 Film Lubrication

There can be no doubt that a much more effective type of lubrication was achieved by practical engineers by mistake, so to speak, long before anyone had any idea how this came about. As it happens, this is one of the few discoveries which can be attributed to two individuals: the story has been told very often, but it is so instructive that it bears brief repetition. In the 1870s a number of investigators had been working on the subject of friction, and almost the only point of agreement between them was that the Coulomb 'laws' were invalid. The Institution of Mechanical Engineers in 1879 appointed a committee to study the subject: in turn the committee gave the task to Beauchamp Tower (1845–1904), who started his investigations in 1882, and submitted his first two (and immeasurably most important) reports in 1883 and 1885.

Tower decided to study friction under lubricated conditions, and to this end built the simple and elegant apparatus illustrated in part in fig. 7.1. In brief, this consists of an axle A (102 mm in diameter, as used on railways) which supports a bearing B (153 mm long and subtending an angle of 157°), which can be loaded through the knife-edge K. When the axle rotates the bearing will rotate slightly with it until the friction torque is balanced by the offset of the load: alternatively, as in the later version illustrated, weights can be added to or taken from the scale-pan to balance this torque, and keep the bearing in the original configuration. Notice particularly the oil-bath O, which Tower provided to give a copious supply of lubricant; he showed that the level of oil in this bath was immaterial, provided that the axle touched it.

With this arrangement, Tower found that the friction torque, far from being directly proportional to the load, was in fact nearly *independent* of it, and therefore that the 'coefficient of friction' was inversely proportional to the load. The actual values he obtained differed with the type of oil and with the temperature, but the minimum values, when the bearing was on the point of seizing, were far far lower than had ever been recorded previously, of the order 0·001 or less, as compared with those of about 0·05 which engineers were accustomed to measure or assume.

Tower tested his bearing with many other systems of lubrication

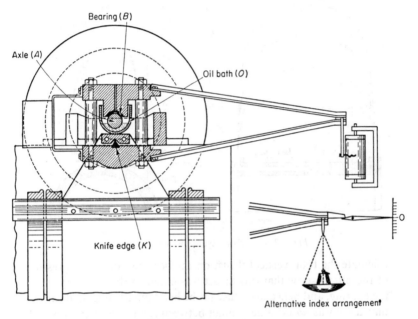

Fig. 7.1 Tower's apparatus

more in accordance with contemporaneous practice, including grooves of many patterns and an oily pad pressed on the shaft surface, but none of these approached in effectiveness the oil bath. In fact, the only major discovery which resulted from these later experiments could fairly be described as a fluke. He had drilled in one bearing a hole for a lubricator as indicated in fig. 7.1, but used this bearing with the oil bath. Oil welled up in this hole and overflowed, causing a mess. Tower therefore inserted a cork to stop this. The cork was pushed out. He tried a wooden plug, with the same result. *He then fitted a pressure gauge* calibrated to 14 bar: when the machine was started the pointer went off the scale. Now the mean load on the bearing was only 7 bar and the oil was being supplied at atmospheric pressure.

Later Tower followed up this astonishing discovery by drilling a few other holes and measuring the pressure developed at them: his results were sufficient to allow him to plot the pressure distribution curves shown in fig. 7.2: even more important, they allowed him to

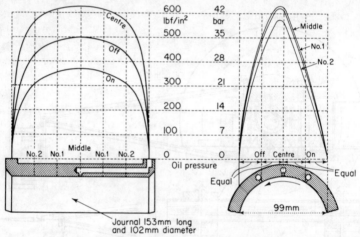

Fig. 7.2 Tower's pressure distribution curves

calculate the total vertical thrust on the bearing due to the pressure of the oil, to show that this thrust was equal to the load applied to the bearing, and hence to deduce (as he had himself suggested earlier) that there was *no* metallic contact between it and the journal, which were completely separated by a film of oil: this condition is described as 'film lubrication'. The importance of this discovery is surely obvious: if film lubrication can be maintained, not only will frictional losses be reduced as compared with boundary lubrication by a factor of the order 50, but all anxiety as to seizure and even as to wear will be eliminated. Under ideal conditions there will be *no* wear. We shall consider shortly what departures from this ideal are likely to occur.

Tower was essentially an experimentalist, and perhaps incapable of pursuing his discoveries much further. Fortunately, however, his results attracted the attention of Professor Osborne Reynolds (1842–

1912), who in 1886, one year later than Tower's second report,[*] which contains the results just described, published a masterly paper[†] which provided a convincing explanation of the results Tower had obtained. The mathematical part of this is beyond the scope of this chapter: we may, however, extract from its eighty-odd pages enough qualitative material for our present purpose.

Reynolds had, a few years earlier, investigated the conditions under which the motion of water in a pipe was 'direct' or 'sinuous', or as we should now say, laminar or turbulent; among many other valuable results he had deduced that the change from one type of motion to the other depended on the value of the quantity $Vd\rho/\eta$ where V is the velocity of the water, ρ its density, η its viscosity, and d the diameter of the pipe. (This quantity is, of course, now universally known as the Reynolds number.) It appeared reasonable to Reynolds, therefore, that a fluid as viscous as oil, in a passage as narrow as the space between the shaft and the bearing (i.e. under conditions of large η and small d), would move even with high velocity in a laminar fashion, and he based his argument on this assumption.

He first considered a layer of fluid between two flat parallel surfaces as in fig. 7.3, a distance h apart, one stationary and the other moving with a speed U. The fluid in contact with each surface will have the same velocity as the surface, and the fluid between them an intermediate velocity as shown, where the displacement of each cross-section of the fluid during a short interval of time is indicated by the difference between the dotted and full lines. There will, of course, be a shearing force or viscous drag uniformly distributed over the upper surface of amount $\eta AU/h$, where A is its area (cf. Introduction, example 0.8).

If now the upper surface has no relative tangential velocity but instead a velocity towards the lower surface, the fluid between the surfaces will have to be expelled. Again the figure (7.4) shows the displacement of various cross-sections of the fluid in a short time: at the centre section no lateral motion will occur, and the maximum fluid velocity will increase proportionately with distance to right and left of this section. As the fluid is viscous, pressure must be developed, and this will clearly have a maximum value at the centre and zero value at each end, roughly as indicated.

[*] *Proc. I. Mech. E.*, 1885, Vol. 36.
[†] *Trans. Roy. Soc.*, 1886, Vol. 177, pt. 1.

This motion could not, of course, persist. But if the upper surface is not strictly parallel to the lower, but slightly inclined as in fig. 7.5, and the lower surface has tangential motion, the result is virtually to provide a combination of the two previous conditions; the oil

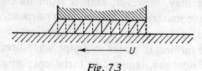

Fig. 7.3

which enters at the right-hand end finds itself between two surfaces which are in effect approaching each other and must escape by flowing outwards in both directions. Pressure is then developed as shown and the upper surface is subjected both to this pressure and to the viscous drag as in fig. 7.3. In this case, since the passage in

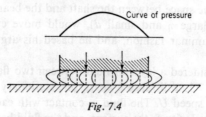

Fig. 7.4

which the oil is flowing outwards is wider at the right-hand (or oil-entry), end, the pressure distribution will not be symmetrical as in fig. 7.4, but asymmetrical as sketched in fig. 7.5. Note that the maximum pressure will not in general coincide with the centre of thrust on the surface.

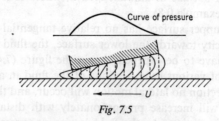

Fig. 7.5

Figs. 7.3, 7.4, and 7.5 have been copied from Reynolds's original paper. He considered qualitatively many variants of this situation, and in particular went on to the part-cylindrical bearing which had been the subject of Tower's experiments. By masterly, but essentially

simple, reasoning he was able to show that if the radius of the
bearing was slightly larger than that of the shaft stable film lubrication was possible, and that only undèr conditions of zero load
would the thinnest point of the film (or point X of closest approach
of the bearing to the shaft, fig. 7.6) be at the middle of the bearing.
He then went on to deal with these situations mathematically,
starting from the basic hydrodynamical equations and solving exactly
or approximately many of them. He achieved an approximate solution for the cylindrical bearing of length large compared with the
diameter, from which he could calculate the 'attitude' of the bearing

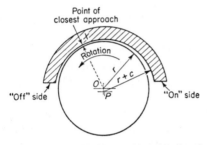

Fig. 7.6 Diagrammatic sketch of journal bearing showing 'attitude'

on the shaft, showing that as the load on the bearing increased the
point X moved towards the 'off' side of the bearing, nearly to its
extremity, and then back again towards the centre. He showed that
a load would be reached at which the oil pressure at the off side
would become negative, and suggested that this would cause 'discontinuity in the oil' and therefore correspond to the maximum safe
load. He calculated the actual difference in radius of Tower's
bearings and shaft – these values being dependent on the temperature,
to a degree which he also worked out – and the pressure at any
point in the oil film. The former figures could not be checked, but
Tower's measured pressures did indeed agree to an extraordinary
degree of accuracy with the Reynolds curve. According to the
Reynolds analysis, the friction torque should vary directly with the
speed: Tower had found a lower rate. Reynolds was able not only
to show that the discrepancy could be qualitatively accounted for
by the rise with speed in the true temperature of the oil in relation
to that of the shaft: he also calculated this difference, carried out a
separate series of experiments himself to determine the effect of
temperature on the viscosity of oil, and demonstrated that the result

was in complete accordance with Tower's experimental findings. Truly research in the grand manner, to which this brief summary comes woefully short of doing justice. The full paper requires to be studied if any appreciation of its quality is to be obtained.

Since Tower and Reynolds, a vast amount of work, both experimental and theoretical, has been carried out on journal bearings, but much of this concerns the specialist only. We may here more profitably give attention to some matters of great importance to every engineer.

What must we do to ensure that journal bearings we design are film lubricated?

First, we have seen that it must be possible for a taper film of reasonable area to form. In the case of a 'complete' bearing (i.e. one extending over 360°) this can be achieved only if there is a finite but small difference between the diameters of journal and bearing. The optimum clearance varies considerably with the conditions, but its order of magnitude is one or two thousandths of the diameter.

Secondly, we must then supply the lubricant at a point remote from the position at which we hope to achieve the high-pressure film. The reason should hardly need elaboration. If it is to support the load the pressure which builds up in the film must necessarily be greater than the mean pressure which the bearing is carrying, probably at least twice as great, and this will normally imply a pressure of the order of hectobars. We have seen that no pressure is *required* in the oil supply: all that is needed is that the oil shall be supplied to the point at which it is needed. In order to ensure this, and sometimes to ensure a flow of oil to cool the bearing, oil is often supplied at a pressure of one or two bars, but this is trivial compared with the film pressure.

Thirdly, it is advisable to provide a smooth entry passage as in the sketch, fig. 7.7a, to ensure that the oil is not scraped off the surfaces, but really does get in between them.

Let us see how we can achieve these simple desiderata. In the case of a uni-directional load, for instance, as shown with greatly exaggerated clearance in fig. 7.7a, we should be able to supply oil at either of the points (x) or (y) as indicated. If the load oscillates, as in fig. 7.7b, (x) is no longer permissible, and it is preferable to supply oil at both (y) and (z). If the load may be applied in a random direction it may be necessary to supply oil at three points (fig. 7.7c). If the load rotates with the shaft it is usually preferable to supply the oil through the shaft itself. Needless to say, the same principle of allowing it to emerge on the side *not* subject to load is unaltered. An oscillating bearing is a more difficult matter. If the load also oscillates, as in the little-end bearing of a connecting-rod, the trouble is not usually too serious, since the oil can get between the surfaces when they separate, but if the load acts in a constant direction seizure is difficult to avoid. One possibility is to supply lubricant to a series of grooves as in fig. 7.7d, so spaced that each part of

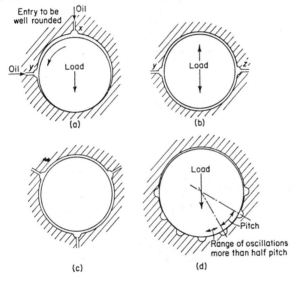

Fig. 7.7 Oil supply to bearings with various loadings

the oscillating member comes into contact with the lubricant. Probably a better solution in this particular case is to abandon a plain bearing in favour of a needle-roller bearing.

It will be noted that this is the first mention of grooves. There is no objection to, and probably some advantage in, a groove which distributes the oil axially, provided it ends within the width of the bearing, but no groove should ever – from the arguments which have been advanced – *cross* an area in which the film pressure is to be high, since the effect is merely to allow the oil to escape, and destroy the lubricating film in this area. In practice, an occasional exception to this rule may be found, as, for instance, when a circumferential groove is made right round the central plane of a bearing and the oil is supplied to this groove. The argument for so doing is that the flow of oil through the bearing is increased, and therefore the cooling effect can be increased. The advantage of greater cooling may outweigh in particular cases the disadvantage of a load-carrying capacity considerably reduced for the reasons indicated in the sketch, fig. 7.8. Again, it has been pointed out that under ideal conditions of film lubrication *no* wear occurs, but such conditions are achieved only when the shaft is running. Now if a shaft can be started and stopped under conditions of light load

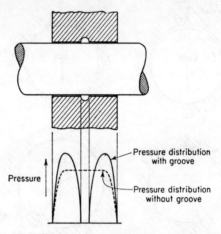

Fig. 7.8 Sketch of pressure distribution in journal bearing with and without circumferential groove

(as, for instance, in the case of the engine of a motor-car) the resulting wear will be small, but if (as, for instance, in the case of the axles of a train) the shaft has to start and stop frequently under full load, the wear which occurs then will be very important, and it may on balance pay to supply the lubricant at a point much nearer the point of maximum pressure than would be otherwise desirable, since this will reduce the time for good boundary lubrication to be established. Here again we see that in practical engineering it is often necessary to adopt compromise solutions which at first appear to be illogical.

After this brief digression on the topic of grooves, let us return to the plain bearing carrying a uni-directional load, and supplied with oil as in fig. 7.7a. We have seen that to obtain film lubrication we must have a finite difference in diameter between the bearing and the journal, and a supply of oil at a point remote from the position at which the film pressure will be generated. Are these necessary conditions also sufficient? We can approach this question analytically or experimentally, or by a judicious mixture of these methods. The purely analytical approach is extremely difficult – far beyond our scope – and no exact treatment deals with the whole problem, which involves the position or 'attitude' which the shaft occupies in the clearance space, the effect of leakage of oil from the sides of a bearing of finite width, and the effect of the variation in viscosity of the oil as it passes through the bearing due to changes in its temperature and its pressure. From an elementary dimensional treatment,* however, we can deduce that the behaviour, for instance, the 'coefficient of friction', of geometrically similar bearings will probably depend on the non-dimensional parameter $\eta\omega/p$, where η is the viscosity of the lubricant, ω the angular speed of the shaft, and

---

* Let us assume, as seems reasonable, that in a series of geometrically similar bearings the coefficient of friction μ under any conditions of film lubrication will depend on some power of

    (i) A linear dimension of the bearing, say l, whose 'dimensions' are L.

    (ii) The pressure on the bearing, say p, whose 'dimensions' are M/LT^2.

    (iii) The speed of running in radians/unit time, say ω, whose 'dimensions' are $1/T$.

    (iv) The coefficient of viscosity of the lubricant, say η, whose 'dimensions' are M/LT.

The dimensions of (i) and (iii) are too obvious to require comment. Pressure is force per unit area, or mass $\times$ acceleration/area or $MLT^{-2} \div L^2$ or $ML^{-1}T^{-2}$. Viscosity (see Introduction, example 0.8) is defined as the shearing stress/rate of shear, and if we consider two plates separated by fluid and in relative motion we see that the stress will have dimensions (like pressure) $ML^{-1}T^{-2}$, and the rate of shear will have dimensions velocity/distance of separation or LT^{-1}/L or T^{-1}. η will therefore have the dimensions $ML^{-1}T^{-2} \div T^{-1}$ or $ML^{-1}T^{-1}$. $\mu (= F/N)$ is non-dimensional.

Our assumption may be written

$$\mu \propto l^a p^b \omega^c \eta^d$$

Now the dimensions on both sides of this equation must be identical,

so $L^0 M^0 T^0 = L^a . M^b L^{-b} T^{-2b} . T^{-c} . M^d L^{-d} T^{-d}$

Equating the coefficients of L, M and T in turn, we have

$$0 = a - b - d$$
$$0 = b + d$$
$$0 = -2b - c - d$$

Whence, by simple algebra, it follows that $a = 0$, $b = -c$, $d = c$, so $\mu \propto (\eta\omega/p)^c$, where c is indeterminable – and not necessarily constant. Or in more general terms $\mu = f(\eta\omega/p)$.

p the load per unit area of the projected area, and there is ample experimental evidence to confirm this. By using this non-dimensional parameter as the independent variable, we can show in one curve the results from bearings of different sizes running at different speeds and under different loads, and using different lubricants. In fig. 7.9, for instance, are shown some typical results from bearings in which the ratio of clearance to diameter was maintained constant. As this ratio is increased the slope of the main part of the friction curve is reduced, but the sharp rise in friction indicated by the dotted line occurs at a larger value of $\eta\omega/p$.

One minor disadvantage of the use of the parameter $\eta\omega/p$ is that

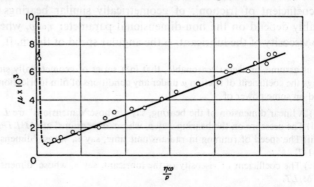

Fig. 7.9 Variation of 'coefficient of friction' with non-dimensional parameter $\eta\omega/p$ for bearings with constant ratio of clearance to diameter

these results are not presented in what might be regarded as a wholly 'natural' form. We normally think of plotting a curve with the independent variable starting at zero at the axis and increasing to the right. This curve is then conventional if we are thinking primarily of speed, or of viscosity, as the single quantity we are varying, but if we are trying to obtain a picture of the effect of changing only the load on a bearing we have to take particular note of the fact that zero load corresponds to $\eta\omega/p = \infty$, and vice versa; in other words, as the load is increased we move along the $\eta\omega/p$ axis from right to left. Thus we see that (as Tower said) as the load on a given bearing is increased the 'coefficient of friction' falls. If the straight line which very satisfactorily represents the mean of the results were to pass through the origin we should have the answer that not the coefficient of friction, but the total frictional torque on a film-lubricated bearing is independent of the load, and this is a very good approximation to the truth. In the curve shown the intercept on the μ-axis is of the order 0·0005; this is a rather lower value than is often obtained, probably because the quality of the bearings used was very high, but it indicates the sort of coefficient which can be regarded as a minimum. Note, however, that it corresponds to a very low value of either speed or viscosity, or a very high value of the load. It is easy to appreciate physically that this implies a very thin film of lubricant (in fact, of the order 1 micron or μm). Now if the surfaces were 'perfect' in finish, and if no distortion occurred, and if the lubricant could be guaranteed to be perfectly clean it might be possible to design for such thin, or even thinner, films. In practice, there would be too small a margin of safety in such a design, and the danger inherent in going to too low a value of $\eta\omega/p$ is illustrated by the way the curve of friction suddenly rises as contact begins to be made between the journal and bearing. In particular, it must be accepted that under normal conditions the lubricant will contain or pick up particles of 'dirt' which are often hard and highly abrasive in nature. Only if the minimum film thickness exceeds the size of these particles will they be able to pass through without causing wear of the surfaces. It is more sensible to use much larger values of $\eta\omega/p$, even though this implies a higher 'coefficient of friction'.

§ 7.5 Thrust Bearings

We have seen that film lubrication of journal bearings must have been achieved (at least occasionally and partially, and, so to speak, by mistake) in practice long before the possibility of its occurrence was appreciated. Given a shaft, a bearing with 'working' clearance, and some oil between the two, some sort of film will automatically form as the shaft rotates. The situation in relation to thrust bearings was very different; there was no correspondingly automatic way in which oil could be dragged into a tapered passage between the elements of such a bearing, which in consequence normally worked under conditions of boundary lubrication, with relatively immensely high coefficients of friction and necessarily larger areas to keep the pressures low enough to avoid disastrous wear. The solution is latent in Reynolds's work, in the sense that he showed that if a tapered film were to exist between two flat surfaces of which one is limited in length, the resultant thrust (cf. fig. 7.5) due to the pressure in the oil would pass not through the centroid of the limited area, but towards its 'off' side. It was, however, nearly twenty years before this knowledge was put to practical use (as it happened, simultaneously and independently by Michell in Australia and Kingsbury in America) by the invention of the 'tilting pad' bearing. The sketch, fig. 7.10, of part of a Michell thrust bearing, is almost self-explanatory. The shaft is provided with a plane collar, but this bears, not on a plane fixed member, but on a series of sector-shaped

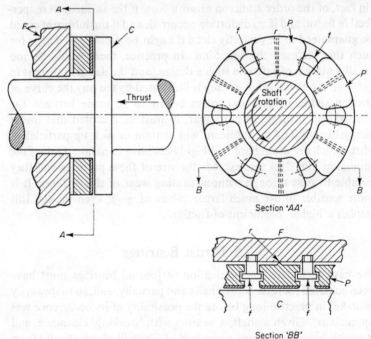

Section '*AA*'

Section '*BB*'

Fig. 7.10 Sketch of Michell thrust bearing

pads *PP*, each of which is restrained from axial or rotational motion by the fixed abutments *f*, *but is free to tilt about the edge rr of its base*, *rr* being radial but passing to the 'off' side of the central axis of the pad. Oil is, of course, supplied to the rubbing face of the collar. When the shaft rotates, this oil is dragged between the surfaces exactly as in fig. 7.5, the pad tilting to provide the essential tapered passage.

By elaborate calculation, it is possible to show (as Michell did) that a stable orientation of the pad and optimum conditions will be reached if the tilting axis is displaced from the central axis by about a tenth of the total width of the pad, and for uni-directional operation this is the order of the displacement used. Physically, it is not difficult to see that if the leading edge of the pad is well rounded to avoid scraping the oil off the collar a considerable margin of error is tolerable, and if the shaft must be able to run in both directions the tilting axis may, at the cost of some increase in frictional loss, be made to coincide with the central axis.

It may occur to the reader that the slight complexity of the tilting-pad bearing could be obviated if the fixed abutment were to be formed to provide tapered passages. This has been tried experimentally, but requires immensely accurate, and therefore expensive, processes, and the 'self-adjusting' feature is lost. It is worth while remarking that the order of magnitude of the taper is only one or two minutes of arc.

The adoption of the tilting-pad bearing effected a major revolution in design. At one step the bearing area required *and* the frictional losses were cut by a factor of the order 20: the benefit to the designer of a ship, in which the whole thrust from a propeller must be transmitted to the hull, or of a hydro-electric alternator, in which a rotating load of many hundreds of tonnes can now be suspended from a single bearing, is not difficult to imagine. Moreover, a well-designed tilting-pad thrust bearing shares with a well-designed journal bearing the desirable property of *running* indefinitely without wear. Wear occurs only at starting and stopping, or if dirt of dimensions greater than the minimum film thickness is admitted to the bearing.

§ 7.6 Lubricants other than Oil

In the paragraph dealing with friction and boundary lubrication some attention was paid to the contaminant or lubricant which, by accident or design, modifies the behaviour of surfaces rubbing on each other, and its importance in such cases must not be overlooked. Once film lubrication has been achieved, however, the *only* quality of a lubricant which matters is its viscosity. Now every fluid, liquid or gaseous, possesses to a greater or lesser degree this property of viscosity: it follows that any fluid may in appropriate conditions be used as a film lubricant. Its suitability in a particular case can be quickly judged in the light of the knowledge that the important parameter is the non-dimensional variable $\eta\omega/p$. Thus, a very viscous liquid, such as a 'heavy' oil, would be suitable for a bearing running at low speed and carrying a heavy load, while a 'light' oil of low viscosity would be more suitable for a high-speed bearing. Water, which has a relatively low viscosity, can be used as the lubricant in special cases, for instance between a steel shaft and a rubber bearing, or in the case of a rolling mill between the steel shaft and bakelite-impregnated fabric bearing, when its high specific heat is helpful in cooling, though grease is normally added to provide adequate boundary lubrication before the film is established. In particular applications it may be essential to keep *any* liquids from the bearing, and a gas, for instance carbon dioxide or air, may be used as the lubricant, but of course these gases provide little or no boundary lubrication, so, to avoid wear, film-lubricated conditions must be established as quickly as possible, and the ratio ω/p must be relatively very high. Moreover, the fact that the film is compressible and has little damping action must be considered in calculations relating to resonant vibrations. In practice, mean bearing pressures of the order 0·1–0·2 bar at starting, and perhaps ten times these values under running conditions, might be regarded as maximum values for gas-lubricated bearings of this type. Higher pressures can be sustained in 'hydrostatic' bearings, to which we shall now devote brief attention.

§ 7.7 Hydrostatic Bearings

In the normal film-lubricated bearing the film can be established only by means of a sufficient relative movement between the parts of the bearing. This type of bearing is therefore frequently described as 'hydrodynamically' lubricated. It follows that when this movement is too small only boundary lubrication at best can obtain between the parts, that a relatively high coefficient of friction is inevitable, and that wear is likely to occur. In particular cases these disadvantages may be intolerable, and special measures may be taken to eliminate them. For instance, in turbo-alternators, where the load on the bearing is due to the dead weight of the rotating parts, a supply of oil may during starting and slow-running conditions be fed to the bearing at such a point and at such a pressure that the journal is lifted by the film which this oil forms in reaching the boundary of the bearing. This type of lubrication may be described as hydrostatic. Needless to say, the primary objective in this instance is the elimination of wear. The same principle may be applied to bearings in which an abnormally low 'coefficient of friction', particularly at running speeds tending to zero, is desirable, and it matters little whether the fluid used is liquid or gaseous. In the latter instance we might object to the prefix 'hydro-', but since it is in any case inappropriate to lubricating oil, we need not worry too much at its further perversion to indicate any fluid. If we wish particularly to distinguish between the use of a compressible and 'incompressible' fluid we can substitute the prefix 'aero-', though this leads to an unpleasant hybrid word.

The sketch, fig. 7.11, shows a simple form of such a journal

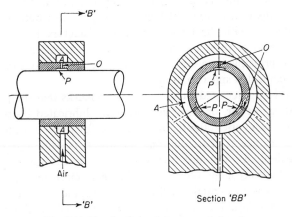

Fig. 7.11 Sketch of simple hydrostatic bearing

bearing in which the lubricant used is air. The air is brought from a mains supply at a suitable pressure – perhaps up to 10 bar, depending on the load to be carried – to the annular passage *A* round the bush. From this bush three or more small orifices *O* lead the air to the clearance space between the bush and the journal, the end at which the air escapes usually being provided with a small flat 'pocket' *P*.

If the journal were to rest against the bush in the neighbourhood of one pocket only a very small quantity of air would be able to escape through the residual clearance space, so nearly the full mains pressure would in this region provide a radial force on the journal. In the neighbourhood of each of the other pockets, however, the greater clearance would allow more air to flow through the orifice, with in consequence a larger pressure drop therein, so a smaller radial force would be exerted on the journal. There is thus a self-centring action, and if the mains pressure and journal are sufficiently large in relation to the external force the journal will, so to speak, float in mid-air. The frictional torque on the journal will, at zero speed, be literally zero. It will rise, due to the normal viscous drag, as the speed rises; but because of the low viscosity of air it will remain extremely low in relation to a normally lubricated bearing.

In more sophisticated designs the orifice can terminate not in a single shallow pocket but in a series of grooves which lead the air, once it has passed the orifice, to a larger area of the journal surface. This provides a much higher self-centring action, but it is clear that if the bearing is to be used for high speeds these distributing grooves may interfere with the formation of a satisfactory hydrodynamic 'film'. Design of these bearings is at present something of an art. Identical principles can even more easily be applied to thrust bearings, or conical or spherical bearings.

It will be appreciated from these brief remarks that the principal applications of air, or more generally gas, lubricated bearings fall into three categories: the hydrostatic type for use when the lowest possible coefficient of friction and the lowest possible rate of wear are required, for instance in instruments and measuring devices, and the hydrodynamic type *either* when extremely high speeds and low friction are essential *or* when, as for instance in some nuclear reactors, no liquid lubricants can be tolerated. Although gas-lubricated bearings have only relatively recently been widely used, they have aroused much interest, and an excellent bibliography* prepared in 1959 lists nearly three hundred items.

§ 7.8 Lubrication in Higher Pairing

The preceding paragraphs have been principally concerned with lubrication between surfaces of considerable area such as obtain in 'lower pairs'; in particular turning and sliding pairs. In chapter 1 it was pointed out that in engineering practice loads had often to be transmitted through 'higher pairs', such as cams and tappets, or gear-teeth, between which only 'line', or even 'point', contact could in general occur. In the very brief discussion in chapter 2 on the strength and wear-resisting qualities of gear-teeth it was re-

* *First International Symposium on Gas-Lubricated Bearings.* Office of Naval Research, Washington, U.S.A., 1959.

marked that such phrases as 'line contact' were not strictly accurate, that since no actual material was infinitely rigid, any two bodies pressed together would if necessary deform, initially elastically, until the area of contact was sufficient to resist the pressure, and that the elastic problem had for one or two simple cases been solved by Hertz. Even when due allowance has been made for this effect, however, it remains true that the pressure between the contact surfaces in higher pairing is in general extremely high and the area of contact extremely small, and it was for long assumed that these conditions would rule out all possibility of film lubrication and that only boundary lubrication was possible.

· In a remarkable pair of papers published in 1949 Grubin* discussed the contact stresses which occur in gearing and the 'fundamentals of the hydrodynamic theory of heavily loaded cylindrical surfaces'. It is well known that the viscosity of oils is very strongly temperature-dependent, falling rapidly as the temperature rises. This, after all, is a matter of common experience. It is less generally realised that the viscosity of oils is similarly pressure-dependent, rising rapidly as the pressure rises. In fact, most oils tend to freeze at atmospheric temperature at pressures of the order of 3 kbar. Grubin took account of the influence of both these factors, as well as of the compressibility of the oil and of the elastic deformations of the cylindrical surfaces in the region of the 'contact', and was able to show that under conditions approximating to practice, with cylindrical surfaces rolling and/or sliding over each other, hydrodynamic lubrication could well be achieved. His solution shows that over most of the area of 'contact' (that is, of very close approach of the solid surfaces, since, of course, if these surfaces are separated by a film of lubricant 'contact' is a misnomer) the film of oil is of sensibly *uniform* thickness so that to a first approximation the distortions of (and therefore the pressures on and stresses in) the solid members are almost exactly the same as those which would occur if the film were not present, that is, as in the Hertzian analysis. We may crudely visualise this situation if we realise that, since the film is very thin, endwise flow of the lubricant (in the direction of the axes of the cylinders) is utterly negligible, so the problem is a two-dimensional one, and even in the direction of the width of the

* A. N. Grubin, Central Scientific Research Institute for Technology and Mechanical Engineering, Moscow, Book No. 30, 1949. English translation D.S.I.R.

'contact' zone the film tends to be thin in relation to this width, so that the flow of the oil, made extremely viscous by the pressure to which it is being subjected, is very restricted, and the oil which has got into this zone is in effect trapped therein. More surprising is Grubin's deduction relating to conditions at the outlet end of this zone, where the oil is, so to speak, released with a sudden reduction in pressure, and hence viscosity, into a rapidly widening channel. This implies a sudden 'disappearance' of the separating film analogous to the situation which would occur if, under static conditions, the cylindrical surfaces were separated by a thin plate of rigid material which ended abruptly *within* the Hertzian zone. At its edge a large stress concentration would undoubtedly occur, and this appears in Grubin's solution as an extremely high-pressure peak. Dowson and Higginson* have amplified Grubin's work, and their calculations show that for steel surfaces lubricated by mineral oil the height of this peak may at certain reasonable speeds greatly exceed the maximum Hertzian pressure. This, then, is a complicating factor which could be important, but it may be doubted whether in practice a peak as sharp as that predicted by such calculations would in fact occur. What has received experimental confirmation is that hydrodynamic lubrication can be, and in good practice is, achieved, and the importance of this fact in relation to wear surely needs no further emphasis. In relation to the stresses in the elements we must accept the conclusion that even hydrodynamic lubrication does *not* reduce the Hertzian values, but could even add greater ones.

Books for Reference and Further Reading

F. P. BOWDEN and D. TABOR. *The Friction and Lubrication of Solids*, rev. edn. Oxford, Clarendon Press, 1954.

F. T. BARWELL. *Lubrication of Bearings*. Butterworth, 1956.

* D. Dowson and G. R. Higginson. Various papers in *J. Mech. Eng. Sci.*, Vol. 1, No. 1; Vol. 2, No. 3; Vol. 4, No. 2.

Answers

Introduction, p. 40

0.1 About 13 km/hr and direction from 53° E. of S.

0.2 2 510 Nm; 9·86 × 10⁵ Nm.

0.3 38 mm.

0.4 37·5 and 2·5 m/s in the opposite directions; 1 875 Nm.

0.5 7·64 m/s at 49°; 22·2 per cent lost to heat.

0.6 (a) ML/T^2; ML/T; ML^2/T^2; ML^2/T^3
     (b) F; FT; FL; FL/T.

0.7 1·013 × 10⁵ N/m², 1·013 bar; 1·016 × 10⁵ N/m², 1·016 bar; 6·90 × 10⁶ N/m², 0·69 hectobar; 15·44 × 10⁶ N/m², 1·544 hectobar; 9·81 × 10⁶ N/m², 0·981 hectobar.

0.8 kg/ms; m²/s; non-dimensional.

0.9 30·5 km/hr; 59°.

0.10 438 N; stable.

0.11 Leans to left causing the front wheel to turn to the left (and so not only making bicycle turn corner but also promoting automatic stability).

0.12 23·0 rev/min. Note that if centripetal acceleration is neglected a slightly lower answer (22·3 rev/min) will be obtained.

If, and only if, the spindle is blunt, the point of contact will be offset from the axis of spin as sketched in fig. 0.22*a*. Slipping will occur and a horizontal friction force F_1 will therefore be exerted on the spindle: the moment of this force about the C.G. of the top, as indicated in fig. 0.22*b*, tends to rotate the axis of spin towards the vertical.

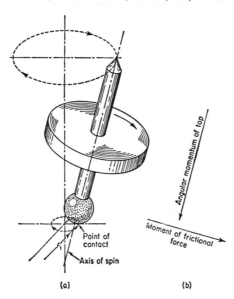

Fig. 0.22.

0.13 5·25 g; 1 135 Nm tending to force nose of aircraft down.

0.15 Range 2·39 m; $F_{\max} = 37\cdot6$ N.

Chapter 1, p. 103

1.1 (*a*) No. The Watt proportions would be very difficult to achieve without either (1) excessive angular displacement of upper link or (2)
excessive length of lower link or (3) lower steering pivot too remote
from tyre–ground contact. Compromise adopted gives very small
lateral movement of this contact point over normal range of movement.

   (*b*) Horizontal displacement of *C* is zero if $AC/OA(1 - \cos\theta) =$

$$BC/PB(1 - \cos\phi); \text{ and } OA\sin\theta \simeq PB\sin\phi, \text{ hence } \frac{BC}{AC} = \frac{OA}{PB}.$$

1.3 (*a*) 19 m/s, 1° right of vertical; 3 100 m/s², 12° right of vertical.
   (*b*) 18 m/s, 1° right of vertical; 3 400 m/s², 4° left of vertical.

1.4 See fig. 1.67.

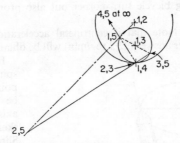

Fig. 1.67.

1.5 3·1 *v* rad/s.

1.6 35 rad/s.

1.7 52 rad/s; -96 rad/s².

1.8 0·44 m/s at 21° to horizontal (i.e. the positive direction of conventional *x* axis).

1.9 0·6 m/s² at 117° to horizontal.

1.10 270 N at 143° to horizontal passing above 0 at perpendicular distance
of 1·7 m; 710 N × 33 mm = 23·4 Nm.

1.11 204 Nm anti-clockwise.

1.12 2 160 m/s² at $-64°$ to horizontal.

1.13 500 rev/min anti-clockwise; zero. Note that these answers can be
written down (cf. § 1.6) but that the example is worth working out.

1.14 10 rad/s; 445 rad/s².

1.15 306 m/s² at 166° to horizontal.

1.16 3·1 m/s to left; 41 m/s² to left.

1.17 8 500 N at 62° to line of stroke, 27 mm along rod from centre of big end; 146 Nm retarding crankshaft.

1.18 4 200 N at 218°; 3 700 N at 254°; 9 600 N at 0°.

1.19 ±1·28 per cent.

1.20 1·76 kg m²; ±3·6 per cent; ±2·6 per cent.

1.21 127°; 26½°; 5 520 kg.

1.22 ±2·9 rad/s. This assumes constant motor torque.

Chapter 2, p. 186

Completion of partially worked Example 2.3.

Line		F	J	K	G	H	E	ω_j/ω_k
1	Fix K	+1	$\dfrac{-92}{28}$	0				
2		+28	−92	0				
3	Fix F	0	−120	−28				+4·29 (1st)
4	Fix F	0	−120	−28	+24·5			
5	Fix G		−144·5	−52·5	0			+2·75 (2nd)
6	Fix G		−144·5	−52·5	0	84·5		
7	Fix H		−229	−137		0		+1·67 (3rd)
2	(K fixed)	+28	−92	0			−12·35	
8	Fix E		−79·65	12·35			0	−6·45 (reverse)

2.1 (a) If belt does not stretch and is not tightened 7·35 kW. If belt is tightened to a mean tension of 1 082 N power can be raised to 9·04 kW.

(b) Speed is too high.

2.2 198 N/cm; 20·8 N/m².

2.3 114 mm²; Yes.

2.4 321 N. (Note that $T_0 = 111$ N does not affect the answer.)

2.5 25·8 m/s; 11·9 kW; 1 112 N.

Data for curves:

2.6 $d = 25$ mm; $R = 100$ mm.

2.8 60·8 mm; 1·37; 141·5 bar; 3 kbar.

2.9 670 mm.

V	5	10	15	20	25	30	40·7	44·7	m/s
for T_{mean} = 1 112 N	2·32	4·62	6·93	9·25	11·5	10·5	0		kW
for T_{mean} = optimum	3·41	6·56	9·20	11·05	11·87	11·35		0	kW

2.10 (*a*) -54 rev/min; (*b*) $-1\frac{13}{37}$ rev/min.

2.11 Clutch locked, gear ratio $= 1$; r.h. brake on, reduction ratio $3\cdot68$; l.h. brake on, ratio $-8\cdot625$ (reduction and reverse).

2.12 413 Nm; 700 Nm; 410 rev/min.

2.13 $-6\cdot23\ T_{\mathrm{i}}$; $7\cdot23\ T_{\mathrm{j}}$.

2.14 -556 rev/min up to $+833$ rev/min; $4\cdot78$ Nm.

Chapter 3, p. 232

3.1 66 N; $58\cdot5$ N; $3\cdot15$ kg at $206°$.

3.2 (*a*) 94 N and 27 N.
(*b*) $0\cdot153$ kg m at $225°$ and $0\cdot0384$ kg m at $308°$
or $1\cdot53$ kg and $0\cdot384$ kg at 100 mm rad.
(*c*) $0\cdot227$ kg m at $220°$ and $0\cdot09$ kg m at $5°$
or $2\cdot27$ kg and $0\cdot9$ kg at 100 mm rad.

3.3 $2\cdot26$ kg, $1\cdot82$ kg, $1\cdot80$ kg; $195°$.

3.4 $11\cdot5$ kg mm at $77°$.

3.5 $0°$ (say), $192°$, $155°$, $348°$ (or $168°$, $205°$, $12°$), $6\cdot09$ tonnes.
$\pm3\cdot88 \times 10^5$ N; $\pm5\cdot60 \times 10^5$ Nm about C/L of crankshaft.

3.6 $\alpha = 34°\ 54'$; $\beta = 52°\ 24'$; $4\cdot58$ tonne.

3.7 Forces balance. Amplitude of primary moment 24 500 Nm, amplitude of secondary moment 13 900 Nm.

3.8 Both engines balanced.

3.9 Forces balanced. Amplitude of primary moment 3 130 Nm.
Amplitude of secondary moment 1 170 Nm.

3.10 2 090 N rotating with crank; secondary force of amplitude 890 N; $0\cdot0306$ kg m; $0\cdot00162$ kg m in each wheel of Lanchester balancer.

3.11 (*a*) Secondary force of amplitude 8 840 N. All moments about centre are zero.
(*b*) An unbalanced moment of 1 483m Nm rotating with crankshaft, where m kg is the effective rotating mass associated with each crank, i.e. the out-of-balance mass at each crank plus a mass equal to the reciprocating mass per cylinder, which can be balanced by adding rotating balancing masses.

3.12 $\dfrac{3}{2} \cdot \dfrac{1}{q} \cdot m\omega^2 r \cos(2\theta + 3\alpha)$, where the crank angle θ is measured from plane inclined at angle α to one of the cylinders.

3.13 Single-cylinder unit: $M\omega^2 r\ \{0\cdot35 \sin\theta + \dfrac{0\cdot68}{q} \sin 2\theta\}$.
Triple unit: Forces balance. Moments are $d\sqrt{3}$ times force from single-cylinder unit. (*N.B.* Crank angles were taken as $\theta + 10°$ and $\theta - 10°$.)

3.14 Single-cylinder unit: Primary force balance. Secondary force $\pm4\cdot63 \times 10^5$ N.
Complete engine: Complete balance.

3.15 Primary forces balanced by a mass with an mr value $0\cdot282$ kg m opposite the crank, Secondary force ±5508 N.

Chapter 4, p. 300

4.1 1·926 Hz.

4.2 2·41 kg m².

4.3 178 mm from centre of small end, $k = 98·1$ mm.

4.4 369 mm from first flywheel, 155·6 Hz.

4.6 21·01 Hz.

4.7 $I_1I_2\omega^4 - \{I_1k_2 + I_1k_3 + I_2k_1 + I_2k_2\}\omega^2 + (k_1k_2 + k_2k_3 + k_1k_3) = 0$

$\dfrac{a_1}{a_2} = +0·732$ or $-2·732$.

4.8 $0·139\sqrt{\dfrac{\lambda}{m}}$ and $0·364\sqrt{\dfrac{\lambda}{m}}$.

4.9 Amplitude of torque $= 7·74$ kN m.

4.10 Linear interpolation suggests 67·6 Hz, more accurate value 67·3 Hz.

4.11 $m_5 = 14·1$ kg. Amplitude of force applied to specimen 17·4 kN.
 Amplitude of force applied to foundation 0·81 kN.

4.12 $m_4 = 13·34$ kg.

4.13 Amplitude of vibration torque applied to $I_4 = 1\,920$ Nm.

4.14 (*a*) and (*b*) 2 054 Hz.

4.15 2 660 Hz.

4.16 4 141 Hz.

4.17 44 Hz.

4.18 $0·0929\sqrt{\dfrac{EI}{ml^3}}$ and $0·618\sqrt{\dfrac{EI}{ml^3}}$.

4.19 249·4 mm.

4.20 Natural frequency by exact solution or assuming linear deflection curve $= 2·34$ Hz.

4.21 (*a*) $0·989\sqrt{\dfrac{EI}{Ml^3}}$; (*b*) $0·997\sqrt{\dfrac{EI}{Ml^3}}$; (*c*) $0·993\sqrt{\dfrac{EI}{Ml^3}}$.

4.22 26·1 Hz.

Chapter 5, p. 362

5.1 (*a*) 4·21 Hz; (*b*) 2·65 kN s/m; (*c*) 0·661.

5.2 (*a*) 4·21 Hz; (*b*) 0·423 N s/m; (*c*) 0·0016.
 Amplitude of force $= 0·112$ N.

5.3 11·4 g.

5.4 $t = 0·00375$ s.

5.5 0·39 per cent ($\omega = 523·6$ rad/s, $\omega_n = 55\,300$ rad/s, $c = 0·00542$).

5.6 0·108 mm.

5.7 0·139 mm.

5.8 Mean torque ± 487 Nm ($\omega_n = 42·9$ rad/s, $c = 0·216$, $\Theta = 0·0139$ rad.

5.9 10·1 kN s/m, 403 N.

5.10 0·138 kg m; $c = 0·0626$.

5.11 $\left(\dfrac{\omega}{\omega_n}\right)^2 = \dfrac{-1 + \sqrt{(1 + 8c^2)}}{4c^2}$

Maximum transmission ratio $= \dfrac{1}{\sqrt{\left(1 - \dfrac{1}{2c^2} - \dfrac{1}{8c^4} + \dfrac{(1 + 8c^2)^{\frac{1}{2}}}{8c^4}\right)}}$

5.12 Length of pendulum $= 22\cdot2$ mm. Maximum torque absorbed $= 149$ Nm.

5.13 $27\cdot9$ N/m.

5.14 $c = 0\cdot634$. (Note that second solution $c = 0\cdot774$ gives ω/ω_n as a negative quantity.) $\omega/\omega_n = 1\cdot43$.

5.15 $c = 0\cdot6, f_n = 20\cdot1$ Hz.

5.16 $\pm1\,120$ N.

5.17 (*a*) 12 680 rev/min; (*b*) 6 340 rev/min.

5.18 2 015 rev/min. (*a*) 1 512 rev/min; (*b*) 8 060 rev/min.

5.19 Considering only the first 5 engine orders as they repeat after this:

For 1st mode:

$\frac{1}{2}$, 1, $1\frac{1}{2}$ and 2 engine order give zero excitation.

$2\frac{1}{2}$ engine order gives maximum excitation with the components from all five cylinders in phase.

For higher order mode:

$2\frac{1}{2}$ engine order gives zero excitation.

$\frac{1}{2}$ and 2 engine order give large excitation. If $\Theta_1 = \Theta_5 = 1$ and $\Theta_2 = \Theta_4 = 0\cdot75$, then vector summation is $2\cdot8$.

1 and $1\frac{1}{2}$ engine order give small excitation, with above values of Θ the vectors are $0\cdot25$.

5.20 It is assumed that the concrete slab can be treated as a concentrated mass at the centre of the beam, as no details are given.

(*a*) Machine mounted rigidly on the floor:

The amplitude will vary with the speed as indicated in fig. 5.12. In the absence of information as to the damping which the floor will provide, it is impossible to estimate accurately the peak value, but the value at the running speed is almost unaffected by the probable damping. If there is no damping the amplitude would be $0\cdot025$ mm, which is unlikely to be objectionable. (Note the natural radiancy is $47\cdot6$ rad/s.)

(*b*) Machine on flexible mounting:

Defining ω_{n1} as the natural radiancy of the machine on its mounting of stiffness λ_1, ω_{n2} as the natural radiancy of the floor of mass m_2, and writing down the disturbing force as $F = R\omega^2 \cos \omega t$, we can by methods very similar to those used in §5.6 derive for the amplitude of the floor

$$X_2 = \frac{R\omega^2 \omega_{n_1}^2 / m_2}{(\omega_{n_1}^2 - \omega^2)(\omega_{n_2}^2 - \omega^2) - \lambda_1 \omega^2 / m_2}$$

which gives a double-peaked curve of the type drawn in fig. 5.17. Again the peak values cannot be calculated, but the upper resonance is closer to the running speed ($\omega = 55$ rad/s), and the amplitude at the running speed is reduced only by a factor of about 4 (with no damping $X_2 = 0.0065$ mm).

It would seem reasonable to recommend that the machines should be installed rigidly in the first instance, flexible mountings to be added only if even the very small vibrations which will result are found to be unpleasant.

In an extreme case (hardly likely to apply to a factory) when all trace of vibration had to be eliminated, a dynamic absorber could be added.

Chapter 6, p. 416

6.1 (a) $\dfrac{x_2}{x_1} = \dfrac{TD}{1 + TD}$, where $T = \dfrac{f}{\lambda}$.

(b) $\dfrac{x_2}{x_1} = \dfrac{T_2(1 + T_1 D)}{T_1(1 + T_2 D)}$, where $T_1 = \dfrac{f}{\lambda_1}$ and $T_2 = \dfrac{f}{\lambda_1 + \lambda_2}$.

(c) $\dfrac{E_2}{E_1} = \dfrac{\{1 + (R_1 C_1 + R_2 C_1)D + R_1 R_2 C_1 C_2 D^2\}}{\{1 + (R_1 C_1 + R_2 C_1 + R_1 C_2)D + R_1 R_2 C_1 C_2 D^2\}}$.

(d) $\dfrac{x_2}{x_1} = \dfrac{c}{a + b} \dfrac{(1 + T_1 D)}{(1 + T_2 D)}$, where $T_1 = \dfrac{b + c}{Kc}$ and $T_2 = \dfrac{a}{K(a + b)}$.

6.2 9·48 per cent, 1·242 s.

6.3 0·485 s.

6.4 Viscous damping reduced to 103·7 Nm s/rad or $C' = 0.3$. First derivative of error compensation $K_1 = 0.0208$ s.

6.5 (a) $K = 28.8 \times 10^{-4}$ Nm/rad, $K_1 = 3.47 \times 10^{-2}$ s.

(b) $S_1 = 0.1041$ s.

6.6 (a) $3.5°$; (b) 7·28 volts.

6.7 $\dfrac{x}{\theta} = \dfrac{K}{1 + TD}$, where $K = a/b$ and $T = \dfrac{(a + b)A}{bk_1}$; level rises 0·297 m.

6.8 Gain margin $= \dfrac{T_1 + T_2 - KT_1T_2}{(T_1 + T_2)}$; K for instability $\geqslant \dfrac{T_1 + T_2}{T_1 T_2}$.

6.9 $\omega = \dfrac{1}{T\sqrt{2}}[\sqrt{(1 + 4K^2 T^2)} - 1]^{\frac{1}{2}}$,

phase margin $= \tan^{-1} \dfrac{\sqrt{2}}{\sqrt{[(1 + 4K^2 T^2)^{\frac{1}{2}} - 1]}}$.

6.10 *Proof:* See fig. 6.40a $\left|\dfrac{\theta_o}{\theta_i}\right| = M = \dfrac{x^2 + y^2}{\sqrt{[(1 + x)^2 + y^2]}}$.

Hence $\left(x + \dfrac{M^2}{M^2 - 1} \right)^2 + y^2 = \left(\dfrac{M}{M^2 - 1} \right)^2.$

Construction: See fig. 6.40b $\sin \beta = \dfrac{M/(M^2 - 1)}{M^2/(M^2 - 1)} = \dfrac{1}{M}.$

Draw OD and find circle which touches OD and frequency response curve and note $OB = OA \cos \beta = 1$, which gives the scale. *Answer:* Gain constant 1·725 and resonant radiancy (i.e. at point of tangency with circle) = 1·1 rad/s.

6.11 Note solution above for **6.10**. Maximum value of $K = 1·39$. Resonant frequency = 0·15 Hz.

6.12 (*a*) Unstable. (*b*) Stabilising. (*c*) Destabilising.

6.13 $K = 7$. This allows for transition curves on the log plot. If transition curves ignored $K = 6·76$.

6.14 Gain margin 6 db or 0·5, Phase margin 16°, Steady state error 0·5 rad.

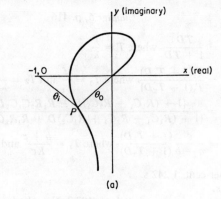

(a)

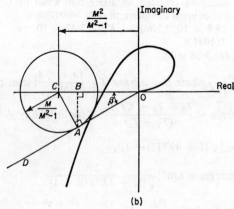

(b)

Fig. 6.40

Index